Electronics and electronic systems

R ED

al n
er

To Gillian

Electronics and electronic systems

George H. Olsen

Butterworths
London Boston Durban Singapore Sydney Toronto Wellington

First published 1987

© G. H. Olsen, 1987

British Library Cataloguing in Publication Data

Olsen, G.H.
Electronics and electronic systems.
1. Electronic apparatus and appliances
I. Title
621.381   TK7870

ISBN 0-408-01369-9

Library of Congress Cataloging in Publication Data

Olsen, George Henry.
Electronics and electronic systems.

Bibliography: p.
Includes index.
1. Electronics.   2. Electronic apparatus and appliances.   I. Title.
TK7816.0414   1987   621.381   86-26400
ISBN 0-408-01369-9

Typeset by Scribe Design, Gillingham, Kent
Printed and bound in Great Britain by Page Bros Ltd., Norwich, Norfolk

Preface

The subject of electronics has changed and expanded very rapidly in recent times; every year sees the invention of new devices or the development of existing ones, and the pace at which electronics is expanding is so rapid that even electronic engineers often feel that they are being overwhelmed. This is especially so in the field of digital electronics where the introduction of microprocessor-based systems has brought its special problems; not the least of these is the introduction of the new jargon terms and phrases associated with this branch of electronics. The people best able to cope with the changes are those who have received a thorough grounding in the fundamentals of the subject. Unpopular though it is in some quarters, a knowledge of physics and mathematics is still a valuable starting point from which the budding engineer can proceed to the study of electronics. The growing dominance of the digital side of the subject has led some to think that fundamental circuit theory and a knowledge of analogue electronics are, if not obsolescent, at least of rapidly diminishing importance. Nothing could be further from the truth. All digital systems must be interfaced with other systems, especially those incorporating a wide variety of transducers. A knowledge of physics is essential in the development and exploitation of transducers for use in industrial and scientific fields, Very often such transducers must be connected to analogue circuitry before the signal can be transferred to digital systems.

This book has been written for first and second year undergraduates reading for degrees in electronics, electronic engineering, physics and allied subjects; those undertaking TEC courses at the higher levels should also find the book useful. Students in subjects other than those mentioned often require a knowledge of electronics in order to equip themselves to deal with the increasing amount of electronic instrumentation being introduced into every field of scientific endeavour. It is hoped that they will find the descriptive parts of the text useful in their work.

The author is indebted to several firms for their willingness to allow publication of data and circuits from their commercial literature: Ferranti Ltd, Texas Instruments Incorporated, Marconi Ltd, Motorola Semiconductor Products Incorporated, Rockwell International Corpn, Burr-Brown, and Analogue Devices have all been helpful in this respect. The firms are, of course, not under any liability for any damage, loss or expense resulting from use of the information supplied, and retain the copyright of their material at all times.

Thanks are also due to Marian Irving who, in typing the manuscript, tackled the problems of a new word processor with great fortitude and stamina. I am also very grateful to my former colleagues W. G. Allen, J. Dunn and R. G. Earl who kindly read the typescript and who made valuable suggestions for its improvement.

Contents

Part 3   Digital electronics

Part 1

Passive components and circuit theory

Chapter 1

Introduction

Few people would deny that the recent recession has been accompanied by enormous changes in our social structures. Millions are now unemployed, and for many of these the prospect of finding paid employment is indeed bleak; our attitudes to work and leisure have therefore undergone remarkable changes in recent years. One of the major causes of such profound changes is the phenomenal increase in our technological knowledge, especially in the field of electronics. The growth of information networks that depend upon digital computers has been almost incredible. Home computers available today are extremely powerful compared with the large industrial computers of only a decade ago, and thus huge data acquisition systems are within easy reach of most of the population. The transmission of messages via inter-computer networks will in future eliminate the need for extensive postal services. The French are already substituting computer terminals for conventional printed telephone directories. The acquisition of data via CEEFAX and ORACLE is now a commonplace use of TV receivers.

Those who are seriously looking for jobs must take account of the fact that for the rest of the decade, and beyond, large sections of industry as we knew it will disappear. Many routine manufacturing jobs are being taken over by machines; office work that once employed tens of thousands will be largely automated and hence need far fewer workers in future. The service industries and professions such as medicine, teaching, retailing, the building trades, welfare and social services, all need more people. Many jobs could be created in these fields if only we were intelligent enough to organize our society so that the wealth created by machines could be used to expand such services. Unfortunately the brilliant advances made in the field of technology are not matched by comparable advances in social and political thinking.

One area in which the future is bright and exciting is electronics. For those with the inclination and willingness to work and train in a subject that is not easy, electronics provides a lucrative and satisfying career. Since all of technology now depends so heavily on electronics it is difficult to see how this industry can ever become redundant. Apart from developments in automation and control of industrial processes, we will see an explosive expansion of electronic apparatus in our homes and offices. Two-way cable television systems will become widespread; high-resolution television and video systems will be developed using flat-screens and flat loudspeakers that can be hung upon a wall, like pictures. Holographic (3-D) television is not now far in the future. Electronic kitchen computers and home appliances will be the norm for the future. It will be possible to work from home, instead of travelling to the office as ever more sophisticated electronic apparatus is developed. The future for those trained in electronics is very good.

Readers of this book are almost certain to be convinced that electronics is a worthwhile career. Those aspiring to a career in electronics should, however, realize that the training path is not an easy one. Although the arrival of ever more complex integrated circuits makes it relatively easy for anyone to assemble a sophisticated electronic system, mere assemblers of such systems are little more than low-grade technicians. They are very vulnerable to changes in technology and are at the mercy of the manufacturers and designers of such circuits. It is hoped that readers of this book will be aspiring to engineering status, and will consequently be looking for an understanding of the principles involved in the design and operation of electronic systems. They will be ready to adapt to changing conditions, and will have the fundamental knowledge to be able to cope with a rapidly changing electronic environment. Their mathematics and physics background will be such that, together with training in electronics, they will be able to cope with change and retraining schemes. It ought to be remembered that electronic apparatus is rarely operated in isolation. In the vast majority of cases such apparatus is connected to transducers and output devices. The transducers convert physical phenomena such as light, heat, temperature, sound, etc., into an equivalent electrical signal; output devices such as digital readouts, servomotors, loudspeakers and indicating devices are required to perform tasks or communicate (usually with human beings). A knowledge of basic physics is very useful in understanding the operation of transducers and output devices. Consider, for example, some of the changes that are taking place in Polaroid's research department. Although this firm was previously concerned only with cameras, electronic techniques now play a very important role in their activities. Several of the developments combine electronics with optics including the complete digitization of the image and subsequent signal processing for use in video camera systems. It will be well known that all important camera manufacturers now incorporate sophisticated electronic systems in their products. Designers of such electronic systems are much more valuable to their firms if they possess enough knowledge of physics to enable them to understand the physical principles involved in the production of the final product.

Work associated with engineering and electronics involves measurement and quantitative results; mathematical skills are therefore essential. We can go a long way in electronics using basic algebra, the j-notation, Laplace transforms and matrix algebra. The text in this book includes what is hoped to be a sufficient explanation of the mathematical techniques used in connection with both the j-notation and Laplace transforms. Although the reader is assumed to have little or no previous knowledge of electronics, some acquaintance with physics and mathematics must be taken for granted. The reader is expected to have heard of Ohm's Law, to be able to manipulate algebraic expressions, and to perform simple differentiation and integration.

It is hoped that a satisfactory balance has been kept between examination-type material and the kind of text required by those who want an introductory account of the subject. The book should therefore be useful to those who are not specialist electronic engineers (e.g. physicists, chemists, mechanical engineers), but who increasingly are required to use electronic systems in their work. Primarily, however, the book is written for first and second year undergraduates reading electronic engineering, physics and physical electronics, and should be useful for those taking TEC courses at levels III, IV and V.

Modern electronics has to a large extent seen the eclipse of discrete component circuitry, and for that reason a great deal of emphasis has been placed on the role of integrated circuits. Inevitably, with the arrival of the microprocessor and microcomputer, digital ICs are assuming an ever-increasing importance. The digital engineer would, however, be at a serious disadvantage if he had little understanding of analogue electronics and circuit theory.

Chapter 2

Passive components

Electric current may be regarded as a flow of electrons. The flow may be along a wire, a carbon rod, through a gas (as in a fluorescent tube), a vacuum, or within a crystal (as in a transistor). Each electron carries a small electric charge $(1.602 \times 10^{-19}$ coulombs) and when many millions of them flow, say along a wire, a charge q passes a particular point during a specified interval of time. At the particular point the wire has a cross-section area. If a small charge, δq, passes through this area in a small interval of time, δt, then $\delta q / \delta t$ is the rate at which charge is passing. In the limit when t approaches zero $\delta q / \delta t$ becomes the differential coefficient dq/dt. This rate is called electric current, i.e.

$$i = \frac{dq}{dt}$$

When the units of charge and time are respectively the coulomb and the second then the unit of current is the ampere. 1 ampere is thus equivalent to the passage of 1 coulomb per second through the cross-sectional area at a given point along the wire.

Electrical engineers are concerned with the fact that electric current can be made to do work. The electron flow (often loosely called the current flow) may be through an electric motor or an electric fire element; the flow may be used to heat a filament to white heat and thus produce light; alternatively the flow may be through transformers in order to effect an efficient transfer of energy from generating stations to schools, factories, homes and offices. In these applications the electron flow is not carrying any special information, and we are then involved in the field of 'heavy' electrical engineering.

Electronics is the science and technology of controlling electron flows that have been suitably modified so as to convey information. Radio, television, telephone systems and high-fidelity sound reproduction are obvious areas in which electron flows convey information. Computers, control systems and robotics are examples in the field of digital electronics.

Waveforms

Information is impressed upon electron flows by varying the size, i.e. magnitude, of the currents. The usual way of doing this is by varying the electric pressure responsible for the movement of electrons through the system. The electric pressure is known as the voltage or potential difference. It is important for the beginner to understand what is meant by the term 'voltage drop'. The voltage, or potential difference, between two points in a circuit is a measure of the amount of work done in moving electric charge from one of the points to the other. Usually we define the potential at one point in a given circuit as being zero; other points in the circuit have potentials different from zero, provided the circuit is energized. The difference in potential between a specified point and the reference point is the voltage between the two points; it is this voltage that moves, or attempts to move, electrons between the two points. The voltage is a measure of the strength of electric field that causes the electrons to drift through the circuit, thus establishing an electric current. (Electrons can also be 'swept' along a conductor by means of a changing magnetic field, as, for example, in a transformer or dynamo. At this stage, however, we shall not consider this method of producing an electric current.)

In order to impress information upon the electron flow we must in some way alter the instantaneous magnitude of the currents with time. The usual way of doing this is to apply a time-varying voltage to the system. Graphs depicting the way in which the currents and voltages vary with time are known as waveforms. Figures 2.1 and 2.2 show some of the waveforms that are commonly encountered in electronic apparatus. Perhaps the most important of these is the sine wave. This is because all periodic waves can be synthesized, i.e. built up, by combining sine waves of differing amplitudes, frequencies, and phase relationships. 'Amplitude', 'frequency' and 'period' are three of the most commonly used terms associated with sine waves.

A sinusoidal voltage is one which varies with a sine wave pattern. The voltage exists across two points in the circuit. If one of the points is maintained at a fixed reference potential – say by connecting it to earth (which is defined as being at zero potential) – then the potential of the other point is continually changing, first in a positive sense and then in a negative sense relative to the earth potential. The maximum difference in potential (i.e. the maximum voltage between the two points) is known as the *amplitude* of the voltage. The *periodic time* (τ) is the time taken for the voltage to change through a complete cycle from zero to its maximum positive value, down through zero to the maximum negative value, and then back again to zero. The number of cycles completed during 1 second is called the frequency of the waveform. Thus if a voltage undergoes 50 complete cycles in a second (i.e. the sinusoidal pattern is repeated 50 times per second) the *frequency*, f is said to be 50 Hz (Hz = hertz, i.e. cycles per second). One cycle is completed in $1/50$ of a second. The periodic time of the wave is therefore $1/50$ second, i.e. 20 ms (1 s = 1000 ms). We see therefore that $f = 1/\tau$.

The signals found in electronic equipment are not often sinusoidal, but they are frequently periodic, i.e. they have a regular repetitive pattern. The value of studying the response of circuits to sine wave signals lies in the fact that a periodic waveform can be considered as a combination of sine waves.

The square wave of Figure 2.1 is of particular importance in logic and computer work, while other waveforms will be seen to be important in connection with oscillators, cathode-ray oscilloscopes, power supplies and audio amplifiers. These waveforms are shown here not only to introduce the reader to their shapes, but also to define various terms and expressions associated with them.

Figure 2.3 shows some examples of traces from an electrocardiograph. By examining the waveforms, the experienced doctor is able to diagnose heart ailments. We can appreciate here the supreme importance of designing electronic equipment so as to avoid distortion of the waveform. Distortion, if it were present to any significant degree, could easily mask the true position and mislead the doctor when he comes to make his diagnosis. In like manner we can diagnose faults in electronic equipment with the aid of a cathode-ray oscilloscope. By examining the waveforms of the voltages and currents throughout the equipment, and comparing them with the waveforms that ought to be present, we may reach conclusions about the performance of the equipment. Obviously the traces on the face of the cathode-ray tube must be a true representation of the waveform of the signal presented to the oscilloscope's input terminal; if unsuspected distortions were to be

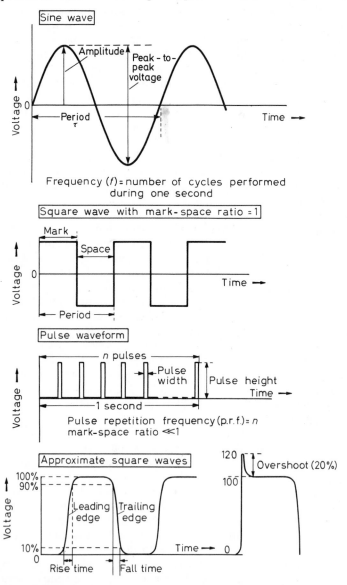

Figure 2.1 Some common waveforms with associated terms

introduced because of poorly designed electronic circuits within the oscilloscope, incorrect conclusions would be reached. We see, therefore, from these two examples alone, why it is so important for the electronics engineer to have a sound knowledge of the behaviour of his circuits, and the effect such circuits have on the

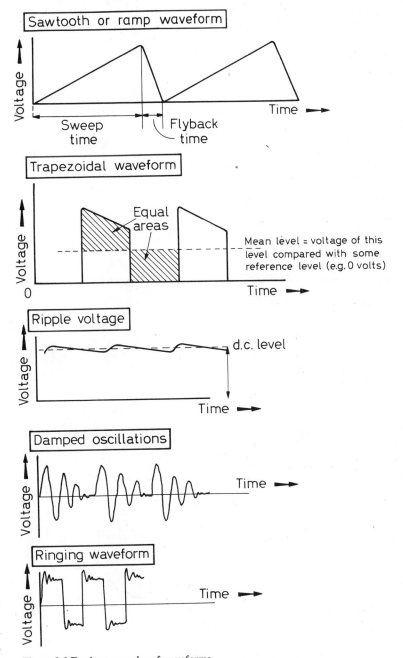

Figure 2.2 Further examples of waveforms

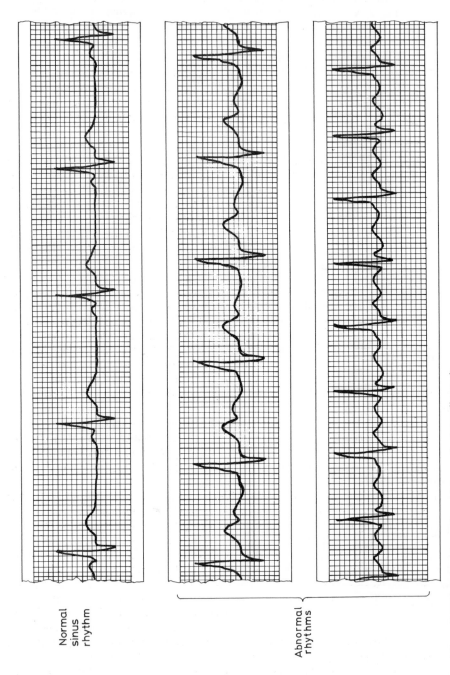

Normal
sinus
rhythm

Abnormal
rhythms

Figure 2.3 Examples of waveforms from an electrocardiograph

information being processed. He must develop the necessary mathematical skills to enable him to analyse in a quantitative way the circuits he designs and uses.

The basic electronic system

Human beings are very restricted in what they can do in the physical world. Their strength and stamina are not very great and their senses are limited. We cannot, for example, detect electromagnetic radiation except over a very small range of wavelengths. Our eyes respond only to that narrow range which we define as visible light; we cannot detect with our eyes ultraviolet or infra-red radiation. Our skin can detect ultraviolet light (and may become sunburnt) or is conscious of infra-red radiation as heat, but the range of detection is extremely small compared with the whole of the electromagnetic spectrum. As a measuring instrument the human body is useless. Nevertheless, we find it necessary in our technological world to measure and control our enviroment; we need to measure temperature, light intensity, strain in structures and radiation levels; we need to sail ships, fly aeroplanes, point guns at our enemies, measure the pH of liquids and so on. Since it is physically impossible to perform these tasks unaided we must invent machines to help us. The incorporation of electronics into these machines has led to the development of very sophisticated apparatus. In all cases the problem associated with these tasks is basically the same. We need some device to move the large gun platform, to control the welding robots, to indicate the temperature, to control the position of a rudder on a large ship, to give us our position at sea during fog, and so on. We also need some detector that can respond to the physical quantities involved (such as temperature, acceleration, position, light intensity, mechanical strain, etc.). Rarely will such detectors be able to supply enough power to operate the working devices directly. It is the role of electronic circuitry to take the output information from the detectors and then supply sufficient power to activate the output devices in the required way. Since electronics is involved, the detectors must be able to convert the physical quantities into an equivalent electrical signal. The signal information is then processed by the electronic circuitry. Detectors able to perform the conversions are called transducers. Microphones, gramophone pickups, strain gauges, thermocouples and light-dependent resistors are common examples of transducers.

Figure 2.4 shows the position in diagrammatic form. The electronic apparatus consists of an assembly of electronic components interconnected to produce a suitable circuit. The electronic components can be divided into active devices and passive devices. Transistors and allied semiconductor devices are classified as active devices because they modify the power supplied to them. The electric power is, of course, supplied from an external source, such as batteries or the electric mains supply. Such power can be modified to produce electrical oscillations or to amplify small signals. Although transistors are the agents by which such modifications are made, they do not actually supply power to the circuit; in spite of this, we often find it convenient in analysing circuits to regard such devices as doing so, while bearing in mind that they are in fact energized from a power supply. Associated with transistors are those components that consume power or otherwise control the flow of energy in some way. These components (resistors, capacitors and inductors) are regarded as passive elements.

The signals that come from the transducers correspond (i.e. are analogous) to the

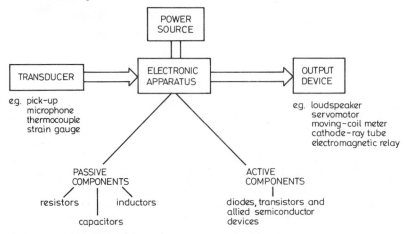

Figure 2.4 The basic electronic system

physical phenomena giving rise to the varying voltage and currents. The corresponding waveforms show a continuous variation with time. If the waveforms (albeit in amplified form) are preserved throughout the system up to the output device stage then we are said to be using analogue circuitry.

Since the processing and storing of information is so conveniently performed by digital computers, nowadays we may convert the analogue signal from the transducer into digital form with an analogue-to-digital converter. The output voltages from the converter are in on/off form, and the subsequent circuitry must be suitably designed to take account of this; such circuitry is said to be digital.

Figure 2.4 shows the electronic apparatus being represented by a rectangle, having an input port (i.e. terminals) into which the transducer signal is fed, and an output port which feeds power to the output device. Within the rectangle the circuitry may be divided into the initial processing stages and the later power stages. The initial stages are required to handle signals of only small amplitude, and analyses of the circuit performance are usually made with the aid of equivalent circuits and general network theory. We shall proceed to discuss these sections of the electronic apparatus in some detail, commencing with a brief description of passive electronic components and the associated network theory.

Resistors

Resistors are used to provide specific paths for electronic currents and to serve as circuit elements that limit currents to some desired value. They provide a means of producing voltages, as, for example, in a voltage amplifier; here variations of transistor current produce varying voltages across a resistor placed in series with the transistor. Resistors are also used in connection with capacitors to build filters and networks of various kinds. Used as feedback elements in connection with integrated circuits they set the gain of an amplifier to some specified value.

Figure 2.5(a) shows the construction and circuit symbols of fixed-value resistors, i.e. resistors whose values are not mechanically adjustable. General purpose resistors are almost invariably made of carbon composition. Resistance values in the range $1\,\Omega$ to $10\,M\Omega$ (megaohms = millions of ohms) are readily available with

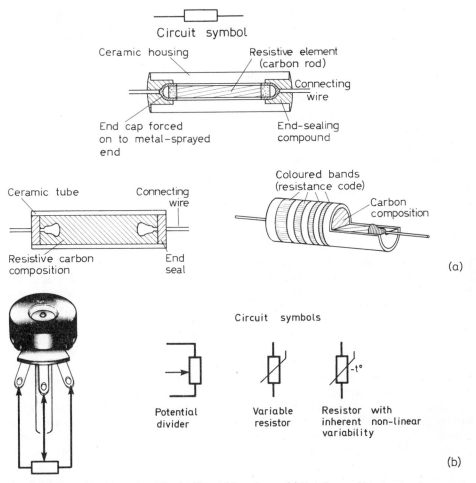

Figure 2.5 Physical appearance of fixed and variable resistors: (a) Two forms of insulated carbon composition resistors; (b) the potential divider: a variable resistor is formed by connecting the slider tag to one of the tags making contact with the end of the track

power ratings of ¼ W to 2 W. The power rating of the resistor is largely determined by the physical size. If this power rating is exceeded the component overheats and is destroyed or performs in an unreliable way.

Heat generation is evidence of the consumption of energy. Work is done in forcing an electric charge through a potential difference. The work done, J, is equal to the potential difference, V, across the ends of the resistor times the charge forced through the component. Since $i = dq/dt$ then the total charge, $Q = \int i\,dt = It$ (for constant current) and $J = VIt$. If volts, amperes and seconds are involved then the work is in joules. Now J/t is the rate of doing work (i.e. the power P) therefore $P = VI$. Since Ohm's Law is operating, $P = VI = V^2/R = I^2R$. The unit of power is the watt.

The manufacture of resistors to cover a whole range is most efficiently achieved by using a system of resistor values based on a logarithmic scale. The ranges are then known as preferred values (see Appendix 1). In order to improve the electrical performance in respect of stability of resistance with temperature, shelf-life and noise, ordinary carbon elements are replaced by metal oxides or special wires such as nichrome and manganin.

For many purposes it is necessary to be able to alter at will the resistance value in a given circuit, e.g. to control the volume of sound from a radio, or to alter the brightness of a TV picture. For these purposes so-called potentiometers are used. The term is not a good one since the measurement of potential is rarely involved; potential divider would be a better term. General purpose variable resistors are manufactured by spraying carbon suspension onto a plastic strip and curing at high temperature. The strips are then formed into an arc subtending an angle of about 300°. Alternatively the track may be on a disc. Electrical connections are made to the ends of the track and brought out to solder tags. A mechanically adjustable wiper arm is connected to a third tag. By moving the position of the arm it is possible to select a suitable fraction of the total resistance. (See Figure 2.5(b).) Potential dividers in which the slider moves linearly are also available.

Various so-called laws are available. A linear potential divider is one in which the resistive track is uniform throughout its length; equal changes of angular rotation of the arm bring about equal changes of resistance between the slider terminal and one end of the track. Logarithmic, inverse log and semi-log laws are also available. With logarithmic potential dividers, equal angular shaft rotations bring about equal changes in loudness when used in audio equipment. In some scientific applications it is convenient to have sine-wave and cosine-wave laws.

Series and parallel connections

Resistors are said to be in series when they are connected to form a chain. If the ends of the chain are connected to a power source such as a battery, the same current passes through each resistor, i.e. the magnitude of the current is the same at every point in the circuit. A voltage must exist across the ends of each resistor; the sum of the voltages across each resistor in the chain is equal to the voltage applied to the ends of the chain, from which we see that the total resistance presented to the voltage source is given by $R = R_1 + R_2 + R_3$.

Resistors are said to be in parallel when each resistor is connected across the same voltage source. Since the same voltage, V, exists across each resistor, the current drawn by (say) R_1, R_2 and R_3 are respectively V/R_1, V/R_2 and V/R_3 from which the total resistance, R, presented to the power source is given by

$$\frac{1}{R} = \frac{1}{R_1} + \frac{1}{R_2} + \frac{1}{R_3}$$

The effective resistance is always less than the lowest single resistance in the parallel circuit.

Since the use of reciprocals often produces inelegant mathematical expressions, it is neater with parallel circuits to use conductances. Conductance, G, is the inverse of resistance; $G = 1/R$. The unit of conductance is the siemen, S. Thus the above relationship may be simply expressed as $G = G_1 + G_2 + G_3$; $(I = GV)$.

Capacitors

Capacitors are components that have the ability to store electric charge. A capacitor consists of two conductors in close proximity separated by an insulator called the dielectric. If a potential difference, V, is established across the dielectric by connecting the conductors to the terminals of a battery or some other generator of steady voltage, a charge, q, is stored. Doubling the voltage increases the charge by a factor of 2; the ratio of charge to potential difference is constant, i.e.

$$\frac{q}{V} = C \text{ (a constant)}$$

The constant C is known as the capacitance. When q is in coulombs and V in volts, C is in farads. The farad, F, is a very large unit of capacitance for electronic purposes so the microfarad, μF, which is one millionth of a farad, is used. The picofarad (pF) is one-millionth of a microfarad so $1 \text{ pF} = 10^{-12} \text{ F} = 10^{-6} \, \mu\text{F}$. Many types of capacitor have their values colour coded in picofarads. The nanofarad is one-thousandth of a microfarad, i.e. $1 \text{ nF} = 10^{-3} \, \mu\text{F} = 10^{-9} \text{ F}$.
Since

$$\frac{q}{V} = C \text{ i.e. } q = CV$$

$$\frac{dq}{dt} = C\frac{dV}{dt}$$

We have already seen that dq/dt (a rate of change of charge with time) is the current flowing, therefore

$$i = C\frac{dV}{dt}$$

The simple Ohm's Law relationship for resistors cannot therefore be applied to capacitors since the current is proportional, not to the applied voltage but to the rate of change of the voltage. An alternative way of expressing this relationship is

$$V = \frac{1}{C}\int i \, dt$$

The physical appearance and properties of a capacitor vary a good deal depending upon the nature of the 'plates' and the dielectric material. Values of capacitance from 1 pF to several thousands of microfarads are readily available. One classification of capacitors depends upon the dielectric used. Hence we have general purpose paper capacitors, mica capacitors, ceramic, electrolytic, polystyrene and polycarbonate types. The main physical constructional features are shown in Figure 2.6.

Capacitors are used to separate steady voltages from varying ones. Used in connection with other components, electric filters can be designed that will pass signals of only a single frequency or desired band of frequencies from the input to output terminals; capacitors are also used in timing circuits. As with resistors, readers will increasingly appreciate the many functions of capacitors as their knowledge of electronics grows.

The choice of a particular type of capacitor depends upon the function to be performed. The nature of the dielectric is often the criterion used in the choice. Where very low leakage is required, say for a timing circuit, polystyrene capacitors would be chosen. Mylar types are popular and inexpensive, but have poor temperature stability. Electrolytic types are used when high-value capacitance values are needed in physically small capacitors; the leakage is high, however, and

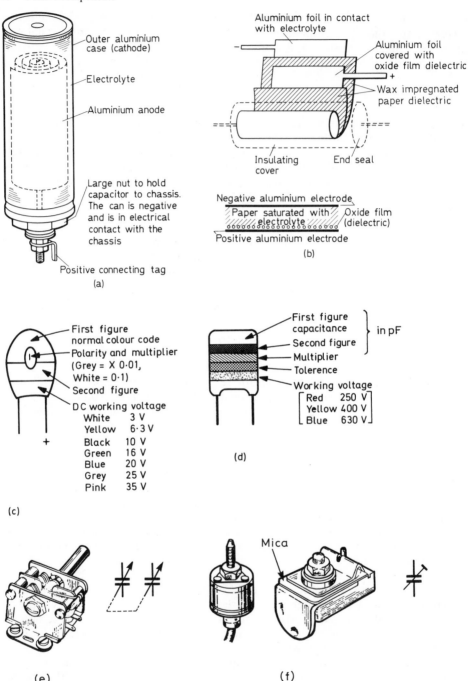

Figure 2.6 Various forms of capacitors: (a) The construction of a 'wet type' of eletrolytic capacitor; (b) Paper capacitor; (c) Tantalum capacitor: with dot facing viewer, positive lead to right; (d) Polyester capacitor; (e) Twin-ganged variable capacitors; (f) Trimmers

their use is usually confined to power supply filter circuits. Being polarized, electrolytic capacitors must be connected into the circuit observing the correct device polarity.

Where the capacitance value is to be adjustable, variable air-spaced capacitors may be used. These types are best known as the tuning agency in radio sets. Where the capacitance value is only to be infrequently adjusted, 'trimmer' types are used (Figure 2.6).

Capacitors in parallel and series

When capacitors are connected in parallel, the same voltage exists across each capacitor. Each capacitor therefore stores a charge proportional to the capacitance. The total charge is the sum of all the charges. The parallel arrangement therefore has a capacitance which is equal to the sum of the individual capacitances, i.e.

$$C = C_1 + C_2 + C_3, \text{ etc.}$$

Connecting capacitors in series results in a reduction of the total effective capacitance to a value less than that of the smallest capacitance in the chain. The total effective capacitance, C, can be found by considering Figure 2.7. If a voltage, V, is applied to a series of capacitors that are originally uncharged then a distribution of charge takes place. The total charge on two interconnected plates,

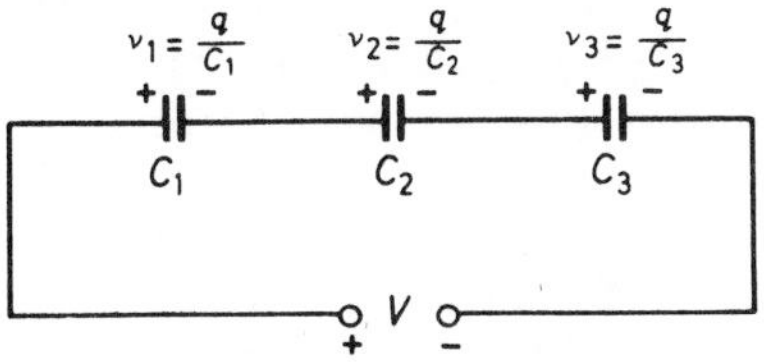

$$v_1 = \frac{q}{C_1} \qquad v_2 = \frac{q}{C_2} \qquad v_3 = \frac{q}{C_3}$$

$$V = v_1 + v_2 + v_3 = q\left(\frac{1}{C_1} + \frac{1}{C_2} + \frac{1}{C_3}\right)$$

$$\therefore \frac{V}{q} = \frac{1}{C} = \frac{1}{C_1} + \frac{1}{C_2} + \frac{1}{C_3}$$

Figure 2.7 Capacitors in series

however, is zero; the electric field causes electrons to leave one plate, and hence that plate becomes positively charged. The plate to which the electrons are attracted becomes negatively charged. The two charges are equal in magnitude, but are of opposite sign. The magnitude of the charge on each capacitor in the chain is therefore the same, say q. The voltages across the respective capacitors, must therefore be q/C_1, q/C_2, etc. The sum of the voltage is V, therefore

$$V = \frac{q}{C_1} + \frac{q}{C_2} + \frac{q}{C_3}$$

Since V/q is $1/C$, where C is the effective capacitance, then

$$\frac{V}{q} = \frac{1}{C} = \frac{1}{C_1} + \frac{1}{C_2} + \frac{1}{C_3}$$

Energy stored in a charge capacitor

The process of charging a capacitor involves the expenditure of energy. As the capacitor is being charged, an amount of work, δJ, must be performed in forcing a small charge, δq, on to the capacitor plates against the voltage v that exists across the capacitor terminals connected to the plates, i.e. $\delta J = v\delta q$. Unlike the process in a resistor, however, the energy is not lost as heat. The energy is stored as electric potential energy in the electric field that exists across the dielectric. Such energy can be recovered during a discharge period. A perfect capacitor does not therefore dissipate energy. The total energy, J_c, stored is measured in terms of the sum of all the increments of work performed during the charging process, i.e.

$$J_c = \int_0^J \mathrm{d}J = \int_0^Q v\mathrm{d}q = \int_0^Q \frac{q}{C}\mathrm{d}q = \frac{Q^2}{2C}$$

Since $Q = CV$, $J_c = QV/2 = CV^2/2$

Inductors

Inductance is that property of an electrical circuit which opposes any change of current in that circuit. Devices having the primary function of introducing inductance into a circuit are called inductors. Inductors usually consist of coils of insulated wire wound onto a suitable bobbin or former. Although the core of the coil may be air, it is more usual to concentrate the magnetic flux in the core by using suitable ferromagnetic substances; magnetically soft iron and ferrites are used. Since inductors oppose changes of current, one of their functions in electronic circuits is to provide a large opposition to the flow of alternating current while simultaneously presenting very little opposition to the flow of steady currents. A common example of this is found in d.c. power supplies.

To understand the way in which inductors present an opposition to alternating currents, but not to direct current, we must recall our ideas on electromagnetic induction. It is not possible here to go into any great detail; standard books on electricity and magnetism are the works to consult.

An electric current flowing through a conductor has a magnetic field associated with it. It is usual to visualize the magnetic field by inventing magnetic flux lines, i.e. the paths that would be taken by a fictitious isolated north pole. Such flux lines are shown in Figure 2.8. Whereas steady currents produce magnetic fields, Faraday

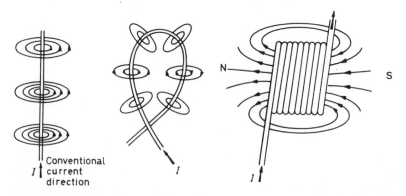

Figure 2.8 Lines of force representing magnetic fields around current-carrying wires in which the current is steady. Steady magnetic fields associated with conductors do not, however, induce currents in the wires. To induce a current, the field must be varying

showed that the converse was not true; to induce a current in a wire, it is necessary for the magnetic field to vary. Consider now the coil represented in Figure 2.8. If no current exists in the coil, and by some means a magnetic field is made to grow in the coil (e.g. by plunging a bar magnet into the coil), an e.m.f. will be induced. The magnitude of the e.m.f. is given by

$$e = -\frac{\mathrm{d}N}{\mathrm{d}t}$$

where N is the magnetic flux and e is the e.m.f., i.e. the e.m.f. is equal to the rate of change of the magnetic flux.

The minus sign arises because the direction of the induced e.m.f. is such as to oppose the change producing it (Lenz's Law). This means that if an external circuit exists, currents will be produced in the coil which themselves give rise to a magnetic field; this latter field is opposite in sense to the magnetic field inducing the current. Instead of plunging a magnet into the coil let us apply a voltage to the ends of the coil. If the voltage is steady the only opposition to current is the resistance of the conductor from which the coil is made. If the voltage is varying, however, there will be a continual change in the current which in turn produces a changing magnetic field. As the magnetic field grows, an increasing magnetic flux is established within the coil. The rate of increase depends upon the frequency of the applied alternating voltage. Such a changing magnetic field induces an e.m.f. in the coil that opposes the applied e.m.f.; there is thus some opposition to the establishment of the magnetic field. When the applied e.m.f. is diminishing the magnetic field collapses. This collapsing field attempts to keep the current at its maximum value. Whether the applied e.m.f. is rising or falling, therefore, the inductor opposes changes of current in the circuit. For a given inductor the opposition increases as the frequency of the applied e.m.f. increases. In a sense, the inductor may be said to have 'electrical inertia'.

The back e.m.f. induced in the coil is proportional to the rate of change of current, therefore

$$e \propto -\frac{\mathrm{d}i}{\mathrm{d}t}$$

where e is the back e.m.f. and i the instantaneous value of the current. Hence

$$e = -L\frac{\mathrm{d}i}{\mathrm{d}t}$$

where L is the constant of proportionality known as the self-inductance of the coil. When i is in amperes, t is in seconds and e is in volts, L is in henries (H). Thus if the current is changing in a coil of inductance 1 H at the rate of 1 A s^{-1} then the back e.m.f. produced is 1 V. Like the capacitor, no simple relationship exists between the applied voltage and the current; the implications of this are discussed later. We shall also discuss the combination of capacitors and inductors to form tuned or resonant circuits. Such tuned circuits enable a signal of one frequency to be isolated from other signals of different frequencies. Radio and television sets use many such circuits. Used in this way inductors must be designed for operation at frquencies much higher than 50 Hz. The losses associated with iron cores are too great to make operation at radio frequencies efficient; the cores of high-frequency inductors are therefore made from ferrites, or iron-dust held in a suitable binding material.

Ferrites are chemical compounds of the form MFe_2O_4 where M is a divalent metal, commonly Mg, Mn, Zn or Ni. This non-metallic material combines reasonable permeability (from several hundreds up to about 1200) with high resistivity. Eddy currents at high frequencies are therefore largely avoided.

Inductors for filter and other purposes are now based on what are called pot-cores (see Figure 2.9). Here the ferrites are cast as cylinders closed at one end. The coil of wire, wound on a suitable former, is placed within two of these cylinders so as to be completely shielded. The magnetic circuit is completed by a central cylindrical core. For a given inductor some variation of inductance is possible by adjusting the position of the central core or by varying the pressure at which the two ferrite cylinders are held together.

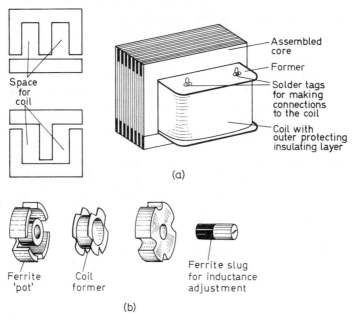

Figure 2.9 Physical appearance of two forms of inductor: (a) Construction of an inductor for power frequencies. The shape of the laminations is shown on the left. Either combination may be used; (b) Essential parts of a 'pot-core' inductor

Energy storage in inductors

The process of establishing a current and associated magnetic field in an inductor involves the expenditure of energy. The total energy stored, J_L, is measured by summing all the increments of work performed during the process of establishing the field, i.e.

$$J_L = \int vi\,dt = \int L\frac{di}{dt}i\,dt = \int_0^I Li\,di$$

$$= \frac{1}{2}LI^2$$

(v is the instantaneous applied voltage, i.e. $L di/dt$.)

As the magnetic field collapses the energy is returned to the source. In a perfect inductor no energy is lost; like the capacitor, an inductor does not dissipate energy.

Mutual inductance

If two coils are placed close to each other so that a varying magnetic field in one coil induces an e.m.f. in the second coil, the two coils are said to be inductively coupled. The changing magnetic flux due to the current in the first or primary circuit must interlink the secondary circuit in order to induce an e.m.f. in the secondary coil. The phenomenon is called mutual induction. The mutual inductance, M, between the two coils is measured in henries and depends upon (a) the number of turns in the primary coil, (b) the number of turns in the secondary coil, (c) the relative position of the coils, and (d) the permeability of the medium between the coils.

$$M = \frac{\text{Induced voltage } (e) \text{ in the secondary coil}}{\text{Rate of change of current in the primary}}$$

When inductors are connected in series the total inductance is calculated in the same manner as that used for resistors in series. Since each inductor contributes to the opposition to change of current the total inductance, L, is equal to the sum of the individual inductors.

$$L = L_1 + L_2 + L_3 \ldots$$

If, however, mutual inductance exists between the individual coils, then the total inductance will depend upon the relative connections of the coils. If the magnetic fields are mutually assisting then the total inductance is given by $L = L_1 + L_2 + 2M$; whereas if the fields are opposing $L = L_1 + L_2 - 2M$.

Inductors in parallel can be represented by a total inductance, L, given by

$$\frac{1}{L} = \frac{1}{L_1} + \frac{1}{L_2} + \frac{1}{L_3}$$

provided no magnetic coupling exists between coils.

Chapter 3

Basic circuit theory

In the past, those studying electronics have found circuit theory to be one of the less attractive aspects of the subject. This has been largely because many tedious algebraic manipulations and arithmetic calculations were involved. Nowadays all colleges and universities are well-equipped with computers, and many students possess their own microcomputers. Much of the tedium of calculation can therefore be avoided by using readily available software packages. Many of the solutions to problems posed later in this chapter, and elsewhere in the book, have been obtained with the aid of a BBC Model B computer. In the field of scientific computers the BBC instrument is very modest; nevertheless it is sufficiently powerful to tackle many of the problems commonly encountered by students embarking on their studies of electronic circuits. Those wishing to write their own programs cannot, of course, do so unless they understand the basic principles of the circuit network analyses.

Electronic circuits can be analysed in one of three ways:

(a) The actual circuit can often be represented by an equivalent circuit that is easier to analyse. The equivalence always depends upon some simplifying assumptions. The individual components are assumed to be linear in their operation. Thus straight-line graphs are obtained when we plot applied voltage against current for resistors, applied voltage against rate of change of current for an inductor, and current against rate of change of applied voltage for a capacitor. The components are assumed to be operating within their specified ratings. The equivalent circuit represents the behaviour of the real circuit only in respect of the signal information; all power supplies and biasing circuits, although not shown, are assumed to be present and that the real circuit is properly energized. Often the frequency range of operation must be restricted to make the analyses valid.

(b) Graphical techniques can be used where small-signal linear parameters are not available. For those devices where the V/I characteristics are not linear, the anaylses are made on graphs of current versus applied voltage; these graphs are called characteristic curves.

(c) Once the theory of operation of devices and circuits is understood we often develop rules of thumb to avoid a lot of theoretical work. Such rules must be soundly based and be the outcome of knowledge and experience.

Throughout this book we shall use all three methods, relying on (c) wherever appropriate.

Foundations

Before discussing in detail the behaviour of analogue circuits that are incorporated into electronic circuits, it will be necessary to revise some basic concepts, *viz.* Ohm's Law, Kirchhoff's Laws, and the theorems of Thévenin and Norton. Initially the discussion will be confined to resistors and time-invariant voltages and currents (i.e. the so-called d.c. cases in which, once the voltages have been switched on, they, and the resulting currents, do not vary with time). To calculate the magnitudes of the currents and voltages that exist in a network we must make use of mathematical equations. This involves an agreed convention that allows us to deduce from the solutions to the equations the physical conditions that exist in the circuit. (A comparable situation arises in choosing a convention when discussing geometric optics.) It is usually the case that less confusion arises in the minds of students if they first establish the equations, using an agreed convention, without thinking about the physical situation within the circuit. Only after the solutions to the equations have been found should the physical conditions be deduced. Examples of this procedure are given later.

Consider Figure 3.1 in which a resistor R is being supplied with current from a battery in which the internal resistance is r and the e.m.f. is E. The current, I (measured by a centre-zero ammeter) is represented by an arrow. This arrow is a mathematical one and is conventionally drawn in a clockwise direction around the loop; the arrow merely enables us to determine the sign ($+$ or $-$) placed in front of the expression in an equation. If, on solving the equations, I turns out to be negative, then the physical (conventional) current will be flowing in the opposite

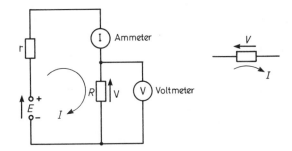

Figure 3.1 Circuit diagram used to establish the mathematical conventions used in circuit analyses

direction to that shown by the mathematical arrow; the needle of the ammeter is then deflected to the left-hand side of zero. For mathematically positive values of I, the conventional physical current flows in the same direction as the mathematical arrow, and the needle is deflected to the right. Voltage polarities are shown in circuit diagrams as arrows next to the component. For mathematical purposes the head of the arrow is positive with respect to the tail. If the solutions to the equations yield a positive value for the voltage, V say, then the arrow on the diagram will indicate correctly the polarity of the physical voltage across the component. Conversely, negative values for V mean that the physical voltage has a polarity in the opposite sense to that indicated by the arrow.

Mesh and nodal analyses

These analyses are based on Kirchhoff's Laws and form a very useful basis for writing computer software. Kirchhoff's Laws are:

(1) The algebraic sum of the currents at a junction point in an electrical network is zero.
(2) The algebraic sum of the potential differences across each circuit component in a closed loop, or mesh, is zero.

The first law follows from the fact that charge is not stored at a junction point in the circuit. Such a junction point is often called a node. All the currents approaching the point must, therefore, leave the point (Figure 3.2).

The second law follows from the fact that the work involved in moving an electric charge from a given point in a closed loop completely around the loop back to the given point must be zero. If this were not so the potential of the given point would

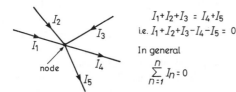

$$I_1+I_2+I_3 = I_4+I_5$$
i.e. $I_1+I_2+I_3-I_4-I_5 = 0$

In general

$$\sum_{n=1}^{n} I_n = 0$$

Figure 3.2 If n current-carrying paths are connected to a node, the algebraic sum of all the currents flowing in the paths is zero

not be unique. Changes in potential around the loop may be due to the voltages across impedances, or the e.m.f.s generated by sources such as batteries. (Since we have not yet discussed capacitors, inductors or the term impedance we shall for the present confine our statements to resistances.) The voltages developed across resistances in a network are the products of current and resistance, viz. IR. It will be seen, therefore, that around any closed loop of an electrical network, the rises in potential due to the generators must equal the falls of potential (IR) across the resistors.

To establish the equation relating to the voltages and e.m.f.s around a closed loop we may refer to Figure 3.3. Initally we consider only a mathematical situation,

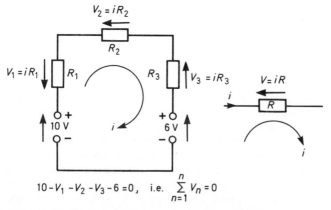

$$10-V_1-V_2-V_3-6 =0, \quad \text{i.e.} \quad \sum_{n=1}^{n} V_n = 0$$

Figure 3.3 Establishing the equation relating the voltages around a closed loop from Kirchhoff's Second Law

temporarily ignoring the physics of the situation. We assume initially that the conventional current, I, is flowing in a clockwise direction. This will then establish mathematically the direction of the 'voltage' arrows and hence the assumed polarity of the voltages across the resistors.

When solving problems the reader should bear in mind two rules: first, when establishing an equation for the sum of the voltages around a closed loop we assume that we move around the loop in the direction of the current arrow. Voltages are counted as being positive if we travel in the direction of a 'voltage' arrow, i.e. from the tail to the head of the arrow. If we travel from the head to the tail of the voltage arrow then the voltage is given a negative sign. Secondly, when a current is passing through a resistor the sense of the voltage produced across the resistor is in the opposite direction to that of the conventional current flow. The position may be made clear by considering the following example:

Example 3.1 Let us be required to find V_3 in Figure 3.4. Once again the arrows are drawn in directions dictated by the convention we have agreed to use. To find V_3 we need first to determine the current I.

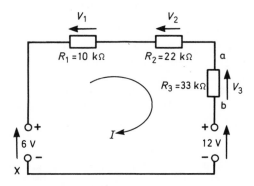

Figure 3.4 Example of finding voltages and currents in a circuit

Starting at the point X and 'travelling' clockwise in the direction of I

$$+6 - IR_1 - IR_2 - IR_3 - 12 = 0$$

Therefore

$$- I (10 + 22 + 33) = 6$$

and

$$I = -\frac{6}{65} = -0.092 \, \text{mA}$$

(It should be noted that when voltages are in volts and resistances are in kilohms then currents are in milliamps.)

Therefore

$$V_3 = IR_3 = -0.092 \times 33$$
$$= -3.04 \, \text{V}$$

Hence the voltage across R_3 is 3.04 V, but the end b is positive with respect to the end a; also the conventional current flows in the opposite direction to that indicated for I in Figure 3.4.

When the network consists of two or more loops the same procedure is adopted, and a set of simultaneous equations is obtained.

Example 3.2. Find the power dissipated in the 10 kΩ resistor of the circuit of Figure 3.5.

In this problem, if we find the current through the resistor then the power is calculated from the I^2R formula from Ohm's Law.

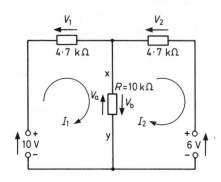

Figure 3.5 Circuit for the problem given in example 3.2, in which the power dissipated in the 10 kΩ resistor is to be calculated

A slight complication exists here because the 10 kΩ resistor is common to both loops. This, however, does not present us with any difficulty. The net current through this resistor is $I_1 - I_2$. If this quantity turns out to be positive then the conventional current flows in the direction of I_1; if $I_2 > I_1$ then $I_1 - I_2$ is negative and the direction of conventional current is that of I_2. If I_2 itself were to be negative than the net current through R would be $I_1 - (-I_2)$, i.e. $I_1 + I_2$. The direction of I_2 would be opposite to that shown for I_2 in the diagram, and the net current through R would be in the direction of I_1 (and I_2). The voltage across R due to I_1, say V_a, will have a sense opposite to that of I_1, and similarly the voltage due to I_2, V_b, will be opposite in sense to I_2. The net voltage across R is $V_a - V_b$. If this quantity is positive then the polarity of the actual voltage across R is such that the end x is positive with respect to the end y.

To solve the problem we must first establish the appropriate equations. For the left-hand loop (around which I_1 is travelling in a clockwise direction)

$$+10 - V_1 - V_a + V_b = 0$$

For the right-hand loop

$$-V_b + V_a - V_2 - 6 = 0$$

Therefore $10 - I_1 (4.7 + 10) + I_2 10 = 0$
and $-I_2 10 + I_1 10 - I_2 4.7 - 6 = 0$

Therefore $\left. \begin{array}{l} 10 - 14.7I_1 + 10I_2 = 0 \\ -6 + 10I_1 - 14.7I_2 = 0 \end{array} \right\}$
and

By using the usual algebraic techniques we obtain $I_1 = 0.75$ mA and $I_2 = 0.1$ mA. Both are positive, therefore the flow directions of I_1 and I_2 are as shown by the original arrows. The polarities of the voltages V_1 and V_2 are also as originally assumed.

The actual current through the 10 kΩ resistor is $I_1 - I_2$, *viz.* 0.75 −0.1, i.e. 0.65 mA. The voltage across this resistor is therefore 6.5V with the end x being positive with respect to the end y. The power dissipated in the $10\,k\Omega$ resistor is $(0.65 \times 10^{-3})^2 \times 10 \times 10^3 = 0.0042\,W$, i.e. $4.2\,mW$.

Nodal analysis depends upon the Kirchhoff's Law that states that the algebraic sum of the currents at a point is zero. Here we still rely on Ohm's Law too, but instead of using the relationship $V = IR$ it is preferable to use the form $I = VG$ where G is the conductance equal to $1/R$. (Later we shall see how to generalize the expression in order to accommodate capacitors and inductors; then $I = VY$ where Y is the admittance equal to $1/Z$, Z being the impedance.) Figure 3.6 indicates the

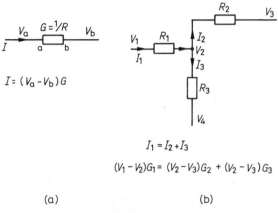

$$I_1 = I_2 + I_3$$

$$(V_1 - V_2)G_1 = (V_2 - V_3)G_2 + (V_2 - V_3)G_3$$

(a)   (b)

Figure 3.6 Conventions for establishing equations in nodal analysis

way in which linear equations can be established. The mathematical arrow representing current can be drawn in any arbitrary direction, but once the direction is chosen, the voltage difference must be expressed as shown in Figure 3.6(a). If the potential at the end a of the component is V_a (relative to some reference line voltage) and that at the other end, b, is V_b, then $I = (V_a - V_b)G$ if the current arrow is drawn in a direction from a to b; if the arrow is reversed in direction then $I = (V_b - V_a)G$. We shall find this a useful convention when, for example, we come to analyse active filters.

The term 'nodal' analysis arises because the algebraic sum of the currents at a given point is zero; the point is therefore called a node.

The Wheatstone bridge

Figure 3.7 shows a circuit diagram of four resistors in an arrangement commonly known as the Wheatstone bridge. The network is one of the most important since, when used in conjunction with an integrated circuit operational amplifier, the combination has extensive applications in the field of analogue measurement. Readers are probably aware that the bridge has long been used in the measurement of resistance values; when the resistors are replaced by capacitors and/or inductors the resulting a.c. bridges are used to measure capacitance and inductance. In instrumentation and control, replacing one of the resistors with a suitable resistive transducer (e.g. a thermistor, strain gauge or light-dependent resistor) results in a very versatile measuring arrangement. From Figure 3.7(a) we can see that the

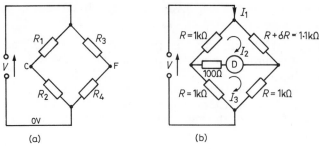

Figure 3.7 The Wheatstone bridge; (b) shows an out-of-balance condition. Establishing a set of simultaneous equations for I_1, I_2, and I_3 enables a calculation of the out-of-balance detector current to be made when δR is not small compared with R

potential at C (relative to the 0 V line) is $R_2 / (R_1 + R_2)$ of the supply voltage, V; similarly the potential at F is $R_4/(R_3 + R_4)$ of V. When these two potentials are the same, a detector connected between C and F will register a null reading; the bridge is then said to be balanced. This occurs when $R_2/(R_1 + R_2) = R_4/(R_3 + R_4)$ whence $R_1/R_2 = R_3/R_4$.

In the measurement of temperature, light intensity or strain, the resistance of the appropriate transducer is not generally such as to bring about the balanced condition. Usually the bridge is so designed that balance occurs at only a single value of physical condition, this value being somewhere near the centre of the range of values likely to be measured. At this mid-value let the resistance of the transducer be R. It is usual to make the resistance of the other three resistors equal to R also. Provided the change in transducer resistance is small compared with R, then the out-of-balance voltage, V_{OUT}, can be calculated as follows:

Suppose R_3 of Figure 3.7 to be replaced by a transducer whose resistance is $R \pm \delta R$ over the range of change in the physical phenomenon. The potential at C is $R_2(R_1 + R_2)$ of V; since $R_1 = R_2 = R$, this potential is $V/2$. At the point F the potential is $R/(2R \pm \delta R)$ of V. The potential difference between C and F is therefore given by

$$V_{OUT} = \frac{V}{2} - \frac{VR}{2R \pm \delta R} = \frac{V}{2}\left(1 - \frac{1}{1 \pm \delta R/(2R)}\right)$$

$$= \frac{V}{2}\left[1 - \left(1 \pm \frac{\delta R}{2R}\right)^{-1}\right]$$

$$= \frac{V}{2}\left[1 - 1 \mp \frac{\delta R}{2R}\right] \text{ when } \delta R \ll 2R$$

$$\doteqdot \pm \frac{V\delta R}{4R}$$

When the change in resistance is too large to permit making the above approximation we could use mesh analysis to establish the necessary equations. As can be seen from Figure 3.7(b) several loops may be identified, and these give rise to a somewhat complicated set of linear simultaneous equations from which the circuit currents and voltages can be calculated. Calculating the out-of-balance current using simultaneous equations can be a tedious process in which it is easy to make careless mistakes. It would be a great advantage if we could find an easier method for solving our problem. Thévenin's theorem is such a method, in which the Wheatstone bridge problem can be solved almost by mental arithmetic. The theorem is, of course, suitable for solving problems associated with many other types of network. The proof of Thévenin's Theorem can be established using the Superposition Theorem. This latter theorem can often prove useful in the analysis of electronic circuits.

The Superposition Theorem

This theorem states that in any linear network of impedances and generators, the current in one branch of the network is equal to the sum of the individual currents in that branch due to each generator taken one at a time with all other generators replaced by their internal impedances. (The term 'impedance' has been used even though a.c. circuits have not yet been discussed. Impedance is a measure of the opposition to current in circuits in which alternating voltages are involved. At this stage the reader may substitute 'resistance' for 'impedance' since we are dealing at this point in the book only with direct voltages and currents.)

In appropriate cases the Superposition Theorem enables the analyst to avoid the tedious processes associated with finding solutions to a set of simultaneous equations. Take, for example, the problem posed in connection with Figure 3.5. The current in the $10\,k\Omega$ resistor may be found by summing the individual currents supplied by the 10 V and 6 V sources separately. With the 6 V source removed and replaced by a short-circuiting link (assuming that is internal impedance is zero), the total current supplied by the 10 V source is 1.27 mA. This current divides in the ratio 4.7:10, the current in the $10\,k\Omega$ resistor being 0.41 mA. Similarly, the separate current from the 6 V supply will be found to be 0.24 mA. Hence with both supplies operating simultaneously the Superposition Theorem tells us that the current in the $10\,k\Omega$ resistor is 0.65 mA.

Thévenin's Theorem

Thévenin's theorem gives us a procedure for establishing a simple equivalent circuit for a complicated network. The theorem states that any network of linear impedances and voltage generators can be replaced for circuit analyses purposes by a simple equivalent circuit consisting of a single voltage source in series with an impedance, as shown in Figure 3.8(a). $V_{o/c}$ is the voltage that exists across the terminals a, b of the network when no load is connected; Z is the impedance 'looking into' the terminals of the network when all the generators have been reduced to zero and replaced by their internal impedances. The current in any load impedance connected to the terminals a, b is therefore the same as it would be if the load were connected to a single voltage generator whose e.m.f. is equal to the open-circuit voltage across a, b and whose internal impedance is the impedance of the network looking into the terminals a, b with all generator voltages reduced to zero and the generators replaced by their internal impedances.

The validity of Thévenin's Theorem can be appreciated by considering Figure 3.8(b) in which a linear network containing generators is connected to a load impedance, Z_L. A voltage generator, with zero internal impedance, is introduced into the load circuit and connected as shown. Its voltage is then adjusted to bring

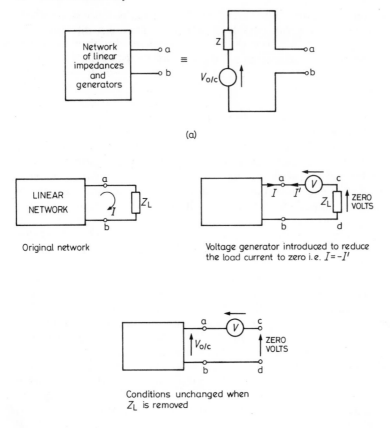

(a)

Original network

Voltage generator introduced to reduce
the load current to zero i.e. $I = -I'$

Conditions unchanged when
Z_L is removed

(b)

Figure 3.8 (a) The Thévenin equivalent circuit; (b) Circuits for proving the validity of Thévenin's Theorem

the load current to zero at which time the voltage is V. Using the Superposition Theorem it is seen that V acting alone must produce a current, I', equal and opposite to I, the current that would flow in the load if V were absent. Thus if Z is the impedance 'looking into' the network when all of the generators are replaced by their internal impedances, then

$$I' = \frac{V}{Z + Z_L} = -I$$

Under these circumstances, since the net current through the load has been reduced to zero, the load may be temporarily removed; thus $V = V_{o/c}$, the open-circuit voltage. These are exactly the results that would be obtained if the actual network were replaced by the equivalent circuit of Figure 3.8(a).

Example 3.3. Let us see how we may apply Thévenin's theorem to the problem of finding the out-of-balance current through the detector, D, of Figure 3.7. We imagine initially that the load (a detector having a resistance of $100\,\Omega$) is removed.

Our first task is to find the voltage across a,b, *viz.* $V_{o/c}$, in the absence of this load. Taking the potential at c as being zero, the potential at d must be 10 V. The potential at a must therefore be 5 V, and that at b must be 1.1/2.1 of 10 V, i.e. 5.24 V. The potential difference across a,b is therefore 0.24 V, which is $V_{o/c}$. Z in this case is the resistance between a and b when the 10V generator (assumed to have zero internal resistance) is shorted-out, i.e. the generator voltage is reduced to zero. Figure 3.9 shows how the network can be redrawn to make the calculation of Z obvious. The two 1 kΩ resistors in parallel are equivalent to 500 Ω, and 1 kΩ in parallel with 1.1 kΩ is 524 Ω; therefore $Z = 1024$ Ω. When we connect the load (100 Ω) to the Thévenin voltage equivalent circuit we see that the current flowing through the detector will be 0.24/1124 A, i.e. 0.21 mA. Readers are invited to compare this method with that which involves solving a set of simultaneous equations.

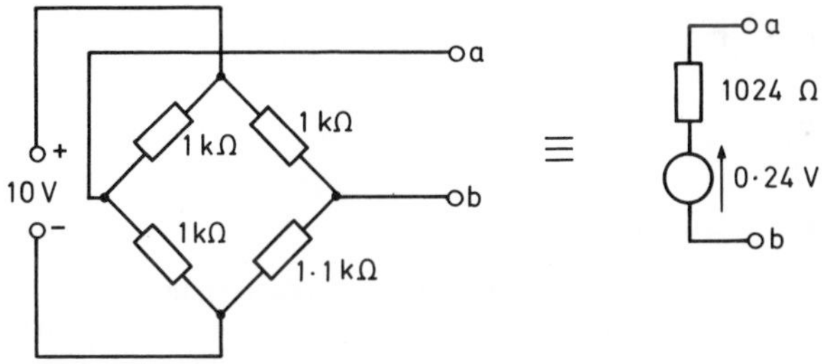

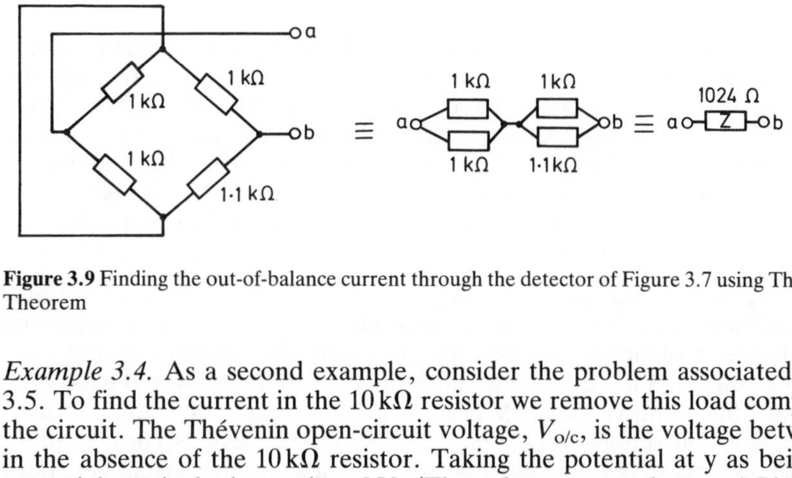

Figure 3.9 Finding the out-of-balance current through the detector of Figure 3.7 using Thévenin's Theorem

Example 3.4. As a second example, consider the problem associated with Figure 3.5. To find the current in the 10 kΩ resistor we remove this load component from the circuit. The Thévenin open-circuit voltage, $V_{o/c}$, is the voltage between x and y in the absence of the 10 kΩ resistor. Taking the potential at y as being zero, the potential at x is, by inspection, 8 V. (The voltage across the two 4.7 kΩ resistors is 4 V hence the potential at x must be 2 V positive with respect to the positive terminal of the 6 V supply.) The equivalent internal impedance is 4.7 kΩ in parallel with 4.7 kΩ, i.e. 2.35 kΩ, the generators having been shorted-out. (Once again the internal resistances of the generators are assumed to be zero.) When we connect the 10 kΩ load resistor to the Thévenin generator the current is 8 V/(2.35 + 10) kΩ, i.e. 8/12.35 = 0.65 mA. Hence the power dissipated in the load is $(0.65 \times 10^{-3})^2 \times 10^4$ W = 0.42×10^{-2} W, i.e. 4.2 mW. This is the same result as previously obtained by solving the simultaneous equations.

Norton's Theorem

Instead of using the voltage equivalent circuit of Thévenin it is sometimes more convenient in circuit analyses to use the current equivalent circuit of Norton. His theorem states that the current in any load connected to two terminals of a network of linear components is identical to that which is produced if the load were connected to a constant-current generator (whose generated current is equal to the current that flows through the two terminals when these are short-circuited) and shunted by an impedance equal to the impedance of the network 'looking into' the two terminals when all of the generator voltages are reduced to zero and the generators are replaced by their internal impedances. Figure 3.10 illustrates the

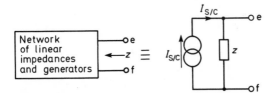

Figure 3.10 The Norton equivalent circuit. $I_{S/C}$ is the current that would flow from e to f if these two terminals were connected by a short-circuiting link. Z is the impedance 'looking into' the terminals e and f when the short circuit is removed, the generator voltages reduced to zero and the generators replaced by their internal impedances

theorem. We see that not all of the current produced by the constant current generator flows in the load circuit; some of the current is shunted through the parallel impedance Z. (This may be compared with the Thévenin voltage equivalent circuit. In this case not all of the voltage produced by the constant voltage generator is available to the load; some voltage will be 'lost' across the series impedance, Z.) Norton's theorem is said to be the *dual* of Thévenin's theorem. Norton's theorem is often used as an alternative to Thévenin's theorem when the network under consideration contains several parallel elements.

Example 3.5. As a first example of the application of Norton's theorem we may consider again Figure 3.5. We need firstly to remove the $10\,k\Omega$ load resistor and find the Norton equivalent for the remaining network. The terminals x and y are shorted and we must first determine the short-circuit current that would pass along the wire connecting x and y. The 10 V generator will supply 10/4.7, i.e. 2.13 mA; the 6 V generator will supply 6/4.7, i.e. 1.28 mA. From previous considerations (and by inspection) these currents must be added. The constant current generator therefore produces 3.41 mA. The shunt impedance is that impedance between x and y with the short-circuit removed and all generators suppressed. In this case the shunt impedance is $4.7\,k\Omega$ in parallel with $4.7\,k\Omega$, i.e. $2.35\,k\Omega$. With the load now connected, the current from the generator will divide, part flowing through the load (I_L) and part through the shunt resistor (I_S). $I_L + I_S = 3.41\,mA$. Also $I_L/I_S = Z/\text{load resistance}$, i.e. 2.35/10 = 0.235.

Therefore

$$I_L + I_L/0.235 = 3.41, \text{ i.e. } I_L + 4.25I_L = 3.41$$

Thus $I_L = 3.41/5.25 = 0.65$ mA. The power dissipated in the load is, therefore, $(0.65 \times 10^{-3})^2 \times 10^4 = 4.2$ mW.

Example 3.6. Find the current that flows through the 1.35 kΩ resistor in the circuit shown in Figure 3.11.

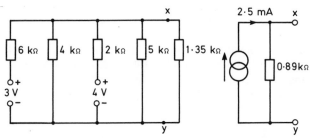

Figure 3.11 Circuit diagram for the problem given in example 3.6. The right-hand circuit shows the Norton equivalent for the given network with the 1.35 kΩ load removed

With the load removed and the generators suppressed, the impedance looking into the terminals x,y is the parallel combination of 6 kΩ, 4 kΩ, 2 kΩ and 5 kΩ. The total conductance is $0.167 + 0.25 + 0.5 + 0.2$, i.e. 1.12 mS. The parallel shunt resistance is therefore 1/1.12, i.e. 0.89 kΩ. To find the current generated by the constant-current Norton generator we short the terminals x,y. This means that the 4 kΩ and 5 kΩ resistors are effectively shorted. The total current passing along the short-circuit link is, therefore, 3 V/6 kΩ + 4 V/2 kΩ, i.e. 2.5 mA. This current will divide when the 1.35 kΩ load is connected, in inverse ratio to the resistance values. Using the symbols I_S and I_L, as before, $I_S + I_L = 2.5$ mA and $I_S/I_L = 1.35/0.89 = 1.52$. We see therefore that $I_L + 1.52 I_L = 2.5$ mA, hence $I_L = 1$ mA practically.

Exercises

1. State Kirchhoff's Laws. Hence, or otherwise, find the current in branch CF of the network shown in Figure 3.12.
2. The total dissipation in the circuit shown in Figure 3.13 is 9 W. Find R_1 and R_2.

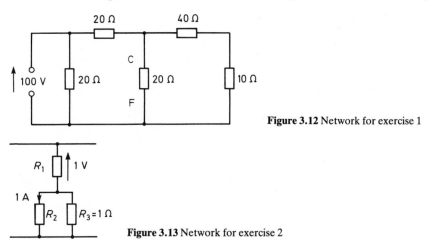

Figure 3.12 Network for exercise 1

Figure 3.13 Network for exercise 2

3. An electric motor takes 20 A at 240 V when driving a pump raising water at 220 gallons per minute to a height of 60 feet. Determine the overall efficiency of the system.

  1 gallon of water weighs 10 lb
  1 horsepower = 550 ft lb of energy per second
  1 foot = 0.3048 m
  1 lb = 0.4535 kg

4. State Thévenin's Theorem. Using this theorem find the current in the 1Ω branch AB of the circuit shown in Figure 3.14. (Before attempting this question make an estimate of the approximate value of the current without using calculators or paper and pencil.)

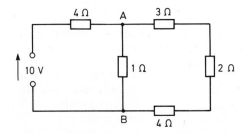

Figure 3.14 Network for exercise 4

(Ans. 1. 2.08 A   2. $R_1 = 0.33\,\Omega$ $R_2 = 2\,\Omega$ 3. 62% [62.15]   4. 1.84 A)

Chapter 4

Passive component networks

Electronic systems that process analogue signals can usually be represented as four-terminal networks as shown in Figure 4.1. The symbols for the voltages and currents are shown in lower case letters to represent the instantaneous values, or sometimes only the signal components. (Capital letters are used for steady voltages and currents, or their r.m.s. values.) v_1 and v_2 are the input and output voltages respectively whilst i_1 and i_2 represent input and output currents. From an external point of view we are usually interested in four properties of the circuit, *viz*:

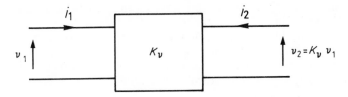

Figure 4.1 'Black-box' representation of a four-terminal network. K_V is a mathematical expression known as the transfer function

(a) the ratio of output to input voltages, v_2/v_1;
(b) the ratio of output to input currents i_2/i_1;
(c) the input impedance as 'seen' by the transducer or input driving circuit, i.e. the load that is connected to the transducer or driving circuit; and
(d) the output impedance. This is the internal impedance of the network when the latter is represented by its Thévenin or Norton equivalent. (The term 'impedance' is defined later in the chapter.)

The output signals depend not only on the properties of the input signals, but also upon the properties of the processing circuit. This fact may be summarized concisely by the equation

$$\text{Output} = \text{Circuit property} \times \text{Input}$$

The simplicity of this expression is so attractive that we attempt to preserve its form whenever possible. For example, $V = RI$ is satisfactory for steady voltages and currents, and for resistors that obey Ohm's Law. When the voltages and currents are sinusoidal, however, and capacitors and inductors are included in the circuit the simple Ohm's Law rules do not apply. Under these circumstances we must use differential equations or preferably develop mathematical techniques (the j-notation) that yield expressions of the above simple form.

Many of the signals encountered in electronic apparatus are not sinusoidal; this is particularly so in digital equipment. Where, for example, square waves, ramp waveforms or triangular waveforms are encountered we must further develop our mathematical skills to enable us to reduce the analyses to elementary algebraic techniques. In these cases we make use of Laplace transforms.

For linear circuits containing only resistors, and where the voltages and currents are time-invariant we can easily solve problems such as the one illustrated in Figure 4.2 in which the output voltage, V_2, must be found. The expression $R_1/(R_1 + R_2)$ is

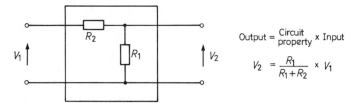

$$\text{Output} = \frac{\text{Circuit property}}{} \times \text{Input}$$

$$V_2 = \frac{R_1}{R_1 + R_2} \times V_1$$

Figure 4.2 Simple potential divider. In this case $K_v = R_1/(R_1 + R_2)$. The resistors obey Ohm's Law, and the voltages are time-invariant

the transfer function that operates on the input voltage to produce the corresponding output voltage. (Upper case letters have been used because the voltages are steady in this case.) Ohm's Law, Kirchhoff's Laws and the theorems of Thévenin and Norton all apply to this type of circuit as discussed in the last chapter. These laws and theorems can also be applied to circuits in which the voltages and currents are sinusoidal, and in which capacitors and/or inductors are incorporated, provided we develop the appropriate mathematical techniques.

Sine wave response in LCR circuits

The study of the application of sinusoidal signals to circuit arrangements of resistance, capacitance and inductance is of great importance, especially when it is remembered that all periodic waveforms may be resolved into a related set of sine waves. (As an example see Figure 4.3.)

The generation of sine waves in a laboratory is usually carried out by a piece of electronic equipment called an oscillator. Such oscillators produce at their output terminals voltages that vary sinusoidally with time. Various forms of electronic oscillators are discussed later in the book. Figure 4.4 illustrates the terms associated with sine waves.

When an alternating voltage is applied to the ends of a resistor, the results are easy to understand. We may draw directly on our experience with direct current and infer that an applied sinusoidal voltage produces a sinusoidal current through the resistor. Furthermore, at the instant when the voltage is a maximum, the current is also a maximum; when the voltage is zero, the current is zero. The current and voltage are said to be in phase (see Figure 4.5(a). By dividing the r.m.s. value of the voltage by the r.m.s. value of the current we calculate the resistance, R. Strictly, we do not need to use r.m.s. values as long as the units are consistent.

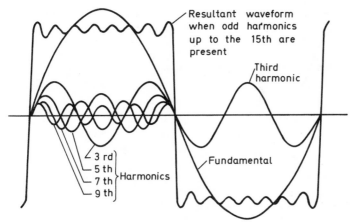

Resultant waveform
when odd harmonics
up to the 15th are
present

Third
harmonic

Fundamental

3 rd
5 th
7 th   Harmonics
9 th

Figure 4.3 A square wave consists of the fundamental plus all the odd harmonics out to an infinite frequency. The sum of the odd harmonics up to 15 is shown together with the appropriate odd harmonics up to the ninth

The majority of instruments used for measuring sinusoidal alternating voltages and currents are, however, calibrated in r.m.s. values.

The application of a sinusoidal voltage to a capacitor results in a somewhat more complicated behaviour than that experienced with a resistor. Since the applied voltage is alternating, a constant ebbing and flowing of electrons takes place between the supply and the capacitor because of the charging and discharging actions. No electrons, of course, actually pass through the dielectric of the capacitor. It is the charging and discharging process that causes the movement of electrons in the leads to the capacitor; a meter placed in a lead thus registers the presence of an alternating current in the lead. The size, or magnitude, of this current depends upon the rate at which the charges are passing a given point, i.e. upon the rate of charging or discharging of the plates of the capacitor. This rate depends, as we have already seen, on the rate of change of voltage ($q = CV$, thus $dq/dt = i = CdV/dt$). If the voltage across a capacitor does not change then the charge held by the capacitor does not alter. If the applied voltage is sinusoidal then the voltage has zero rate of change when the voltage is at its peak or trough. It is at such times that the charging/discharging process is temporarily halted. We see, therefore, that the current in the leads to a capacitor is zero when the applied voltage is a maximum or a minimum. The current in the leads is a maximum when

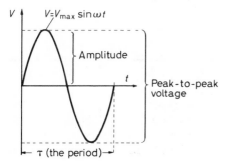

$V = V_{max} \sin \omega t$

Amplitude

t

Peak-to-peak
voltage

τ (the period)

Figure 4.4 Terms associated with sine waves

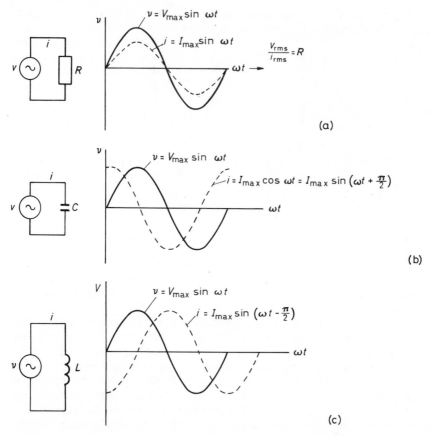

Figure 4.5 The phase relationships between voltage and current for the basic passive components: (a) Current and voltage in phase; (b) Current leads voltage by 90°, i.e. current reaches its maximum value a quarter of a period before the voltage reaches its maximum; (c) Current lags voltage by 90° (alternatively voltage leads current by 90°)

the applied voltage is changing most rapidly; for a sinusoidal voltage the maximum rate of change ($\mathrm{d}V/\mathrm{d}t$) occurs when the instantaneous value of the voltage is zero. We see that the current and voltage do not reach their maximum values simultaneously; they are out of step in a time sense, and are said to be out of phase.

The facts may be summarized mathematically as follows;

$$i = C\frac{\mathrm{d}v}{\mathrm{d}t}$$

where i is the instantaneous current value, C the capacitance and v the instantaneous value of the applied voltage. If a sinusoidal voltage is applied then $v = V_{max} \sin \omega t$ and

$$i = C\frac{\mathrm{d}(V_{max} \sin \omega t)}{\mathrm{d}t}$$

$$= \omega C V_{max} \cos \omega t \tag{4.1}$$

Now cos ωt = sin (ωt + $\pi/2$) so the waveform of the current is sinusoidal. However, there is a phase displacement of $\pi/2$ or 90°, which tells us that the maximum current flows at the time the applied voltage is zero. Figure 4.5(b) shows the relevant waveforms. The current reaches its maximum at some time t, say, and a quarter of a period later the voltage reaches its maximum; when a sinusoidal voltage is applied to a capacitor, the current leads the voltage by 90°. The time t was deliberately selected as being some time when the voltage and current had achieved their regular waveforms, i.e. after the so-called 'steady-state' conditions had been established. Initially, of course, if the voltage is applied at time $t = 0$ the current and voltage waveforms do not follow the patterns already discussed, but go through a transient state before the steady-state condition is reached. These initial transients are usually of very short duration, and can be analysed by a technique described later in the chapter.

Equation 4.1 can be rearranged to give

$$\frac{V_{max} \cos \omega t \text{ (voltage)}}{i \quad \text{(current)}} = \frac{1}{\omega C} = \frac{1}{2\pi f C} = X_C$$

The term $1/(2\pi f\, C)$ is called the capacitative reactance of the capacitor and is often given the symbol X_C. The reactance of a capacitor is analogous to the resistance of a resistor since it is a measure of the opposition to the flow of alternating current. This opposition to flow depends upon the capacitance (the larger the capacitance the less is the opposition to the flow of current) but, unlike the resistor, the opposition also depends upon the frequency of the applied voltage. The greater the frequency the less is the opposition to the current. Conversely, when the frequency is zero, i.e. when a steady voltage is applied to the capacitor, X_C is infinitely large and no current passes. This accords with our previous knowledge that direct current is prevented from flowing through a capacitor by the dielectric. When f is in hertz and C is in farads X_C is measured in ohms: for example, if an r.m.s. voltage of 10 V at a frequency 500 Hz were applied to a capacitance of 20 μF then the r.m.s. current that would flow in the lead to the capacitor would be V/X_C, i.e.

$10/[1/(2\pi\ 500 \times 20 \times 10^{-6})] = 2\pi \times 10^{-1} = 0.628$ A (r.m.s.)

The application of a sinusoidal voltage to an inductor results in a steady-state current that lags the voltage by 90° (see Figure 4.5(c)). If the current through the inductor is given by $i = I_{max} \sin \omega t$, then

$$v = L\frac{di}{dt} = L\frac{d(I_{max} \sin \omega t)}{dt} = \omega L I_{max} \cos\omega t = \omega L I_{max} \sin (\omega t + \frac{\pi}{2})$$

Once again we can regard these statements as representing the alternating current version of Ohm's Law. The voltage, v, is equal to the product of a current term, $I_{max} \sin(\omega t + \pi/2)$ and a term that is a measure of the opposition to current flow in the circuit. This latter term, ωL, is known as the inductive reactance and is given the symbol X_L. (The subscript L is often omitted when there is no ambiguity.) We see that the opposition to current flow is proportional to the frequency and to the inductance involved.

When resistors, capacitors and inductors are interconnected to form a required network, the analysis of the network's performance appears at first sight to be complicated. Consider a simple series arrangement as shown in Figure 4.6. The sum of the instantaneous voltages across each component must be equal to the

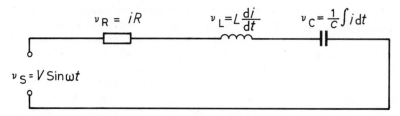

Figure 4.6 A simple series circuit; v_S is the driving voltage, assumed to be sinusoidal

instantaneous supply voltage, i.e. $v_S = v_R + v_L + v_c$. Since a simple Ohm's Law relationship between current and voltage holds only for the resistor, we must analyse the circuit with the aid of differential equations.

For this circuit, assuming a sinusoidal driving voltage,

$$V \sin \omega t = iR + L\frac{di}{dt} + \frac{1}{C} \int i dt$$

$$\therefore \omega V \cos \omega t = R \frac{di}{dt} + L\frac{d^2 i}{dt^2} + \frac{i}{C}$$

If we consider only the steady-state solution, then the voltage $V \sin \omega t$ is assumed to have been applied for a long time, and to continue to be applied for a long time; we are not therefore concerned with the initial conditions applying to the differential equation, or with the initial transient behaviour of the circuit. The solution of the equation may then be assumed to be of the form $i = I \sin (\omega t + \phi)$ where I is the current amplitude and ϕ the phase difference between the voltage and current waveforms.

Since the circuit is such a simple one it is a pity that the analysis of the circuit's behaviour depends upon the relatively complicated business of solving a second-order differential equation. Such a complication can be avoided provided we restrict ourselves to sinusoidal excitations. The sophisticated way of doing this is to transform the differential equation into an algebraic one using a suitable technique. The resulting algebraic equation can then be solved using elementary procedures as, for example, were used in the previous chapter. Since it is the aim of the text to lead the reader as gently as possible from familiar to less familiar concepts, we shall first consider another method of solving network problems before discussing the transformation technique.

Phasor diagrams

When resistors, capacitors and inductors are connected to form various series and parallel circuit arrangements, it is still possible to represent the alternating waves graphically by plotting their instantaneous values against time. Such graphs give us the mutual relationships between the various currents and voltages in the circuit, but they are tedious to draw. A much better method of representing the alternating quantities and their mutual relationships is to use what is called a 'phasor diagram'. Some books use the term 'vector diagram', but as the expressions 'vector' and 'vector analysis' have definite and well-defined meanings in physics and mathematics, it is better to avoid these terms. We may use phasors, which have the same mathematical properties as vectors, and thus avoid the physical anomaly of representing scalar quantities such as voltage and current by vectors.

In a phasor diagram, voltages and currents are represented by lines (called phasors), the lengths of which are proportional to the maximum values of the waves involved. The angles between the lines represent the phase angles between the waveforms. The phasors are, by definition, supposed to rotate about a fixed point, in an anticlockwise direction, at a constant angular velocity, ω. Figure 4.7 should make the position clearer. Here we have a voltage and a current represented by the lines OA and OB, respectively. The length of OA represents the amplitude, or maximum value, of the voltage, while OB represents the amplitude of the current.

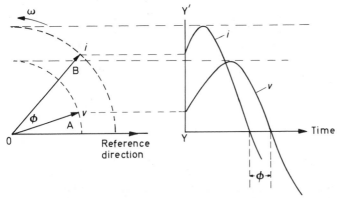

Figure 4.7 The representation of voltage and current graphs by means of phasors. ϕ is the angle representing the time or phase displacement between the two waveforms

The angle AOB is the fixed phase angle between the voltage and the current and is given the symbol ϕ. As both phasors rotate about O at a constant angular velocity, ω, the projections of OA and OB on YY' represent the instantaneous values of the voltage and current. If we plot the projections at subsequent times we regain the original graphs which are to be superseded by the phasor diagram. It is customary to adopt the trigonometrical convention and take our reference direction along the positive x-axis from O.

To be able to construct and understand phasor diagrams there are three basic configurations to remember; these are given in Figure 4.8. In the case of the resistor, the angle between the phasors representing v and i is zero, showing that the current and the voltage are in phase. The diagram associated with the capacitor shows the current leading the voltage by 90°; for the inductor the diagram shows the current lagging the voltage by 90°. It is now much easier to represent the relationships between the phase and the amplitudes of the various currents and voltages in a circuit containing combinations of resistance, capacitance and inductance.

Consider, as an example, a simple series combination of a resistor and a capacitor; Figure 4.9(a) shows the related phasor diagram. The total opposition to current as 'seen' by the generator is the supply voltage, v_s, divided by the supply current, i_s. This total opposition is known as the impedance of the circuit and is usually given the symbol, Z. We have thus retrieved our simple Ohm's Law type of relationship

$$v_s = i_s Z$$

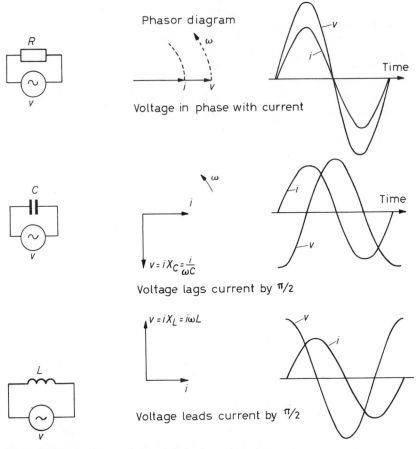

Figure 4.8 Phasor diagrams for the three basic passive components

Although a simple mathematical expression has been obtained it must not be forgotten that the equation contains information about both the magnitudes of various voltages and currents as well as phase information. From Figure 4.9(a), using Pythagoras' Theorem,

$$|Z| = \sqrt{\left[R^2 + \left(\frac{1}{\omega C}\right)^2\right]} \text{ and } \phi = \tan\frac{1}{\omega CR}$$

Readers may care to show that the impedance of a series arrangement of resistance, capacitance and inductance is given by

$$|Z| = \sqrt{[R^2 + (X_L - X_C)^2]}$$

where $X_L = \omega L$ and $X_C = 1/(\omega C)$

Then $v = iZ$

We have thus recovered a simple Ohm's Law relationship for this circuit, too.

For the circuit shown in Figure 4.9(c) we may be asked to find expressions for the current through the capacitor, i_C, the current through the resistor, i_R, the

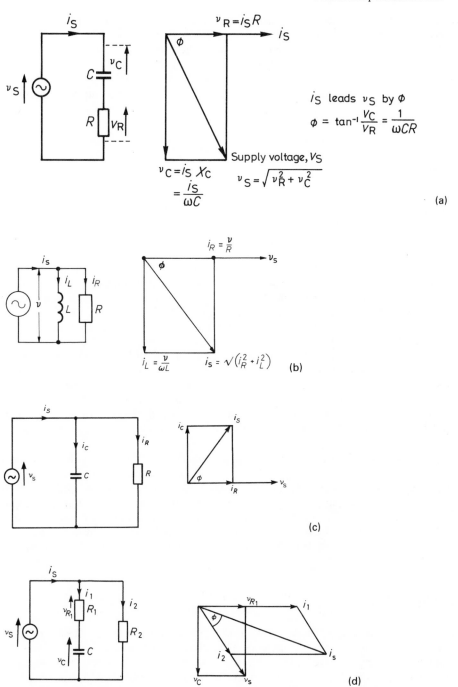

Figure 4.9 Phase relationships for simple circuits; (a) Phase relations in a simple CR circuit: (b) Phase relations in a simple parallel LR circuit; (c) See text; (d) Circuit for the phasor diagram for example 4.1

impedance of the parallel combination and the phase angle between the supply voltage and the supply current.

Since we have a parallel combination we take the supply voltage as our reference phasor. $i_R = v_S/R$ and $i_C = v_S\omega C$. From the diagram we see that i_c leads the supply voltage, v_S, by 90° and is, therefore, drawn in the 'northerly' direction. By completing the rectangle and using Pythagoras' Theorem, i_S, the supply current, is given by

$$i_S = \sqrt{(i_R{}^2 + i_C{}^2)} = v_S\sqrt{\left(\frac{1}{R^2} + \omega^2 C^2\right)}$$

Therefore the impedance, Z, is given by

$$Z = \frac{1}{\sqrt{\left[\dfrac{1}{R^2} + (\omega C)^2\right]}} = \frac{R}{\sqrt{[1 + (\omega CR)^2]}}$$

The phase angle, ϕ, is $\tan^{-1}(\omega CR)$.

Example 4.1. Draw the phasor diagram showing the phase relationships for the various voltages and currents associated with the circuit of Figure 4.9(d).

Although this circuit is a comparatively simple one, the construction of a phasor diagram and obtaining therefrom expressions for ϕ and Z are not straightforward. From the phasor diagram (Figure 4.9(d)) the reader will see that the use of complicated trigonometrical ratios are involved. In this case we take i_1 for our reference phasor. v_{R1} is in phase with i_1, while v_c lags i_1 by 90°. The phasor sum of v_{R1} and v_c is v_s, the supply voltage. i_2 must be in phase with v_s. If we now select suitable lengths for the phasors i_1 and i_2, the phasor sum of these currents is the supply current i_s. ϕ is the phase angle between the phasors for v_s and i_s. Once again it should be emphasized that the arrows on the circuit diagram are for mathematical purposes; the various voltages and current involved are, of course, alternating quantities.

Since the phasor diagram for such a simple circuit is not easy to draw and analyse, it can be seen that such diagrams are of limited use in a.c. analyses. Fortunately, these diagrams (from which we could recover the sinusoidal graphical information) can be replaced by using algebraic techniques to obtain analytical expressions. Before such expressions are derived and explained, however, we need to extend our algebraic techniques to include what is known as the j-notation.

It is left as an example for the reader to establish the relationships shown in Figure 4.9(b).

The j-notation

Since the construction of phasor diagrams is not easy for any but the simplest of circuits, a much more convenient way of analysing a.c. circuits is to replace the phasor diagram by an algebraic technique. The possibility of doing this was first shown by Steinmetz (1893) who recognized the similarity between phasor diagrams and the Argand diagrams used by mathematicians to represent diagrammatically what are called complex numbers.

Having established the fundamentals of manipulating algebraic expressions that include j-terms, it is possible to analyse complicated a.c. circuits. The associated phasor diagrams can readily be replaced by algebraic expressions. Such expressions conform to the standard rules of elementary algebra, and are as easy to handle as those involving Ohm's Law. All of the theorems that have been previously discussed for the d.c. case (Kirchhoff, Thévenin, Norton, etc.) are valid for a.c. circuits, and the same rules apply. The only additional rules that must be remembered are:

(1) j^2 in expressions may be replaced by -1;
(2) the opposition to current for a capacitor (analogous to resistance, R, for a resistor) is given by $X_C = 1/(j\omega C)$;
(3) the corresponding quantity for an inductor is $X_L = j\omega L$.

X_C and X_L can now be treated as simple terms from an algebraic point of view; the j factors automatically contain the phase information. An example may clarify the position.

Example 4.2. For the circuit shown in Figure 4.10 find the voltages across the capacitor and resistor; determine the supply current and the phase angle between the supply current and the supply voltage.

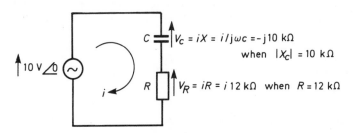

Figure 4.10 Circuit for example 4.2. It is given that $X_C = 10\,k\Omega$ and R = 12 KΩ

It should be noted that when a value for X_C is given we do not need the value of C or the frequency. Had we been told that, say, $C = 0.1\,\mu F$ and $f = 159\,Hz$ then it would first have been necessary to calculate X_C from $1/2\pi fC$. The 'opposition to flow' would then be $X_C/j = -jX_C$. The value of X_C is $10\,k\Omega$ in this case, assuming that $2\pi.\,159 = 1000$.

We now apply the same rules as were established in Chapter 3. The mathematical arrows (which must not be confused with the physical currents and voltages involved) enable us to establish the Kirchhoff equation

$$10 \angle 0 - V_C - V_R = 0$$

$10 \angle 0$ tells us that the supply voltage is to be taken for our reference phase direction.

Hence

$$(10 + j0) - i(-j10k) - i12k = 0$$

Therefore $10 - i(12k - 10kj) = 0$

and $i = 10/(12k - j10k)\,A$
    $= 10/(12 - j10)\,mA$

Rearranging i into the form $a + jb$ involves the rule given for the division of complex quantities.

$$\text{Hence} \qquad i = \frac{10(12 + 10j)}{12^2 + 10^2} = 0.49 + j0.41$$

The phase angle between the supply current and the supply voltage is $\tan^{-1}$ (quadrature component/ordinary component), i.e. $\tan^{-1} 0.41/0.49 = \tan^{-1} 0.84$. The current, therefore, leads the supply voltage by 40°. The magnitude of the current is $(0.49^2 + 0.41^2)^{\frac{1}{2}} = 0.64$ mA; hence $V_R = 7.68$ V and $V_C = 6.4$ V. (As a check the phasor sum of V_R and V_C should equal the supply voltage, i.e. the magnitude of the supply voltage is $\sqrt{(7.68^2 + 6.4^2)}$. This latter quantity should equal 10 V.)

Power in a.c. circuits

The power dissipated in a resistor in which a steady current is flowing because of the application of a steady voltage is calculated in a straightforward way from VI, V^2/R or I^2R. In the a.c. case the instantaneous power, p, is given by $p = vi$ where $v = V_{max} \sin \omega t$ and $i = I_{max} \sin \omega t$. The angles are the same because the voltage and current are in phase. The average power dissipated over a complete cycle is given by

$$P = \frac{1}{2\pi} \int_0^{2\pi} vi \, \mathrm{d}(\omega t)$$

which means that

$$2\pi P = V_{max} \, I_{max} \int_0^{2\pi} \sin^2 \omega t \, \mathrm{d}(\omega t)$$

$$P = (V_{max} \, I_{max})/2$$

We may rewrite this as $P = (V_{max}/\sqrt{2})(I_{max}/\sqrt{2})$ which is equivalent to $P = V_{r.m.s.} I_{r.m.s.}$. We see therefore that for resistors the average power dissipated in the a.c. case is the product of the r.m.s. values of the voltage and current. This is the simple a.c. equivalent of the d.c. case, hence $P = V_{r.m.s.} I_{r.m.s.} = V^2_{r.m.s.}/R = I^2_{r.m.s.} R$.

When circuits contain combinations of resistors, inductors, and capacitors, however, a phase difference exists between the supply voltage and the supply current. The average power supplied by a generator to an a.c. circuit is therefore given by

$$P = \frac{1}{2\pi} \int_0^{2\pi} V_{max} \sin \omega t . I_{max} \sin (\omega t + \phi) \, \mathrm{d}(\omega t)$$

where ϕ is the phase shift between the supply voltage and supply current. Since $\sin A \sin B = \frac{1}{2} [\cos(A - B) - \cos(A + B)]$,

$$2\pi P = V_{max} I_{max} \frac{1}{2} \int_0^{2\pi} [\cos(-\phi) - \cos(2\omega t + \phi)] \, \mathrm{d}(\omega t)$$

As $\cos -\phi = \cos \phi$ this integral yields

$$P = V_{r.m.s.} I_{r.m.s.} \cos \phi$$

Cos φ is known as the power factor of the circuit. When large amounts of power are involved precautions are taken to ensure that cos φ is close to unity (i.e. φ → 0). For a given power, when φ is greatly different from zero, the corresponding r.m.s. value of the current must be large. With large inductive electric motors, for example, it is usual to use a parallel capacitor so that the combination has a power factor as near to unity as is practical. The supply cables need not then be unduly large and expensive.

In the limit when φ = π/2 the power must be zero irrespective of the values of the supply voltage and current. This is the case when the load is purely reactive. When the load, for example, is a pure capacitor the supply current leads the supply voltage by 90°. There can, therefore, be no dissipation of power in a perfect capacitor. All of the energy supplied during the charging process is returned to the source during the discharging process. An r.m.s. ammeter in the leads to the capacitor would, of course, register the average r.m.s. current. Nevertheless no net power is absorbed by the capacitor. For similar reasons no net power is supplied to a pure inductor. This is why a transformer used for a front door bell, for example, can be permanently connected to the mains. For much of the time no load is connected to the secondary circuit. The transformer therefore acts as an inductor presented to the mains and hence no net power is taken (provided the transformer is of reasonable quality). Only when the bell push is operated will power be taken to energize the bell.

In a load which contains resistors and reactive components, only the resistors dissipate power. If the power factor is known, then the power dissipated in the resistive part of the load is $V_{r.m.s.}$ $I_{r.m.s.}$ cos φ. This expression is not always a convenient form since the voltage, current and power factor all need to be known. Often, in a circuit analysis, only one of the quantities is known. For example, we may obtain a general expression for the current in the load, of the form $a + jb$. We can easily calculate the power absorbed by a load $R + jX$ when such a current is flowing from the formula $P = |i|^2 R$, i.e. we square the modulus of the current and multiply this by the resistive part of the load. (As explained previously, no power can be dissipated by the reactive component jX.)

Polar and exponential forms

The expression $a + jb$ is not always the most convenient way of expressing a general (or complex) number. The expression can be illustrated by an Argand diagram based on the ordinary Cartesian (x, y) coordinates. If we use polar coordinates it is possible to represent a point in a plane by using the distance, r, from the origin and an angle, $θ$, between the line joining the point to the origin and a reference direction (usually drawn out to the 'east' i.e. along the positive x-axis.)

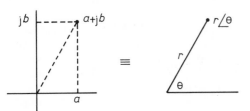

Figure 4.11 Diagram showing the equivalence of $a + jb$ and $r\angle θ$

Thus $r\angle\theta \equiv a + jb$ since both expressions locate a given point in a plane. Resolution, as in Figure 4.11, shows that $a + jb = r(\cos\theta + j\sin\theta)$ because $a = r\cos\theta$ and $b = r\sin\theta$. The multiplication process is somewhat easier in polar coordinates since $r_1\angle\theta_1 \times r_2\angle\theta_2 = r_1 r_2 \angle(\theta_1 + \theta_2)$. (The distances are multiplied and the angles added.) Division using polar coordinates is also easier than handling the $a + jb$ forms; $r_1\angle\theta_1 \div r_2\angle\theta_2 = r_1/r_2 \angle(\theta_1 - \theta_2)$.

The exponential form often makes mathematical manipulation easier since $a + jb = re^{j\theta}$. This result can be seen to be true by remembering the power series expansions, *viz.*

$$e^x = 1 + x + \frac{x^2}{2!} + \frac{x^3}{3!} \cdots$$

$$\therefore e^{j\theta} = 1 + j\theta - \frac{\theta^2}{2!} - j\frac{\theta^3}{3!} + \frac{\theta^4}{4!} \cdots$$

Since

$$\cos\theta = 1 - \frac{\theta^2}{2!} + \frac{\theta^4}{4!} \cdots$$

and

$$\sin\theta = \theta - \frac{\theta^3}{3!} + \frac{\theta^5}{5!} \cdots$$

then

$$e^{j\theta} = \cos\theta + j\sin\theta$$

$$\therefore re^{j\theta} = r(\cos\theta + j\sin\theta)$$

Summarizing

$$a + jb = \sqrt{(a^2 + b^2)}\angle\tan^{-1}(b/a) = r\angle\theta = re^{j\theta}$$

Frequency response

When a complex signal is presented to the input terminals of a given network, the output waveforms depends upon the properties of that network. The complex signal waveform may be regarded as being a synthesis of many sinusoidal waveforms of differing amplitude, phase and frequency (e.g. see Figure 4.3). If the waveform is to be preserved, the network must be able to respond adequately to signals of widely differing frequencies, i.e. the frequency response must be wide. Alternatively there are occasions when some frequency components must be eliminated from the signal, as we shall see when dealing with filters later in the book.

For any network there is a range of frequencies over which the performance is intended to be satisfactory. The range is called the mid-frequency range, or, in the case of filters, the pass band; the actual range depends upon whether the network is designed for low-frequency (l.f.), radio frequency (r.f.) or very high frequency (v.h.f.) operation. At frequencies above and below the mid-frequency range, the performance is likely to be unsatisfactory in some respects. The frequency performance is often illustrated graphically by what is termed a frequency response curve. Frequency response curves use logarithmic rather than linear scales for both the 'horizontal' and 'vertical' axes; such scales enable an enormous arithmetic range to be covered in a satisfactory way, and often interpret physical phenomena better than linear scales. Many physical phenomena are fundamentally logarithmic, and such phenomena can be represented by straight-line graphs when logarithmic axes are used.

When comparing the frequency performance of two networks (e.g. two amplifiers) on a single set of axes, the vertical scale should be a relative logarithmic one. The logarithmic scale prevents the cramping of information displays, and, if it is also a relative logarithmic scale, direct comparisons of performance can be made. The relative logarithmic scale may be divided into a linear scale of units called decibels.

The decibel scale

The bel scale was introduced originally by telephone engineers to express the ratio of two powers, and was named in honour of Alexander Graham Bell, the pioneer in telephone work. The bel is, in practice, too large a unit to use, and so the decibel (dB) was introduced, 1 dB being one-tenth of a bel. Hence:

$$\text{Power ratio (in dB)} = 10 \log_{10}\frac{P_1}{P_2}$$

This is the fundamental and defining equation for decibels. Thus if $P_1 = 2P_2$ then P_1 is $10\log_{10}2$, i.e. 3dB greater than P_2. We often say that P_1 is 3dB up on P_2. If $P_1 = 0.5P_2$, then P_1 is $\log_{10} 0.5 = -3$dB compared with P_2. Here we often say that P_1 is 3dB down on P_2.

To obtain a frequency response curve the output power is observed at different frequencies with the input voltage maintained at a constant r.m.s. value. Since the output power is being developed in the same resistance, R, at different frequencies, it is possible to modify the above equation. If the power at the first frequency is P_{f1} and that at the second frequency P_{f2} then the power gain or loss, N, is given by

$$N = 10 \log_{10} P_{f1}/P_{f2}$$

Now $P_{f1} = V^2{}_{f1}/R$, where V_{f1} is the output voltage at the first frequency. Similarly $P_{f2} = V^2{}_{f2}/R$.

$$\text{Therefore} \quad N(\text{dB}) = 10 \log_{10} \frac{V_{f1}^2/R}{V_{f2}^2/R} = 10 \log_{10} \left(\frac{V_{f1}}{V_{f2}}\right)^2$$

$$= 20 \log_{10} \left(\frac{V_{f1}}{V_{f2}}\right)$$

This alternative equation is convenient because it is easier to measure voltages than powers. However, it must be stressed that this alternative equation is only possible because the resistance across which the measurements are taken is the same at both frequencies.

It is often possible to read or hear about the voltage gain of an amplifier being so many decibels. Strictly, it is incorrect to make this kind of statement because the input and output voltages are being developed across different resistances. However, it is so attractive to express the gains in this way that by convention many people use this type of statement. A straightforward statement such as 'the voltage gain of this amplifier is 60 dB' is taken to mean that $60 = 20 \log_{10} V_{out}/V_{in}$, i.e. that the output voltage is 1000 times greater than the input voltage.

To obtain the frequency response of a network the input terminals are energized by sinusoidal signals of constant amplitude, but varying frequency. The magnitude of the input signal must be satisfactory for the network under examination; for example, if the network were an amplifier then the input signal must not be large enough to cause overloading or excessive distortion. The magnitude of the output voltage is noted at frequencies within the operating band. The output voltage at each frequency is compared with the output voltage at some frequency; often the reference frequency is at the centre of the mid-band range. The ratio between the output voltage at a given frequency and the output voltage at the reference frequency, expressed in decibels, is the response of the network at the given frequency. It is then usual to use log/linear graph paper, and plot the response (in decibels) on a vertical (i.e. y-direction) linear scale against the frequency on the logarithmic (x-direction) scale.

A good deal about the electrical behaviour of the network may be deduced from an examination and analysis of the frequency response. By way of an example let us examine the simple network shown in Figure 4.12. The examination is quite general and the method may be used with any network.

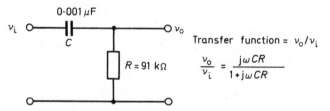

Transfer function = v_o/v_i

$$\frac{v_o}{v_i} = \frac{j\omega CR}{1 + j\omega CR}$$

Figure 4.12 Simple CR high-pass filter

A sketch of the frequency response may be made without any actual output voltages being measured provided we know, or can derive, the transfer function of the network. The transfer function is the mathematical expression for the ratio of the output voltage to the input voltage, i.e. v_o/v_i. Strictly, this function should be in terms of the Laplace operator, s, and the circuit constants. However, if we restrict ourselves to sinusoidal signals we may replace s by $j\omega$, and hence the ratio v_o/v_1 can be expressed in terms of j, ω and the circuit constants. In Figure 4.12, for example, C and R form a potential divider and hence $v_o/v_i = R/(R + 1/j\omega C) = j\omega CR/(1 + j\omega CR)$.

Previously, the input and output voltages have been described mathematically by expressions which are functions of time; we are then said to be working in the *time domain*. In Figure 4.12

$$v_i = V \sin \omega t = \frac{1}{C} \int i(t)\mathrm{d}t + Ri(t); \quad v_o = Ri(t)$$

where i is some functions of time; in this case i has a sinusoidal waveform. When we use the transfer function $j\omega CR/(1 + j\omega CR)$ the time variable has been deleted completely and the expression is a function of frequency; such an expression defines the behaviour of the circuit in what is said to be the *frequency domain*. The ratio v_o/v_i can then be plotted as a function of frequency, and thus the frequency response is obtained.

To sketch the frequency response four main steps are involved:

(1) obtain the transfer function;
(2) consider an approximate expression for the transfer function at very low frequencies, i.e. when $\omega \rightarrow 0$;
(3) consider an approximate expression for the transfer function at very high frequencies, i.e. when $\omega \rightarrow \infty$;
(4) note any special properties of the network that are revealed by an examination of the transfer function (e.g. maximum or minimum values).

When these steps have been taken the sketch can be completed.

With reference to the example shown in Figure 4.12 the transfer function has been shown to be $j\omega CR/(1 + j\omega CR)$. At very low frequencies $\omega \rightarrow 0$ therefore $j\omega CR \ll 1$, hence $v_o/v_i \simeq j\omega CR$. For a fixed r.m.s. value of v_i it is seen that v_o is proportional to the frequency. Each time the frequency is halved the output voltage is also halved. When the frequency is halved it has been reduced by an octave (a term borrowed from musicians). A halving of the output voltage means that v_o has dropped by 6 dB because $20 \log_{10}(1/2) = -6\,\text{dB}$. The slope of the frequency-response curve at low frequencies is constant and is equal to a rate of fall of 6 dB per octave. (This is equivalent to a fall of 20 dB per decade; i.e. when the frequency is reduced by a factor of 10 the output voltage falls by the same factor, i.e. 20 dB.)

At high frequencies $j\omega CR \gg 1$ and therefore $v_o/v_i \simeq 1$, i.e. 0 dB. The frequency response curve is shown in Figure 4.13. It will be noticed that since the transfer function has no discontinuities a smooth transition takes place between the response at high and low frequencies. The frequency at which the slope of the response curve changes from 6 dB per octave to zero is not well defined. For this

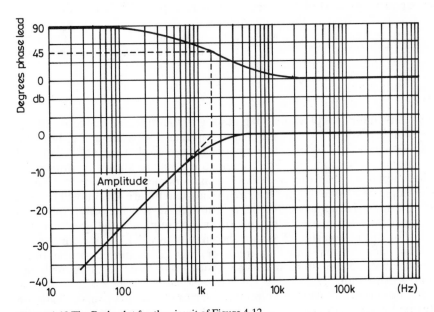

Figure 4.13 The Bode plot for the circuit of Figure 4.12

reason we define this turnover frequency, f_t, as being that frequency at which the response is 3 dB down. If we define f_t in this way it is easy to derive a simple expression for f_t in terms of the circuit constants. At high frequencies the modulus of the transfer function is 1 (i.e. $v_o/v_i = 1$ which corresponds to 0 dB). At the turnover frequency the transfer function has fallen by 3 dB and hence the modulus of $v_o/v_t = 1/\sqrt{2}$. This means that

$$\frac{\omega_t CR}{\sqrt{[1 + (\omega_t CR)^2]}} = \frac{1}{\sqrt{2}}$$

where $\omega_t = 2\pi f_t$

By squaring each side of this equation, and rearranging the terms, we see that $\omega_t CR = 1$ and hence $f_t = 1/(2\pi CR)$.

Phase response

A knowledge of how the phase shift between v_i and v_o varies with frequency is often important, especially when the network is associated with feedback systems. Much of the fundamental work discussed in this part of the book will be required when we consider the stability of operation of integrated circuit amplifiers.

The phase shift can be calculated from the transfer function for any frequency. For the network shown in Figure 4.12 we see that at low frequencies $v_o/v_i \simeq j\omega CR$. The presence of the j term shows that there is a phase shift of $+90°$ at very low frequencies, i.e. there is a phase lead of 90°. We can appreciate this when we consider that $v_o = iR$ and hence the phase of v_o is the same as the current i. At very low frequencies the capacitor is the dominant component because its reactance is much higher than R when $\omega \to 0$. Since the current through the capacitor leads the voltage by 90° we see that i must lead v_i by an angle that approaches 90° as $\omega \to 0$.

At high frequencies $v_o/v_i = 1$, hence there is zero phase shift and the output voltage is practically equal to the input voltage.

At the turnover frequency $v_o/v_i = j/(1 + j)$ because $\omega CR = 1$. By rationalizing the denominator we see that $v_o/v_i = (1 + j)/2$, hence the phase shift is $\tan^{-1} 1$, i.e. 45° (leading), at the turnover frequency.

Bode plots

It is often convenient to plot the magnitude of the transfer function versus frequency as one curve, and the phase characteristic as a separate curve, but with the same frequency axis. Log or semilog paper is used so that several decades of frequency can be accommodated. The combined curves are known as *Bode plots* and find wide application in the analysis of amplifiers and servo systems. Figure 4.13 is an example of a Bode plot for the circuit shown in Figure 4.12. It will be seen only that the high-frequency signals are transmitted satisfactorily; the low-frequency signals are attenuated, and the lower the frequency the greater is the attenuation. For this reason the *CR* circuit shown in Figure 4.12 is an example of a high-pass filter, meaning that only high-frequency signals are passed satisfactorily. Figure 4.14 shows the circuit of one type of low-pass filter together with the corresponding Bode plot.

It should be noted that the amplitude versus frequency curve may be approximated by drawing two straight lines which intersect at the turnover

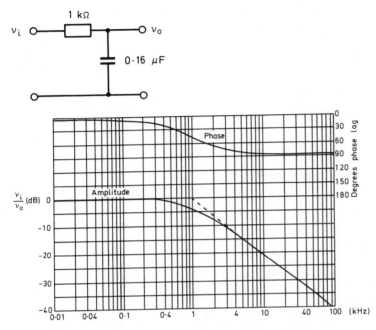

Figure 4.14 The amplitude and phase response of a low-pass filter

frequency. The turnover frequency can then be referred to as the *corner* or break frequency since it is that frequency at which there is a distinct change in the slope of the amplitude characteristic. As previously stated, the actual response curve undergoes a smooth transition at frequencies around f_t, but the errors in using the straight line approximation are not excessive. The greatest error is at f_t, where there is a 3 dB difference. At an octave above, or below, f_t the error is only 1 dB. This can be appreciated by considering the expression for the low-pass filter, for example.

$$\frac{v_o}{v_i} = \frac{1}{1 + j\omega CR} = \frac{1}{1 + j\frac{\omega}{\omega_t}} = \frac{1}{1 + j}$$

at the break frequency when $\omega = \omega_t = 1/CR$. Here the magnitude of $v_o/v_i = 1/\sqrt{2}$, i.e. 3 dB down. When $\omega = 2\omega_t$ then, remembering that $\omega_t CR = 1$.

$$\frac{v_o}{v_i} = \frac{1}{1 + 2j}, \text{ i.e.} \left|\frac{v_o}{v_i}\right| = \frac{1}{\sqrt{5}}, \text{ i.e. 7 dB down}$$

Since the straight line approximation gives 6 dB down, the error is seen to be only 1 dB.

Example 4.3. Obtain a Bode plot for the circuit shown in Figure 4.15. In this example $v_o/v_i = R_2/(R_2 + Z)$ where Z is the parallel arrangement of R_1 and C, i.e. $Z = R_1/(1 + j\omega CR_1)$. The transfer function is therefore

$$\frac{v_o}{v_1} = \frac{R_2(1 + \omega CR_1)}{R_1 + R_2 + j\omega CR_1 R_2}$$

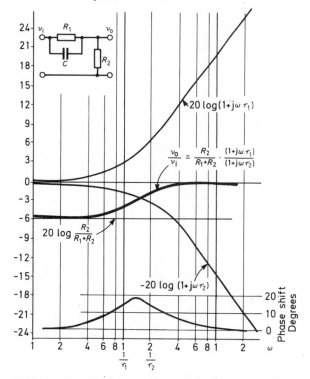

Figure 4.15 A Bode plot displaying the amplitude and phase characteristics of the network, shown inset, as functions of frequency. Since the transfer function can be expressed in decibels as $20 \log (R_2/R_1 + R_2)$ $+ 20 \log (1 + j\omega\tau_1) - 20 \log (1 + j\omega\tau_2)$ the overall response in decibels is the algebraic sum of the individual response curves. (To obtain the above curves it was assumed that $R_1 = R_2$)

At very low frequencies $\omega \to 0$ hence $v_o/v_i \simeq R_2/(R_1 + R_2)$; we have a simple resistive potential divider, and the capacitor can be ignored at very low frequencies. The response at low frequencies is constant and equal to $20 \log_{10} R_2/(R_1 + R_2)$ dB.

At high frequencies $v_o/v_i \simeq j\omega CR_1R_2/j\omega CR_1R_2$, i.e. 0 dB. The frequency response curve is therefore a straight line parallel to the frequency axis at both high and low frequencies; at high frequencies the straight line is at the 0 dB level, while at low frequencies it is at a lower level. Clearly there must be two turning frequencies, which is not surprising since two time constants are involved in the transfer function. This can be seen by rearranging the function slightly as follows:

$$\frac{v_o}{v_i} = \frac{R_2(1 + j\omega CR_1)}{R_1 + R_2 + j\omega CR_1R_2} = \frac{R_2}{R_1 + R_2} \cdot \frac{(1 + j\omega CR_1)}{[1 + j\omega CR_1R_2/(R_1 + R_2)]}$$

$$= \frac{R_2}{R_1 + R_2} \frac{(1 + j\omega\tau_1)}{(1 + j\omega\tau_2)}$$

where $\tau_1 = CR_1$ and $\tau_2 = CR_1R_2/(R_1 + R_2)$. The first time constant τ_1 is simply the product of C and R_1; the second time constant is the product of C and the resistance equivalent to the parallel arrangement of R_1 and R_2. By inspection $\tau_1 > \tau_2$, and each time constant is associated with a turning frequency (f_1 and f_2, respectively). This can be seen by considering v_o/v_i in decibel units.

$$\frac{v_o}{v_1} = 20 \log_{10}[R_2/(R_1 + R_2)] + 20 \log_{10}(1 + j\omega\tau_1) - 20 \log_{10}(1 + j\omega\tau_2)$$

At very low frequencies the response curve is flat at $20 \log_{10}[R_2/(R_1 + R_2)]$. At very high frequencies

$$\frac{v_o}{v_1} = 20 \log_{10}[R_2/(R_1 + R_2)] + 20 \log_{10}(\tau_1/\tau_2) = 0 \text{ dB}$$

An advantage of expressing v_o/v_i in decibels is that the effect of each logarithmic term can be considered separately; the overall response is obtained by adding the individual responses. For the given example, at intermediate frequencies (between very high and very low frequencies) the second term, $20 \log_{10}(1 + j\omega\tau_1)$ will be responsible for a 3 dB rise when $\omega\tau_1 = 1$. The overall response will not be 3 dB up, however, unless the two break frequencies are widely separated. In Figure 4.15, as an example, $R_1 = R_2$ and hence the break frequencies are quite close. The contribution of the third term at the first break frequency point cannot be ignored. As the frequency rises the effect of the third term becomes more significant until at a frequency where $\omega\tau_2 = 1$ the response is 3 dB lower than it would have been in the absence of the third term. The response then flattens out as shown in Figure 4.15. For the sake of illustration it has been assumed that $R_1 = R_2$, hence the response is 6 dB down at low frequencies.

The phase shift of the network can be computed from the transfer function, and depends upon the values of the circuit constants at intermediate frequencies. At very high and very low frequencies the transfer function is independent of frequency and hence there is zero phase shift at these frequencies.

$$\frac{v_o}{v_1} = \frac{R_2}{R_1 + R_2} \cdot \frac{(1 + j\omega\tau_1)}{(1 + j\omega\tau_2)}$$

thus $\phi = \tan^{-1}\omega\tau_1 - \tan^{-1}\omega\tau_2$. If, for example, $R_1 = R_2$ then $\tau_1 = 2\tau_2$. At the high-frequency break point (when $\omega\tau_2 = 1$)

$$\phi = \tan^{-1}2 - \tan^{-1}1 = 18.5°$$

At the low-frequency break point $\omega\tau_1 = 1$, i.e. $\omega\tau_2 = \frac{1}{2}$

$$\phi = \tan^{-1}1 - \tan^{-1}0.5 = 18.5°$$

By finding $d\phi/d\omega$ and equating this to zero the maximum phase lead angle will be found to be about 19.5°. An approximate phase response is shown in Figure 4.15; for a more precise curve ϕ must be evaluated at several values of ω.

A bandpass filter

By combining a low-pass filter with a high-pass filter in a single network, as shown in Figure 4.16, then both high- and low-frequency signals are discriminated against; the arrangement passes satisfactorily only those signals whose frequency lies within a specified band. To obtain a Bode plot for this network is easy enough in principle, although the algebraic expressions may be tedious to manipulate. We must be careful to note that C_2 and R_2 act as a load upon C_1 and R_1.

The transfer function may be obtained easily using matrix algebra; the Bode diagram of Figure 4.16 was, however, obtained with the aid of a BBC computer and

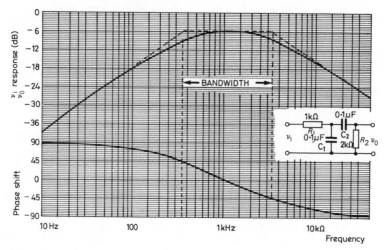

Figure 4.16 Simple bandpass filter

circuit analysis software. The bandwidth is the range of frequencies over which the response is not less than 3 dB below the response at the midband frequencies; it is, in effect, the band of frequencies between the two break, or turning, frequencies.

Resonant circuits

The phenomenon of resonance is well-known in nature. Any system capable of oscillating exhibits resonance; energy storage and release is always involved. As an example we may consider a mass suspended on a spring. When the mass is first attached to the spring, and carefully released, the spring extends. The extension produces an upwards force which balances the weight of the mass (mg) when the system is at rest. Energy is stored in the spring equal to mgh where h is the extension involved. If the mass is then pulled down by an amount x, additional energy (mgx) is stored in the spring. Upon releasing the mass this energy is used to accelerate the mass upwards; such energy is converted into kinetic energy, the latter reaching its maximum value when the mass is at its original stationary position. At this point the mass has its maximum velocity. Inertia carries the mass upwards for a distance x, in a loss-free system. The mass then falls again and the cycle is repeated. If no losses of any kind are involved, the mass oscillates about a fixed position with simple harmonic motion. The natural frequency of oscillation is determined by the physical properties of the system.

In practice some losses do occur, and continuous oscillations can be maintained only by supplying energy to make good the losses. The mass can, of course, be forced to oscillate at any practical frequency by the application of suitable forces. In general, the amplitude of forced oscillations is not large and the energy involved is quite large. If however, the energy is supplied at the natural frequency of oscillations the amplitude becomes large and very little energy is needed to maintain the oscillations. The system is then exhibiting resonance and is said to be oscillating at its resonant frequency.

An electrical circuit is said to be resonant when a sinusoidal driving voltage with a frequency equal to the natural frequency is applied to the circuit. The energy storage components are capacitors and inductors, and when oscillations occur there is a continual interchange between the energy stored as an electrical field in the capacitor and that stored as a magnetic field in the inductor. Resistance in the circuit dissipates the energy in the form of heat.

In practice a number of definitions of resonance exist. It is often convenient to define a circuit as being resonant when the net reactive component is zero and the circuit behaves as a pure resistance. This implies that the magnitude of the inductive reactance is equal to the magnitude of the capacitive reactance. Analyses involving this practical definition of resonance yield a slightly different formula for the resonant frequency from that which would be obtained by using the definition of resonance given in the previous paragraph. For resonant circuits that are usually encountered in electronic equipment the difference is small enough to be ignored; this implies that the resistive losses are small.

For oscillations to occur, both inductance and capacitance must be present; only then can there be an interchange between electric field energy and magnetic field energy. Circuits containing only capacitance and resistance, or only inductance and resistance, cannot therefore be resonant circuits.

Series resonance

A resonant circuit that consists of a series arrangement of its electronic components exhibits what is termed *series* resonance. In the circuit of Figure 4.17 the impedance, Z, is given by

$$Z = R + j \left[\omega L - \frac{1}{\omega C} \right]$$

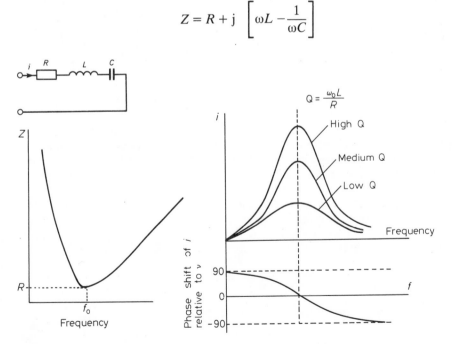

Figure 4.17 Series resonance

The reactive (or j) term is zero for that special frequency that makes $\omega L = 1/(\omega C)$. The special frequency is the series resonant frequency. At this resonant frequency $\omega = \omega_o$ and hence

$$\omega_0 L = \frac{1}{\omega_0 C}, \text{ i.e. } \omega_0 = \frac{1}{\sqrt{(LC)}}$$

from which

$$f_0 = \frac{1}{2\pi\sqrt{(LC)}}$$

At this series resonant frequency $Z = R$ and is hence purely resistive. At frequencies above and below the resonant frequency $|Z| > R$. A curve of Z versus frequency is shown in Figure 4.17 and is an example of a resonance curve. The 'sharpness' of the curve is determined by the value of resistance compared with X_L or X_C. The lower the resistance the greater will be the change of Z with frequency about the resonant frequency. If, theoretically, $R = 0$ then Z would also be zero at the resonant frequency.

The resistive losses are associated mainly with the coil, since practical capacitors are usually almost loss-free. The higher the quality of the coil the smaller are the resistive losses. The quality factor (denoted Q-factor or merely the Q) of the coil largely determines the sharpness of the resonant curve. The practical consequence of this is that a resonant circuit can select a signal operating at the resonant frequency, and discriminate against signals at all other frequencies.

Q-factor

The term Q-factor usually arises whenever oscillatory waveforms are being considered. The Q of a component is a measure of its energy-storing ability. Since resistance in a circuit dissipates energy as heat, then inductors and capacitors that have large resistive losses have poor energy-storing ability; their Q-factor is small. The concept of Q can be expressed quantitatively in the following way. The Q of a circuit is defined as

$$Q = 2\pi \times \frac{\text{Energy stored in circuit}}{\text{Energy dissipated in the circuit during one cycle}}$$

In this general definition the circuit is usually taken to be a resonant circuit. The stored energy is transferred back and forth between the inductive and capacitive parts of the circuit. At some instant all of the stored energy is in the inductance; this occurs when the current in the coil is a maximum. The stored energy at this instant is $LI_{max}^2/2$ where I_{max} is the maximum current achieved during the oscillatory cycle. Half a period later all of this energy is stored in the capacitor. The energy dissipated in the resistor is given by power $\times$ time $= i_{r.m.s.}^2 R\tau$ where τ is the period of oscillation equal to $1/f_o$ (where f_o is the resonant frequency). Since $i_{r.m.s.} = I_{max}/\sqrt{2}$ then the energy dissipated during one cycle is $I_{max}^2 R/2f_0$. Hence

$$Q = 2\pi \frac{LI_{max}^2/2}{I_{max}^2 R/2f_0} = \frac{2\pi f_0 L}{R} = \frac{\omega_0 L}{R}$$

For a series resonant circuit $\omega_o L = 1/\omega_o C$, hence

$$Q = \frac{\omega_0 L}{R} = \frac{1}{\omega_0 C R}$$

The Q of a circuit can be defined in terms of the bandwidth of the resonance curve. A definition based on bandwidth is a very convenient way of describing the steepness of the curve, and hence the ability of the circuit to select a wanted signal from those signals operating at adjacent frequencies. The derivation of a suitable expression for Q depends upon the fact that for a series resonant circuit the current, and hence the power, is a maximum at the resonant frequency; this follows from an appreciation that Z is a minimum at the resonant frequency, so that, for a constant r.m.s. driving voltage the current must be a maximum. At each of two other frequencies, f_1 and f_2 (where $f_1 < f_0 < f_2$) the power dissipated is half of that dissipated at resonance, i.e. the current is $I_{max}/\sqrt{2}$. The range of frequencies between f_1 and f_2 is arbitrarily defined as the bandwidth, Δf, i.e. $\Delta f = f_2 - f_1$. Figure 4.18 illustrates these facts.

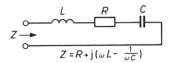

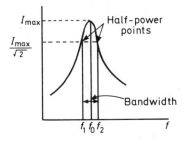

Figure 4.18 Series resonant circuit and associated resonance curve, showing how the current in the circuit varies with frequency

The magnitude of Z at f_1 and f_2 is $\sqrt{2} \times$ impedance at resonance, i.e. $|Z| = \sqrt{2}.R$, hence

$$R^2 + \left(\omega_1 L - \frac{1}{\omega_1 C} \right)^2 = 2R^2 = R^2 + \left(\omega_2 L - \frac{1}{\omega_2 C} \right)^2$$

where ω_1 and ω_2 are $2\pi f_1$ and $2\pi f_2$ respectively.

$$\omega_1 L - \frac{1}{\omega_1 C} = \pm R$$

Since below resonance $\omega_1 L < 1/(\omega_1 C)$, $\dfrac{1}{\omega_1 C} - \omega_1 L = R$.

This can be rearranged into the standard quadratic equation form

$$\omega_1{}^2 + \frac{R}{L}\omega_1 - \frac{1}{LC} = 0$$

Since $Q = \omega_0 L/R$ and $\omega_0^2 = 1/(LC)$, the practical solution to the quadratic equation is

$$\omega_1 = \omega_0 \left[\sqrt{\left(1 + \frac{1}{4Q^2}\right)} - \frac{1}{2Q} \right]$$

(We reject the negative root as having no real physical meaning.)

By using a similar technique, and remembering that $\omega_2 L > 1/(\omega_2 C)$ above the resonant freqency, it is readily shown that

$$\omega_2 = \omega_0 \left[\sqrt{\left(1 + \frac{1}{4Q^2}\right)} + \frac{1}{2Q} \right]$$

Hence $\omega_2 - \omega_1 = \dfrac{\omega_0}{Q}$, i.e. $f_2 - f_1 = f_0/Q$

Since $f_2 - f_1$ is the bandwidth Δf then

$$Q = \frac{f_0}{\Delta f}$$

For large values of Q (say > 10) the term $1/4Q^2$ can be ignored, hence

$$f_1 = f_0 \left(1 - \frac{1}{2Q}\right) \text{ and } f_2 = f_0 \left(1 + \frac{1}{2Q}\right)$$

Since $Q = 2\pi f_0 L/R$ then

$$f_1 = f_0 - \frac{R}{4\pi L} \text{ and } f_2 = f_0 + \frac{R}{4\pi L}$$

We see, therefore, that for high-Q circuits the resonance curve is symmetrical about the resonant frequency, within the bandwidth.

Parallel resonance

When the coil and capacitor are connected in parallel across a generator, parallel resonance is exhibited. The impedance of such an arrangement is small at frequencies below and above the resonant frequency. At the resonant frequency the impedance becomes very large.

An expression for the resonant frequency of a given parallel circuit can be derived by considering the expression for the impedance of the circuit. Initially, the mathematical manipulations are simpler if we consider admittance. With reference to Figure 4.19 $Y = 1/Z$ and is given by

$$Y = j\omega C + \frac{1}{R + j\omega L}$$

$$= j\omega C + \frac{R}{R^2 + \omega^2 L^2} - j \frac{\omega L}{R^2 + \omega^2 L^2}$$

The j-term is zero at a frequency that makes $\omega C = \omega L/(R^2 + \omega^2 L^2)$. Since the j-term is zero for Y it must also be zero for Z; the impedance then has only a resistive component, and the circuit is resonant at the frequency involved. Hence

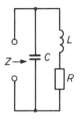

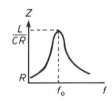

Figure 4.19 Parallel resonant circuit

$$C(R^2 + \omega_0^2 L^2) = L$$

$$\omega_0^2 = \frac{1}{LC} - \frac{R^2}{L^2}$$

$$\text{i.e. } f_0 = \frac{1}{2\pi} \sqrt{\left(\frac{1}{LC} - \frac{R^2}{L^2}\right)}$$

We see, therefore, that the parallel resonant frequency for a given coil/capacitor pair is slightly different from the series resonant frequency. In practice, however, R^2/L^2 is usually much smaller than $1/LC$, and hence can be ignored. For high-Q coils, therefore, the series and parallel resonant frequencies are practically the same.

At the resonant frequency the impedance is given by

$$Z = \frac{R^2 + \omega_0^2 L^2}{R} = R + \frac{L^2}{R}\left\{\frac{1}{LC} - \frac{R^2}{L^2}\right\}$$

$$= \frac{L}{CR}$$

This is called the *dynamic impedance* of the circuit.

Equation transformation

We mentioned earlier the possibility of using a more sophisticated way of transforming differential equations into algebraic form. Having done this we can then use well-known and simple techniques to analyse a.c. circuits.

Let us now reconsider the series LCR circuit of Figure 4.6. The related differential equation describing the behaviour of the circuit is

$$Ri + L\frac{di}{dt} + \frac{1}{C}\int i dt = v(t) \tag{4.2}$$

where $v(t)$ is some function of time.

For the moment we shall restrict ourselves to sinusoidal functions. In general we could express the driving function as $v(t) = \sqrt{2}\, V\, e^{j\omega t}$ where $V = Ve^{j\theta_1}$, V being the r.m.s. value, and θ_1, some arbitrary phase angle. ($\sqrt{2}V$ is the amplitude and corresponds to r in the polar expression $re^{j\theta}$.) The expression for i must also be sinusoidal and therefore $i = \sqrt{2}I\, e^{j\omega t}$ where $I = Ie^{j\theta_2}$, I being the r.m.s. value of the current. Differentiating and integrating the expression for i enables us to transform Equation 4.2, since

$$R\sqrt{2}.Ie^{j\theta t} + j\omega L\sqrt{2}.Ie^{j\omega t} + \frac{1}{j\omega C}\ \sqrt{2}.Ie^{j\omega t} = \sqrt{2}.Ve^{j\omega t}$$

i.e. $\sqrt{2}.Ie^{j\omega t}\left(R + j\omega L + \frac{1}{j\omega C}\right) = \sqrt{2}.Ve^{j\omega t}$

Thus

$$V = I\,Z \qquad\qquad\qquad (4.3)$$

where $Z = R + j\omega L + \dfrac{1}{j\omega C}$

We have thus recovered the simple Ohm's Law expression for an *LCR* circuit. We see that Z is a complex quantity equal to $Ze^{j\phi}$ where

$$Z = \sqrt{\left[R^2 + \left(\omega L - \frac{1}{\omega C}\right)^2 \right]}\ \text{and}\ \phi = \tan^{-1} \frac{\left(\omega L - \dfrac{1}{\omega C}\right)}{R}$$

Z is called the *complex impedance*.

The expression $V = IZ$ contains a good deal of information. Remembering if $a + jb = c + jd$ then $a = c$ and $b = d$ (the ordinary or real parts are equal, as are the quadrature components, or so-called imaginary parts). In this connection it should be noted that if the voltage across an inductor is expressed as $j\omega Li$ then the voltage is 90° out of phase in some respect to some reference voltage; this is why it is called a quadrature component. The voltage is real enough in a physical sense, but when $j\omega Li$ is part of a mathematical analysis the term is often referrred to as the imaginary part of a whole expression. The terms 'real' and 'imaginary' are derived from the pure mathematician's vocabulary when dealing with the theory of complex numbers. Since $a + jb$ can be expressed in the form $r\angle\theta$, then if $r_1\angle\theta_1 = r_2\angle\theta$ then

$$r_1 = r_2 \text{ and } \theta_1 = \theta_2$$

Since $V = IZ$

then

$$Ve^{j\theta_1} = Ie^{j\theta_2} . Ze^{j\phi}$$

From these results

$$V = IZ$$

i.e. the r.m.s. voltage is the product of the r.m.s. current and the modulus of the complex impedance; additionally, $\theta_1 = \theta_2 + \phi$. If $\theta_1 = 0$, as is often the case when the phasor representing the voltage is the reference phasor, then $\theta_2 = -\phi$; $-\phi$ is the phase angle between the current and the voltage.

Because

$$\phi = \tan^{-1}\frac{\left(\omega L - \dfrac{1}{\omega C}\right)}{R}$$

the sign of $-\phi$ depends upon the relative values of ωL and $1/(\omega C)$. When $\omega L > 1/(\omega C)$ the circuit is predominantly inductive. The phase angle of the current relative to the voltage is negative, i.e. the current lags the voltage. For negative values of ϕ (when $1/(\omega C > \omega L$) the phase angle between the current and voltage $(-\phi)$ is positive, i.e. the current leads the voltage by an angle ϕ.

Laplace transforms

We have seen that by representing sinusoidal functions by rotating phasors, and then describing the latter mathematically by expressions such as $v = \sqrt{2}.Ve^{j\omega t}$ we can transform differential equations into algebraic form. We therefore move from the time domain (in which expressions such as $L\,di/dt$ or $1/C.\int idt$ are used) to what is called the frequency domain in which $L\,di/dt$ is transformed to $j\omega Li$, etc. In the frequency domain we are limited to sinusoidal functions, and the solutions obtained so far describe the situation only when the steady state obtains, i.e. only after the currents and voltages become truly sinusoidal. We learn nothing about the initial transient period before the steady state is reached, say on switching on the apparatus. In addition, the methods so far discussed do not allow us to find solutions when the network is energized by non-sinusoidal functions, such as square waves, ramp and step voltages. When presented with these problems we are once again faced with the difficulty of developing a suitable mathematical technique to help us. Fortunately such a technique has been developed, and depends upon Laplace transformation. By using Laplace transforms we may once again convert differential equations into algebraic form, but in addition we can obtain the transient as well as the steady-state solution; we are no longer restricted to sinusoidal functions.

In the transformation we have previously discussed, the sinusoidal time-domain graphs were replaced by rotating phasors of constant length (representing constant amplitude). This led to the symbolic representation $\sqrt{2}.Ve^{j\omega t}$. A more versatile expression is $\sqrt{2}.Ve^{(\sigma + j\omega)t}$ where σ is some real number. Since $\sqrt{2}.Ve^{(\sigma + j\omega)t} = \sqrt{2}.Ve^{\sigma t}e^{j\omega t}$ this latter expression includes sinusoids of varying amplitude. If σ is a positive number then we have an increasing amplitude; when $\sigma = 0$ we are back to sinusoidal waves (of constant amplitude); and if σ is negative then we have damped sinusoidal waveforms. e^{st} where $s = (\sigma + j\omega)$ thus allows a consideration of a wider set of electrical systems and behaviour than $e^{j\omega t}$. A number of integral transforms have been based on the use of these exponential expressions. The Laplace transform is one mathematical device which is useful for solving the types of problems we are likely to encounter in electronic systems. If a function is specified for all positive values of time from time $t = 0$, we can, for the cases we are likely to encounter, write down a related quantity known as its Laplace transform; conversely if by some analysis the Laplace transform is obtained, the corresponding time function can be deduced. The method provides a powerful way of analysing circuits in which both the transient and steady-state solutions are obtained, and where the forcing functions are not necessarily sinusoidal.

Space limitations do not permit a discussion here on the mathematical background and principles underlying the derivation of Laplace transforms. Suufice it to say that differential equations can be transformed into algebraic ones by replacing the first and second derivatives by s and s^2 respectively; in general the nth derivative is replaced by s^n. $\int dt$, is replaced by $1/s$. The appropriate initial conditions must also be included. In simple cases this can be achieved by using the appropriate equivalent circuit as shown below; the more general expressions are given in Table 4.1.

Table 4.1 Some useful Laplace transforms

Definition: $L[f(t)] = \int_0^\infty e^{-st} f(t) \, dt = F(s)$

$f(t)$	$F(s) = L[f(t)]$
K	K/s
t	$1/s^2$
t^n	$n!/s^{n+1}$
e^{-at}	$1/(s+a)$
$\sin \omega t$	$\omega/(s^2+\omega^2)$
$\cos \omega t$	$s/(s^2+\omega^2)$
$t \sin \omega t$	$2\omega s/(s^2+\omega^2)^2$
$t \cos \omega t$	$(s^2-\omega^2)/(s^2+\omega^2)^2$
$\sinh \omega t$	$\omega/(s^2-\omega^2)$
$\cosh \omega t$	$s/(s^2-\omega^2)$
$\sin \omega t - \omega t \cos \omega t$	$2\omega^3/(s^2+\omega^2)^2$
$\delta(t)$	1
$H(t)$	$1/s$
$H(t+T)$	e^{-sT}/s
$f(t-T)H(t-T)$	$e^{-sT}F(s)$
$e^{-at}f(t)$	$F(s+a)$
$\dot{y}(t)$	$s\bar{y}(s) - y_0$
$\ddot{y}(t)$	$s^2\bar{y}(s) - sy_0 - \dot{y}_0$
$d^n y/dt^n$	$s^n\bar{y}(s) - s^{n-1}y_0 - s^{n-2}\dot{y}_0 - s^{n-3}\ddot{y}_0 \ldots \left(\dfrac{d^{n-1}y}{dt^{n-1}}\right)_{t=0}$
$\int_0^t y(t) \, dt$	$\dfrac{1}{s}\bar{y}(s)$

The general procedure for solving differential equations is to replace the differentials by the appropriate operator (s, s^2, etc.) and then find an algebraic expression for the Laplace transform of the quantity under examination. The solution then involves a manipulation of the algebraic expression, using elementary algebraic techniques, until the right-hand side of the expression yields recognizable transforms. The time functions are then obtained from the inverse of the recognizable transforms. Examples given later may serve to clarify the situation.

Circuit equivalents

The reader will recall that the object of developing a mathematical technique is to enable us to reduce the circuits under consideration to simple forms. For time-invariant voltages and resistors, it was shown that Ohm's Law, Kirchhoff's Law and theorems such as Thévenin's and Norton's applied. These simple laws can be extended to the a.c. case for sinusoidal variations by using the expressions $j\omega L$ and $1/j\omega c$ as the 'resistance' equivalents for inductors and capacitors respectively; in moving into the frequency domain the j-terms automatically take into account the phase information. Let us see now how we can develop circuit equivalents, using Laplace transforms for the passive components so as to be able to use simple d.c. laws when the voltages and currents involved are not sinusoidal. Furthermore, we wish to obtain both transient and steady-state information. We shall then be moving from the frequency domain to what is called the s-domain, or complex frequency domain.

The resistor case is easy. Resistors present the same resistance, R, to current irrespective of the waveform involved. The opposition to flow is therefore represented by R in the d.c. case, the time domain, frequency domain and the s-domain.

For inductors the voltage across the component is given by

$$v(t) = L\frac{di}{dt}$$

For non-sinusoidal variations, and where transient information is required we shall use the Laplace transformation technique outlined above. Therefore

$$\int_0^\infty e^{-st}\, v(t)\, dt = L \int_0^\infty e^{-st}\frac{di}{dt}\, dt$$

i.e.

$$v(s) = L \int_0^\infty e^{-st} \cdot \frac{di}{dt}\, dt$$

Integrating by parts

$$\int_0^\infty e^{-st}\frac{di}{dt}\, dt = [e^{-st}i]_0^\infty + s \int_0^\infty e^{-st}\, i\, dt$$

$$= -i_o + si(s)$$

where i_0 is the value of i at time $t = 0$, and $i(s)$ is the Laplace transform of $i(t)$. Hence

$$v(s) = sLi(s) - Li_o$$

When $i(t) = 0$ at time $t = 0$, $v(s) = sLi$, i.e. $Z(s) = v/i$, the simple Ohm's Law form. Note also that $j\omega$ in the frequency domain has been replaced by s in the s-domain.

The equivalent circuit must not assume that $i_o = 0$ at time $t = 0$; the circuit must therefore represent the fact that the applied voltage is opposed by two voltages $sLi(s)$ and $-Li_o$. The circuit is therefore as shown in Figure 4.20(a).

It is important to note that the direction of the voltage arrow representing Li_o is always in the same direction as the initial current. Some care is needed in specifying the direction of the arrow associated with sLi. The conventions used in d.c. cases should be observed. The circulating currents in the closed loops of the networks should be drawn in a clockwise direction; the voltage across sL will then be represented by an arrow pointing in the opposite direction to the current direction. Depending upon the problem, the voltage arrow representing the voltage across sL will sometimes point in the opposite direction to the Li_o arrow, as shown in Figure 4.20(a); at other times it may be in the same direction.

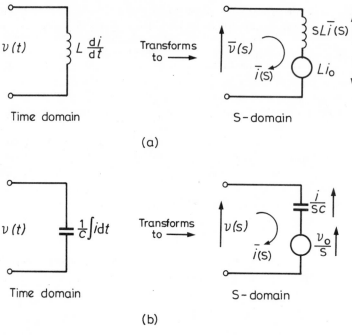

Figure 4.20 Equivalent circuits in the s-domain for inductors and capacitors

For capacitors we commence in the time domain with the expression relating voltage and current, i.e.

$$i = C\frac{dv}{dt}$$

Hence

$$\int_0^\infty e^{-st} i \, dt = \int_0^\infty e^{-st} C\frac{dv}{dt} \, dt$$

i.e.

$$i(s) = C\int_0^\infty e^{-st} \frac{dv}{dt} \, dt$$

Integrating by parts

$$\int_0^\infty e^{-st} \frac{dv}{dt} \, dt = [e^{-st} v]_0^\infty + s\int_0^\infty e^{-st} v(t) \, dt$$

Therefore

$$i(s) = -Cv_0 + sCv(s)$$

If at time $t = 0$ there is no charge on the capacitor, $v_0 = 0$ and

$$v(s) = \frac{i(s)}{sC}$$

$1/(sC)$ is the 'impedance' of the capacitor in the s-domain, equivalent to $1/(j\omega C)$ in the frequency domain. The full equivalent circuit, including the initial conditions, is obtained from

$$v(s) = \frac{i(s)}{sC} + \frac{v_0}{s}$$

and is given in Figure 4.20(b).

It is important to note that the direction of the arrow showing the polarity of the initial voltage v_0/s, is always in the same direction as the actual voltage on the capacitor at $t = 0$. As with the inductor, care must be taken with the arrow representing the voltage across $1/(sC)$; the direction of this arrow may or may not be in the same direction as the v_0/s arrow. The actual direction will be determined by the direction taken for $i(s)$. In Figure 4.20(b) both arrows are in the same direction. If at time $t = 0$, however, the capacitor were to be switched into, say, a resistive network, and the arbitrary direction of $i(s)$ were to be changed, then the direction of the voltage arrow representing $i(s)/(sC)$ would be opposite to that of the arrow representing v_0/s.

Before proceeding to some examples we will need to be familiar with what is called the First Shifting Theorem, which states that if $y(s)$ is the transform of $y(t)$ and a is a constant then $y(s + a)$ is the transform of $e^{-at}y(t)$. The simple proof can be seen from

$$\int_0^\infty e^{-st}[e^{-at} y(t)]dt = \int_0^\infty e^{-(s+a)t} y(t) \, dt$$

i.e. the transform of $e^{-at}y(t)$ is equal to the transform of $y(t)$ in which s is replaced by $(s + a)$.

The reader should now be in a strong position to solve network problems. With the theoretical background outlined above and in previous chapters, problems may be solved by recognizing that they fall into obvious categories.

(a) When steady voltages and currents are evident in resistive circuits then the so-called d.c. theorems are invoked (Ohm's Law, Kirchhoff's Law, Norton, Thévenin, Superposition, mesh and nodal analysis). All of these are simple enough in concept and involve only elementary algebraic techniques.

(b) When time-variant voltages and currents are involved and are sinusoidal (and this implicity excludes transient phenomena) then all the 'd.c. laws' may be applied to the a.c. case provided the 'resistance equivalents' of capacitors and inductors are expressed as $1/(j\omega C)$ and $j\omega L$ respectively. The solutions will be complex and of the form $a + jb$, $r<\theta$ or $re^{j\omega t}$; from these solutions steady-state information may be extracted involving magnitudes and phase. The equivalent circuit diagrams in the frequency domain look the same as those in the real or time domain except that X_L $(=\omega L)$ is replaced by $j\omega L$, and $X_C(=1/\omega C)$ is replaced by $1/(j\omega C)$. Simple algebraic techniques are then all that are required.

(c) When non-sinusoidal quantities are involved, and when information about the transient period is required, we move into the s-domain. The real circuits are then replaced by their s-domain equivalent circuits. All of the 'd.c. laws' may once again be invoked and only elementary algebraic techniques are required. The solutions will be functions of s, the Laplace operator, and from these solutions the time functions can be obtained from the 'inverse' Laplace transform.

Let us see how to apply the s-domain equivalent circuits to problems posed in the following examples.

Non-sinusoidal and transient response in LCR circuits

Example 4.4. Figure 4.21 shows a series arrangement of a capacitor and resistor. The switch is closed at time $t = 0$. Find expressions for the voltages across the resistor and capacitor as functions of time. The capacitor is uncharged at time $t = 0$.

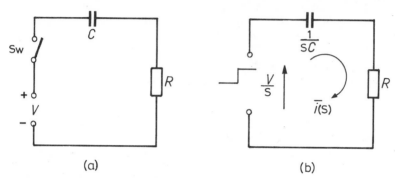

Figure 4.21 Charging of a capacitor via a resistor: (a) Original circuit; (b) Transformed circuit in s-domain

Figure 4.21(b) shows the transformed circuit in which V transforms to V/s because V is a constant. We first need to obtain the Laplace transform for i, because then the voltage across R, $\bar{v}_R$ (s) is given by $\bar{i}R$; similarly the voltage across C is given by $\bar{v}_C = \bar{i}/(sC)$.

$$\frac{V}{s} = \bar{i}\left(R + \frac{1}{sC}\right) \therefore \bar{i} = \frac{VC}{1 + sCR} = \frac{V/R}{s + 1/(CR)}$$

(Note that where no ambiguity exists, $\bar{i}(s)$ may be written as $\bar{i}$.)

To find the inverse transform, i.e. $i(t)$, we make use of the First Shifting Theorem, thus

$$i(t) = \frac{V}{R} e^{-t/CR}$$

Hence

$$v_R(t) = iR = Ve^{-t/CR}$$

Since the voltage across C is $V - v_R$ then the corresponding expression for $v_C(t)$ is

$$v_C(t) = V(1 + e^{-t/CR})$$

Figure 4.22 shows the position in graphical form. If at time $t = 0$ the switch is moved to A and allowed to stay there, the accompanying graphs show the resultant voltages across C and R at any subsequent time. The capacitor is assumed to be uncharged initially. The sum of the capacitor voltage, v_C, and the voltage across the resistor, v_R, must equal the supply voltage, V. After a time equal to about six time constants the charging current will be practically zero; as a result $v_R = 0$, and the

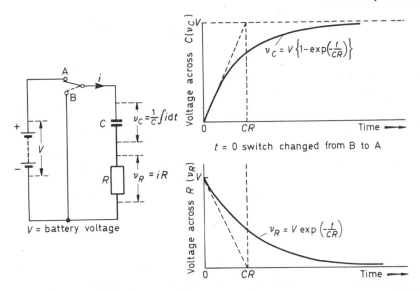

Figure 4.22 Voltage waveforms associated with C and R during charging. $v_C + v_R = V$. $CR =$ time constant, i.e. the time it would take for the voltage across the resistor to become zero if the original rate of charging the capcitor had been maintained. When $t = CR$, $v_R = V/e$, where e is the exponent of the natural logarithms, equal to 2.718. An alternative definition for the time constant is therefore the time it takes for the voltage across R to fall from V to about 37 per cent of V. In the same time the charging of the capacitor will have reached a stage where the voltage across the capacitor has become $[1 - (1/e)]\,V$, i.e. about 63 per cent of V

whole of the supply voltage appears across the capacitor. On moving the switch to B, the capacitor will discharge through the resistor; v_C therefore falls to zero. Since now $v_C + v_R = 0$, the voltage across the resistor will be $-V$ initially and subsequently will approach zero as the capacitor becomes discharged. The position is shown in Figure 4.23.

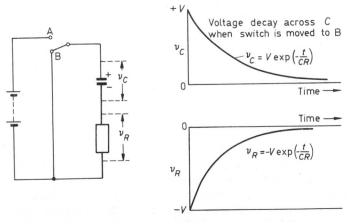

Figure 4.23 Voltage waveforms associated with C and R when the switch is moved to B. C then discharges through R. $v_C + v_R = 0$

Time constant

The time concept of a time constant is very important in electronics. By examining the mathematical expression for v_R we see that theoretically this voltage never reaches zero; we note also that the voltage across the capacitor in Figure 4.22 never quite reaches V. These statements are true irrespective of the values chosen for C and R. How then are we to compare the behaviour of different CR combinations, since the time for v_C and v_R to reach V and zero volts, respectively, is infinite in all cases? Readers will be aware that a corresponding problem exists in radioactivity, in which it is seen that all radioactive substances, no matter what their rate of decay, take an infinite time for the radioactivity to fall to zero, i.e. the life of every radioactive material is theoretically infinite. To compare the rates of decay of different radioactive substances we arbitrarily introduce the concept of 'half-life', i.e. the time taken for half of the radioactive atoms to disintegrate, and hence for the radioactivity of the element to diminish to half its initial value.

When we wish to compare different CR combinations during the charging process, we could choose to compare the times taken for the voltages across the resistors to fall to half the initial value (or alternatively for the voltage across the capacitor to reach half its final value). The choice of half in this case, however, would not lead to simple mathematical expressions. We therefore arbitrarily define the time constant as the product CR so that after the elapse of this time (i.e. $t = CR$) then $v_R = Ve^{-1}$ and $v_C = V(1 - e^{-1})$. These simple mathematical expressions enable us to define the time constant as the time taken for the voltage across R (during the charging process) to decay to Ve^{-1} (that is, 36.8 per cent) of its initial value. After the elapse of six time constants the voltage across R is below 0.3 per cent of its initial value. The initial rate of fall of the voltage across R can easily be found by differentiating the expression for v_R with respect to time and evaluating this rate at time $t = 0$. It will be found that if this initial rate of fall had been maintained during the charging process, then the voltage across R would fall to zero after one time constant had elapsed, i.e when $t = CR$. This gives us an alternative definition for the term time constant (see Figure 4.22).

If we imagine the switch shown in Figure 4.22 to operate continuously, staying as long in position A as in position B, the resulting waveform presented to the CR circuit will be rectangular with a mark-space ratio of 1. The waveforms across C and R depend upon the period of the square wave and the time constant CR. Figure 4.24 shows the resulting waveforms when a square wave with a period of about CR is applied to a capacitor-resistor combination. It is clear that the waveforms across C and R, respectively, are different from the input waveform. The concept of time constant, and especially its comparison with the period of the input waveform is therefore of great importance to designers of electronic circuits. Often the change in waveform is highly undesirable, but in some cases changes are a necessary consequence of performing some special operation on the input signal; differentiation and integration are two cases in point.

From the waveforms shown in Figure 4.24 it will be seen that there is a transient period of about, say, four or five time constants, after which the steady-state conditions obtain. The input waveform has a steady component ($V/2$ in this case) which is loosely called the d.c. level. The square waves are symmetrical about this level. The capacitor, of course, cannot transmit the steady (or d.c.) level and hence in the steady state, the square waveform across the resistor varies about the zero voltage level; the steady component of the input waveform appears across the

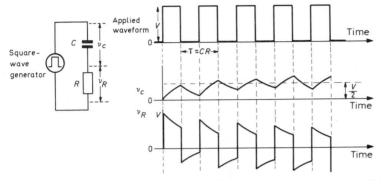

Figure 4.24 Waveforms resulting from the application of an input square wave with a period of about CR

capacitor when steady-state conditions have been established. The actual waveform levels can be calculated by considering each period separately as the switch in Figure 4.22 is moved from A to B and back again. Apart from the transition at time $t = 0$, the capacitor will have a charge upon it, and this must be taken into account when calculating the value of v_R during the next half period of the input waveform. (This means that v_o/s is not zero for all calculations except the first.)

The differentiating circuit

A simple differentiating circuit is formed when the time constant of a CR combination is much smaller than the period of the applied waveform. Figure 4.25 shows the associated waveforms. A small time constant results from the choice of small-value capacitors and resistors. Under these circumstances the effect of the resistor on the instantaneous value of the current in the circuit is small; nearly all of the applied voltage appears across the capacitor. For small values of C and R, therefore, the current, i, is given to a close approximation by the expression $i = Cdv/dt$ because v_C is almost equal to v. If we take the output voltage to be that

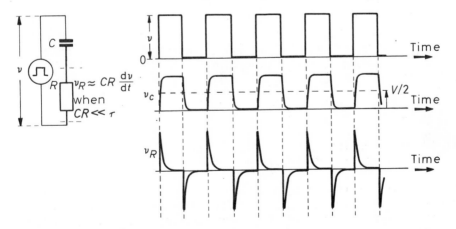

Figure 4.25 A simple differentiating circuit is formed when small values of C and R are used, i.e. CR « the period of the applied waveform

across R this will be iR, i.e. $CR\mathrm{d}v/\mathrm{d}t$. Hence v_R is proportional to the differential coefficient of the input voltage with time.

The integrating circuit

A simple integrating circuit is formed when the time constant of a CR circuit is much longer than the periodic time of the input waveform. Figure 4.26 shows the associated waveforms. In this case when the value of C is large, the voltage developed across the component is not very great compared with the voltage across R. Nearly all of the signal voltage, v, appears across the resistor, and it is this

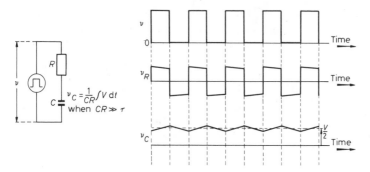

Figure 4.26 The simple integrating circuit. The output voltage across the capacitor is approximately proportional to the time integral of the input voltage when $CR \gg$ period of the applied waveform

large-value component that effectively controls the current. To a close approximation therefore (for large time constants compared with a given input period) the current in the series combination is given by $i = v/R$. Since the voltage across the capacitor is given by $v_C = (1/C) \int i\,\mathrm{d}t$ then $v_C = (1/CR) \int v\,\mathrm{d}t$. If we take the output voltage to be that across the capacitor then this output voltage is proportional to the time integral of the input voltage.

The coupling circuit

The circuit in Figure 4.27 is used extensively to connect one stage of an amplifier to the next. The usefulness of this arrangement will become increasingly apparent as the reader progresses in his studies of electronics. The object of this circuit is to transmit only the alternating component of the signal waveform from one part of

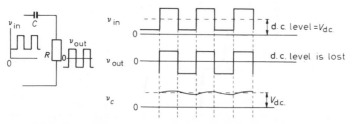

Figure 4.27 The coupling circuit. Here C and R are made large so that CR is a good deal greater than the period of the lowest frequency waveform it is desired to handle

the circuit to the next, blocking out any steady component that the input signal contains. Provided the time constant is large, say greater than five or six times the period of the input signal, the voltage across the resistor has a waveform almost identical with that of the varying part of the input signal. It will be remembered that the voltage across a given capacitor depends upon charge stored. If that charge does not alter, the voltage across the capacitor will not change. We can arrange that very little charging or discharging of the capacitor takes place by making R large. For example, in a coupling circuit to be used with signals whose frequency is as low as 50 Hz, the value of R may well be 20 kΩ with a capacitance of 10 μF. The resulting time constant, CR, is $10 \times 10^{-6} \times 20 \times 10^3 = 0.2s$, which is long compared with the period of a 50 Hz signal, i.e. 0.02s.

Figure 4.28 further illustrates the point about the transmission of a variable voltage by a capacitor. It must be emphasized that the function of the capacitor in a coupling circuit is to block out the d.c. level and to transmit only variations of the input signal. This it does by *not* charging or discharging. In so far as charging and discharging occur, the capacitor is failing to transmit the variations properly. We shall see in later chapters that some charging and discharging is inevitable, but we must arrange for this to be a minimum.

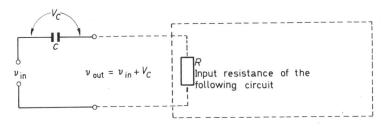

Figure 4.28 By keeping the charge on the capacitor, q, constant, V_C is constant. v_{out} will vary only if v_{in} varies. If there is no leakage in C then the variations are transmitted without change in magnitude or waveform. Only the d.c. level is lost. In practice, the following circuit inevitably allows some change in the charge, q, but this change is usually kept to a minimum by using a large value of C and having a large value of input resistance in the following circuit

When a square wave is passed through a coupling circuit, the top of the waveform droops unless the time constant is very long compared with the period of the square wave. This droop can be calculated in terms of the fundamental frequency of the square wave and the lower 3 dB point for the network, as shown in the chapter on amplifiers. (See under 'sag, in square waves'.)

Inductance-resistance (LR) circuit

Inductance in a circuit resists changes of the current in that circuit; if a series LR circuit is supplied with square waves, the waveforms across the inductor and resistor can be deduced by using Laplace transforms; Figure 4.29 shows the relevant waveforms.

Example 4.5. In the circuit shown in Figure 4.30 the switch, after being in position A for a long time, is switched into position B at time $t = 0$. Obtain an expression for the subsequent current through the inductor, as a function of time.

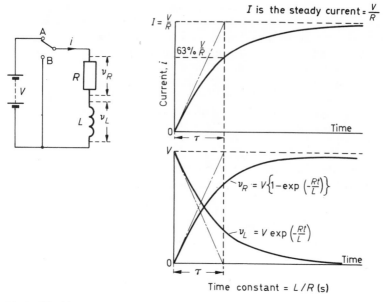

Figure 4.29 (a) The establishment of a steady current (V/R) is shown in the upper graph. Once a steady current is flowing there will be a steady magnetic field associated with the inductor. The lower graph shows the corresponding voltages across the two components

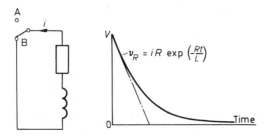

Figure 4.29 (b) Once the current in the circuit in (a) is steady the voltage across the inductor is zero. The inductor stores energy ($\frac{1}{2}LI^2$) as a magnetic field. If the switch is now moved to B the field collapses, and the voltage across R decays exponentially

When the switch is in position A the current taken from the 40 V source is 1 A. The initial current through the inductor is therefore 0.5 A. When the switch is changed to position B the 30 Ω and 20 Ω resistors are in parallel, and the equivalent resistance is 12 Ω. The circuit of Figure 4.30(b) is therefore the s-domain equivalent. The value of i is given by $i = Li_o/(R_1 + R_2 + sL)$; therefore

$$i = \frac{0.15}{32 + 0.3s} = \frac{0.5}{s + 107}$$

Using the First Shifting Theorem

$$i(t) = 0.5 \, e^{-107t}$$

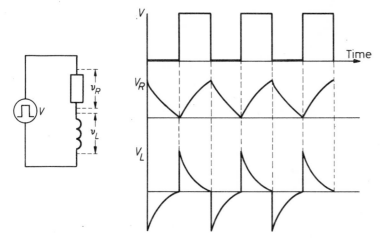

Figure 4.29 (c) Waveforms obtained after steady-state conditions have been achieved. The time constant is less than the period of the applied square wave

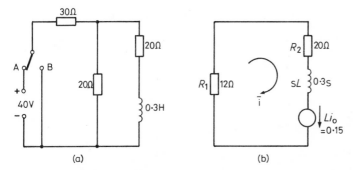

(a)   (b)

Figure 4.30 Circuits for example 4.5: (a) Circuit diagram; (b) equivalent circuit in the s-domain for all positive values of time. Switch is in position B, therefore $R = 30\,\Omega$ in parallel with $20\,\Omega$ i.e. $12\,\Omega$

Example 4.6 A single ramp waveform is applied to the circuit shown in Figure 4.31(a). Obtain an expression for the output voltage, v_{out}, as a function of time, and sketch the output waveform.

Up to time $t = 0$ the input voltage is zero, and therefore we may assume that at the moment the ramp voltage is applied, the capacitor is uncharged. The time equation for the ramp is given by $v_i = mt$ where m is the slope, equal to 1 V per millisecond, i.e. 10^3V.S^{-1}. For the duration of the ramp the equivalent circuit in the s-domain is as shown in Figure 4.31(b). Consulting the table of Laplace transforms (Table 4.1) we see that mt transforms to m/s^2. The initial problem is to find an expression for i from which an expression for the transform of the output voltage is obtained, (since $\tilde{v}_{\text{out}} = \tilde{i}/(s.10^{-7})$).

$$\frac{m}{s^2} = \tilde{i}\left(R + \frac{1}{sC}\right) \quad \therefore \tilde{i} = \frac{mC}{s(1 + sCR)}$$

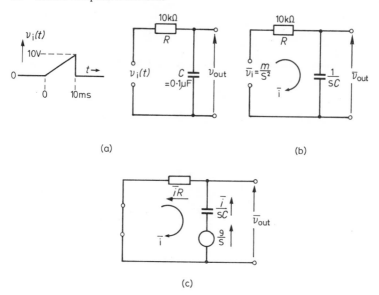

(a)   (b)

(c)

Figure 4.31 Transformations for the problem posed in example 4.6: (a) Circuit diagram; (b) Transformed circuit in the s-domain for the period $t = 0$ to $t = 10$ ms; (c) Transformed circuit for $t >$ 10 ms

The output voltage $\bar{v}_{\text{out}}$ is, for the period $t = 0$ to 10 ms, given by

$$\bar{v}_{\text{out}} = \frac{\bar{i}}{sC} \quad \therefore \quad \bar{v}_{\text{out}} = \frac{m}{s^2(1 + sCR)}$$

Using the method of partial fractions

$$\bar{v}_{\text{out}} = m \left[\frac{A}{s^2} + \frac{B}{s} + \frac{D}{1 + sCR} \right] = \frac{m}{s^2} - \frac{mCR}{s} + \frac{mC^2R^2}{1 + sCR}$$

The right-hand side is now in recognizable forms of Laplace transforms; note that $mC^2R^2/(1 + sCR) = mCR.1/(s +1/[CR])$. Therefore

$$v_{\text{out}}(t) = mt + mCR \left(e^{-t/CR} - 1 \right)$$

Since CR in the example is 10^{-3} and $m = 10^3$ Vs^{-1}, the voltage across the capacitor at $t = 10$ ms is given by $v_{10} = 10 + e^{-10} - 1 = 9$ V (near enough). After 0.5 ms v_{out} (t) will be approximately 0.1 V.

For times greater than 10 ms, the s-domain equivalent circuit is as shown in Figure 4.31(c). The input voltage $v_i(t)$ has dropped to zero, hence $v_i(s)$ must also be zero. At time $t = 10$ ms there is a charge on the capacitor that makes the voltage across the capacitor equal to 9 V. This is our initial condition for an analysis for times subsequent to $t = 10$ ms.

We now start again with a new $t' = 0$ corresponding to the old $t = 10$ ms. From Figure 4.31(c)

$$\bar{v}_{\text{out}} = \frac{\bar{i}}{sC} + \frac{9}{s}$$

but

$$-\tilde{i} - \frac{\tilde{i}}{sC} - \frac{9}{s} = 0$$

(This is derived by using the usual convention as explained for the d.c. case in Chapter 3.)

$$\tilde{i}\left(R + \frac{1}{sC}\right) = -\frac{9}{s}$$

$$\text{i.e. } \tilde{i} = \frac{-9C}{1 + sCR}$$

Thus

$$\tilde{v}_{out} = \frac{-9}{s(1 + sCR)} + \frac{9}{s} = \frac{9}{s}\left(1 - \frac{1}{1 + sCR}\right)$$

$$= 9\left(\frac{CR}{1 + sCR}\right) = 9\left(\frac{1}{s + \frac{1}{CR}}\right)$$

Hence

$$v_{out}(t) = 9e^{-t'/(CR)}$$

Where t' represents all times later than $t = 10\,\text{ms}$. The required sketch is shown in Figure 4.32.

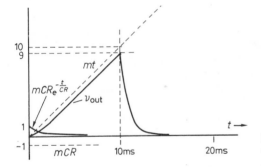

Figure 4.32 Sketch of the output voltage for example 4.6

Exercises

1. Show that when a steady voltage is applied at time $t = 0$ to a series LR circuit which had previously not been energized, that the current for times $t>0$ is given by

$$i = I_o\left[1 - \exp(-Rt/L)\right]$$

2. A voltage, v_i, having a ramp waveform ($v_i = mt$) is applied to the input terminals of a high-pass CR filter. Show that the output voltage developed across the resistor is given by

$$v = mCR\left[1 - \exp\left(-t/CR\right)\right]$$

3. Show that the two-terminal network illustrated in Figure 4.33 has an impedance that is resistive at all frequencies provided $R = \sqrt{(L/C)}$.

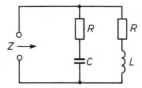

Figure 4.33 Circuit for exercise 3

4. Obtain the Thévenin equivalent circuit for the network shown in Figure 4.34. Hence, or otherwise, determine the power dissipated in a load of $(25 + 50j)\ \Omega$. Assume that in this case the generator voltage is 10 V r.m.s.
   (Ans. 0.4 W)

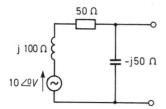

Figure 4.34 Circuit for exercise 4

5. Figure 4.35 shows the input circuit of a high-grade oscilloscope represented by an input resistance of 1 MΩ in parallel with a capacitor of 30 pF. A passive probe consisting of a series resistor R in parallel with a compensating capacitor C is connected to the input terminal of the oscilloscope; the passive probe is designed to attenuate the input signal voltage by 20 dB.
   (a) Calculate the bandwidth of the probe/oscilloscope arrangement in the absence of C. (Ans. 5.9 kHz.)
   (b) Calculate the value of C that gives optimum frequency response. (Ans. 3.3 pF.)

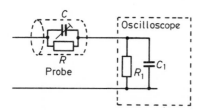

Figure 4.35 Circuit for exercise 5

6. Figure 4.36 shows the circuit of a four-terminal network. A 10 V step is applied to the input terminals at time $t = 0$. Assuming that all of the initial conditions are zero, obtain an expression for the output voltage in the time domain. Sketch the graph of output voltage as a function of time.
   (Ans. $v_o(t) = 6.45 + 3.07 \exp(-590.5t)$)
7. Sketch the output waveforms for the circuits shown in Figure 4.37.

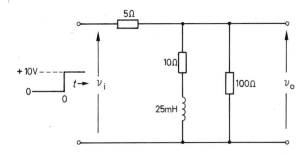

Figure 4.36 Circuit for exercise 6

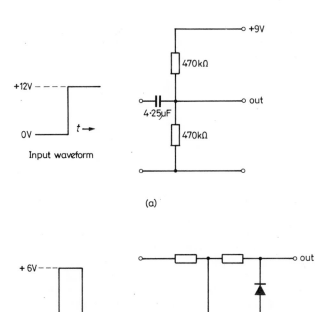

(a)

(b)

Figure 4.37 Circuit for exercise 7

Part 2

Analogue electronics

Chapter 5

Semiconductor devices

Since the announcement of the invention of the transistor in 1948, by the American physicists Bardeen and Brattain, the developments that followed have been far-reaching, and have changed the whole outlook of the science and technology of electronics. The overwhelming number of semiconductor devices in use today depend upon the controlled diffusion of selected impurities into the element silicon.

Silicon is a tetravalent element, the outermost electrons of which are held in covalent bonds. At temperatures above absolute zero, and particularly at normal ambient temperatures, some of these bonds are broken; the freed electrons are injected into the body of the crystal. Such electrons, being charge carriers, are available for conduction. The energy (usually from thermal sources) that 'digs out' the electron leaves behind a hole. The hole, or deficit, produced by removing an electron from the valence-bond structure of the crystal is conveniently regarded as a separate entity having a positive charge equal in magnitude to the charge on the electron. The vacancy, or hole, may be filled by an adjacent electron that has insufficient energy to leave a covalent bond. The hole in effect has moved to the site previously occupied by the adjacent electron; thus the hole can move about the whole volume of the semiconductor material. The movement of a hole is actually a more complicated movement of electrons, just as the movement of a bubble in water is actually a more complicated movement of water molecules. Provided the holes and electrons are not subjected to an electric or magnetic field, they wander throughout the crystal lattice in a random fashion. The presence of mobile charge carriers in the crystal gives rise to the possibility of establishing an electric current. An electric current is a net movement of charges along a system. This may be achieved in one of three ways:

(a) The charges may be swept through the crystal by a varying magnetic field. Except for transformer action this method is unusual in electronics and need not be considered further here.

(b) In the presence of an electric field a steady drift is superimposed upon the random motion of the charge carriers. This drift is due to the continuous electrostatic force acting on the charge carriers. In intrinsic material (i.e. material which is free from impurities) the electric field may be created by the application of external voltages to electrodes on the crystal surfaces. The conduction is termed intrinsic, and in so far as it occurs, degrades the performance of semiconductor devices. Fortunately, in silicon, intrinsic conduction is small, which is one reason why silicon replaced germanium as the basic element used in the manufacture of the majority of semiconductor devices. Intrinsic conduction, though small, is too large to enable silicon to be classed as an insulator; on the other hand the conduction is many orders less than that found in conductors. Silicon is therefore classed as a semiconductor.

As we shall see later in the chapter the conductivity may be increased by adding specified impurities to the silicon; so-called p-type and n-type silicon can be formed. When both types are present in a single crystal, an electric field can be established within the crystal. The consequences of this are described in the section on pn rectifiers.

The drift velocity of the charge carriers is usually small compared with the thermal random velocity of the carriers; it is proportional to the strength of the electric field and depends also upon the mean free path between collisions. The mobility, μ, of the charge carriers is related to the drift velocity, dx/dt by the relation

$$\frac{dx}{dt} = \mu E$$

where E is the electric field strength and x is the distance along the crystal from a specified surface.

(c) A third, and important, way of establishing a current in crystals is by diffusion. Whenever the distribution of electrons and holes in a given volume of semiconductor material is not uniform, diffusion occurs. The movement of charge carriers constitutes a current, and is analogous to the movement of gas molecules in a given volume. In fact the mathematical analyses associated with the Kinetic Theory of Gases are also used in developing theories concerning the currents and charges found in semiconductors. It is a matter of common knowledge that if a canister containing a concentrated gas is opened in a closed room, the gas molecules diffuse so as to fill every part of the room. If sufficient time is allowed for the diffusion to occur, eventually the density of the gas will be uniform throughout the room. The rate of diffusion depends upon the particle velocity and hence on the average thermal energy $3kT/2$, where k is Boltzmann's constant and T is the absolute temperature in K. The number of molecules passing through a small area in unit time is the flow rate, F, which is proportional to the difference in concentrations of the molecules on each side of the area, i.e. proportional to the density gradient (say dn/dx) where n is the concentration and x, as before, is the distance from some arbitrary surface.

Hence

$$F \propto \frac{dn}{dx} \quad i.e. \quad F = D \ \frac{dn}{dx}$$

where D is the diffusion coefficient. The minus sign arises because the concentration n diminishes as the distance x increases.

In an analogous manner the number of electrons passing through unit area per second (i.e. the electron current) is given by $-D_n \, dn/dx$ where D_n is the diffusion constant for electrons and dn/dx the electron density gradient. Similarly, $-D_p \, dp/dt$ represents the position for holes. By analogy again with kinetic theory it can be shown that

$$D_n = \frac{kT\mu_n}{e} \text{ and } D_p = \frac{kT\mu_p}{e}$$

where μ_n and μ_p are the mobilities of electrons and holes respectively, and e is the electron charge. Replacing the molecular flow rate above (F) with the current and remembering that the charge on the electron is negative we obtain

$$I_n = e \, D_n \, \frac{dn}{dt} \text{ for electron current}$$

$$\text{and } I_p = -e \, D_p \frac{dp}{dt} \text{ for hole current}$$

In practice the transition from idealized kinetic theory to charge movement in semiconductor material requires certain empirical modifications to be made to the formulae. Mobilities, for example, shown to be inversely proportional to temperature in kinetic theory are more nearly approximated in silicon by the expression $\mu \propto T^{-3/2}$.

Impurity condition

One way of increasing the conductivity of a crystal is to add a controlled amount of certain impurities. The conduction that occurs is then called impurity conduction and is of paramount importance in the operation of semiconductor devices.

When small amounts (1 part in 10^7 approximately) of a pentavalent impurity such as arsenic or phosphorus are added during crystal formation, the impurity atoms lock into the crystal lattice, since they are not greatly different in size from a silicon atom, and the crystal is not unduly distorted. Four of the impurity valence electrons form valence bonds with adjacent silicon atoms, but the fifth is very easily detached at normal room temperature, and becomes free within the crystal. The pentavalent impurity donates an electron to the crystal and is therefore called a donor impurity. The material is then known as n-type silicon. The crystal as a whole is of course still electrically neutral, but the number of available charge carriers has been considerably increased. Since only about 0.045 eV is required at room temperature to release the fifth electron into the conduction state, all the donor centres are ionized.

Instead of adding a pentavalent impurity to an otherwise pure crystal, we may include a trivalent impurity such as boron or aluminium. The three valence electrons enter into convalent bonding, but a region of stress exists where the fourth bond ought to be, and any stray electron in the vicinity can be trapped by the impurity centre, which then becomes negatively charged. The region from which the electron came is positively charged. In effect the trivalent impurity atom has injected a hole into the crystal. Such an impurity is called an acceptor impurity since it accepts electrons into the impurity centre. Silicon doped in this way has many more holes than free electrons as charge carriers and is therefore termed a p-type semiconductor.

In an n-type semiconductor, electrons are in the majority and are called majority carriers. Some holes exist owing to the formation of electron-hole pairs at room temperature; these are called minority carriers. Alternatively, in a p-type semiconductor, holes are the majority carriers and electrons the minority carriers. The number of charge carriers available determines the resistivity of the material.

The pn junction rectifier

If, during manufacture, a crystal is doped so as to be p-type in one region and n-type in an adjacent region, a very important practical semiconductor device is obtained. It is the concern of the technologist to ensure that the change from p-type to n-type material occurs abruptly at a well-defined boundary, called a pn junction. After formation the electrons diffuse from the n-type material and fill some of the acceptor centres. The result is that a layer of negative charge lies towards the junction on the p-side. A corresponding diffusion of holes to the n-side creates a layer of positive charge towards the junction on the n-side. The final result (Figure 5.1) is the creation of a potential gradient across the junction. The electric field in the junction region is between fixed charged centres, *viz.* the acceptors, which are negatively charged, being on the p-side, while the fixed donor centres, positive because of the loss of an electron, are on the n-side. The potential gradient increases with diffusion until eventually it is great enough to prevent any further migration of electrons from the n-side and holes from the p-side. A dynamic equilibrium is then established. The potential gradient is often represented, as in Figure 5.1, by a small internal fictitious battery with the polarity as shown.

The application of a potential difference across the whole crystal can be considered by referring to Figure 5.2. If the positive pole of the external e.m.f. source is connected to the p-side, and the negative pole is connected to the n-side,

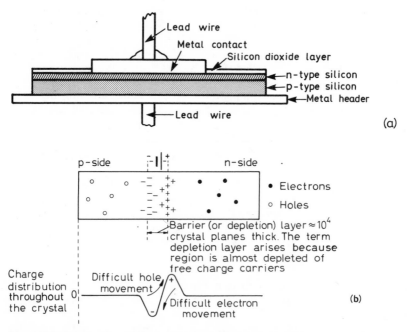

(a)

(b)

Figure 5.1 (a) A planar epitaxial pn junction rectifier; (b) the pn junction. Upon formation the electrons diffuse from the donor impurities leaving the centres positively charged. The electrons are captured by the acceptor centres. Two layers of charge are formed along the junction boundary. Diffusion ceases when the barrier potential (represented by a fictitious internal battery) is high enough to prevent holes diffusing from the p-side, and electrons from the n-side

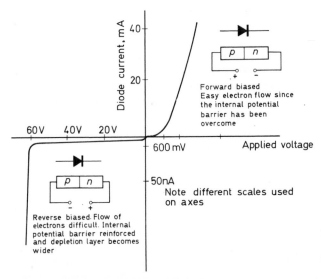

Figure 5.2 Rectifying action of a junction diode

the electrons and holes will be drawn away from the junction and the potential barrier that exists there will be lowered. When the external voltage exceeds a few hundred millivolts, the internal 'battery' is totally overcome. Electrons and holes then flow freely across the junction, and the whole crystal is then a good conductor of electricity. When, however, the polarity of the external voltage source is reversed, so that the negative terminal is connected to the p-type region and the positive terminal is connected to the n-type region, then the potential barrier within the crystal is increased. Electrons are repelled towards the junction on the p-side and holes are repelled towards the junction on the n-side. A comparatively thick barrier layer is formed in which there are very few free charge carriers. Such a layer, known also as a depletion layer, is a very good insulator, and thus very little current flows through the crystal. The pn junction device is therefore a rectifier, conducting electricity well in one direction, but not in the other.

A strong electric field exists across the depletion layer of a reverse-biased diode. If a covalent bond in the depletion layer is ruptured, the resulting electron and hole are swept to the n- and p-sides of the junction respectively (Figure 5.3). The movement of these charge carriers constitutes a leakage current. In addition, minority carriers from the p-side (electrons) and minority carriers from the n-side

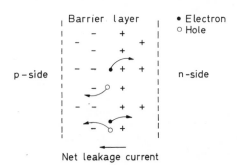

Figure 5.3 At room temperature electron-hole pairs are formed in the depletion layer constituting the junction. The holes are swept into the p-side and the electrons to the n-side. The net leakage current is from the n-region to the p-region

(holes) also move across the junction. The total reverse current, which is very small in silicon devices ($\approx 10^{-14}$ A) is known as the reverse saturation current; it is almost independent of the applied reverse voltage, but is temperature-dependent. In practice it is found that the empirical value of the reverse current is a good deal higher than the figure quoted above. For diodes intended to pass forward currents of one or two amperes the measured reverse current is of the order of 10^{-8} A. This is because most of the observable reverse current is due to 'seepage' around the surface of the device. In manufacture, therefore, a good deal of attention must be paid to surface preparation and packing.

Point contact diodes

Historically, point contact diodes were in use long before pn junctions. Many readers will know of the 'cat's whisker' era in the 1920s.

Radio was becoming increasingly popular and many amateurs built their own crystal receiving sets. To demodulate the radio waves it was necessary to use a metal-to-semiconductor rectifying diode. The crystal involved was a lead sulphide crystal (impure galena). The sharp point of a brass wire was brought into contact with the crystal, the contact being maintained by bending part of the wire into a spring – hence the name 'cat's whisker'.

Although such rectifiers are not used today, it is still necessary to use the principle involved when radio signals of high frequency have to be detected. The ordinary pn junction has quite a large capacitance associated with it when reverse-biased. Charge is stored in such a capacitor. The capacitance is too large for the efficient detection of very-high-frequency signals. At such high frequencies the reactance of even a small capacitance is quite low, being perhaps only tens or hundreds of ohms. This means that during the half-cycle when no current should be flowing there is a significant charging current. The rectification properties are therefore largely nullified.

The difficulty can be overcome by using a metal-to-semiconductor diode. Such an arrangement involves holding the end of a metal wire in contact with p-type or n-type silicon or germanium. A diode of this type cannot store charge, and in any case the effective rectifying area is very small. No disturbing capacitance effects are evident even at high frequencies.

Because of the small rectifying area involved, the current-carrying capacity of a point-contact diode is small. Although this is of no consequence when using the device as a signal detecting device, such a limitation would be severe when rectifying large alternating currents. For power supplies, therefore, pn junctions are needed.

Zener diodes

When a pn junction diode is biased in the reverse direction, the majority carriers (holes in the p-side and electrons in the n-side) move away from the junction. The barrier, or depletion, layer becomes thicker, and the absence of charge carriers is manifest as a very high resistance region across which current transfer becomes difficult. The presence of minority carriers, and the electron-hole pairs formed within the depletion layer, account for the very small leakage current. This leakage current remains very small for all reverse voltages up to a certain critical value. Once this value has been exceeded there is a sudden and substantial rise in the reverse current. Over the operating range of reverse current it is found that the voltage across the diode remains almost constant. For this reason the most common application of Zener diodes is in voltage regulation and for use as voltage reference elements.

The term Zener diode arises because it was thought that the mechanism responsible for the behaviour of the diode was similar to that proposed by Dr Carl Zener in his paper 'A Theory of Electrical Breakdown of Solid Dielectrics' (Proc. Roy. Soc. 1934). It is now believed that for operational voltages in excess of about 8 V the mechanism can be described as an avalanche effect in which, at some critical voltage depending upon the doping levels, electrons are accelerated and acquire sufficient energy to dislodge valence electrons which in turn dislodge others, and so on. Since most Zener diodes have operating voltages in excess of 8 V they should more correctly be termed 'avalanche' diodes, but the expression 'Zener' is still widely used.

True Zener diodes operate at voltages below about 5 V, and are produced by using high doping levels. The resulting depletion layer is extremely thin and the bulk of the crystal has comparatively low resistivity. Nearly all the applied voltage exists across the very narrow depletion layer, and this results in very high electric field strengths within the depletion layer. In this very intense field electrons are excited from the valence band into the conduction band. The excitation is a quantum mechanical process in which electron-hole pairs are formed; this is the Zener effect. Diodes operating at voltages from about 5 V to 8 V use a mixture of both the Zener and avalanche effects simultaneously.

Zener diodes are available at present in 5, 10 and 20 per cent voltage tolerances. The same preferred numbers for their nominal operating voltages are used as those already in use for resistors, e.g. 4.7 V, 15 V, 22 V, etc.; the range up to 75 V is readily available, but units are manufactured with operating voltages up to about 200 V. Power ratings as high as 60 W are available, but for electronic purposes wattage ratings of 400 mW, 500 mW, 1.3 W, 5 W and 20 W are common. Typical characteristics are shown in Figure 5.4.

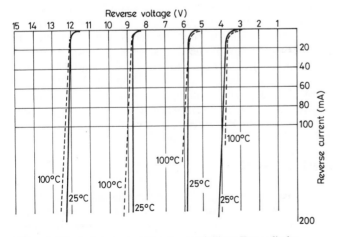

Figure 5.4 Typical characteristics of a set of silicon Zener diodes

Temperature coefficients of diodes that use the avalanche effect are positive while those using the Zener effect are negative. We see from Figure 5.5 that a zero temperature coefficient diode is possible that operates at just over 5 V. The curve shown in Figure 5.5 is for one value of diode current. This means that the user would have to accept the voltage, but more importantly, since the temperature coefficient depends also on the diode current, he would have to ensure that the optimum current was used. A more economical solution, as well as a little flexibility in voltage, can be achieved if a 5.6 V Zener (which has a positive temperature coefficient) is combined with a normal silicon diode having a temperature coefficient of −2.1 V/°C. The normal diode is operated in the forward-biased mode and therefore has a voltage drop of about 600 mV; the combined diode and Zener diode therefore forms a voltage reference source of 6.2 V nominal. The IN821 has a temperature coefficient of only 0.01%/°C while the IN827 has a temperature coefficient of only 0.001%/°C at an operating current of 7.5 mA.

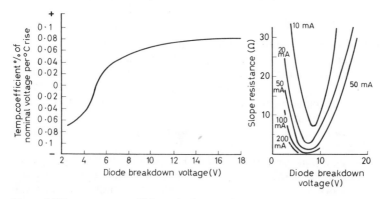

Figure 5.5 Temperature coefficient of voltage and the variation of slope resistance with reverse current for a typical series of silicon Zener reference diodes

Apart from the working voltage and temperature coefficient, the potential user should consider the slope resistance. The a.c. slope resistance is related to the working voltage and the reverse current. It is, in fact, the slope of the reverse voltage/reverse current characteristic at constant junction temperature. When the characteristic is plotted slowly enough for the junction temperature to become steady between each measurement the slope is the d.c. slope resistance. For zero temperature coefficient conditions the a.c. and d.c. slopes are identical. The lower the slope resistance the more constant is the operating voltage with changes in current. Graphs or sets of typical values are available from manufacturers for the diodes they manufacture. Figure 5.5 shows the temperature coefficient curve and slope resistance curves for a range of silicon Zener reference diodes.

The silicon controlled rectifier (SCR)

This device, also known as a thyristor, can be used to control the large amounts of power involved in the operation of electric motors, furnaces and high-power lamps. The resistance of an SCR in the forward or conducting state is generally only a fraction of an ohm; the voltage drop across the device is therefore low, being in the region of 1 volt. Powers of tens of kilowatts can easily be controlled by an SCR that is physically quite small.

The main constructional features of a thyristor are shown in Figure 5.6. The thyristor is a three-terminal device consisting of four layers of semiconducting material in a pnpn sandwich. There is continuity of the crystal structure throughout the device, and therefore three pn junctions exist between the layers.

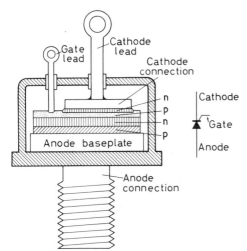

Figure 5.6 Constructional features and circuit symbol of a silicon controlled rectifier (thyristor)

When the device is connected in series with a load resistance and d.c. source, no current can pass through the system if the supply polarity is such as to make the anode negative with respect to the cathode. Under these circumstances both the cathode and anode junctions are reverse-biased. Upon reversing the polarity of the supply voltage, so that the anode is now positive with respect to the cathode, the device remains non-conducting because the gate junction is now reverse-biased. In this condition both the cathode and anode junctions are forward-biased and hence nearly all of the applied voltage appears across the gate junction. The field across the depletion layer at this junction is high because the voltages that in practice are applied are large (e.g. 10 V or so up to several hundred volts). If in this condition the gate is driven sufficiently positive to inject a pulse of current into the device the thyristor will be 'fired' into conduction. The gate current to initiate firing must exceed a certain critical value, and may be regarded as an injection of holes into the gate junction. These holes neutralize the negative space charge on the p-side of the gate junction. Because of the presence of a positive space charge on the n-side of the gate junction, the injected holes migrate along the junction (Figure 5.7). Although the gate current pulse is relatively small (10 mA of gate current may be all that is needed to switch 20 A of load current) the density of holes between the gate and cathode is high enough not only to neutralize the local space charge, but also to induce the injection of a large number of compensating electrons from the cathode. These electrons are attracted to the positive space charge on the n-side of the gate junction. They gain sufficient energy to cause avalanche breakdown (as discussed in the section on Zener diodes). This avalanche breakdown spreads right along the gate junction. The thyristor now carries a large forward current. The gate can now no longer influence the conduction process; it cannot stop conduction because it

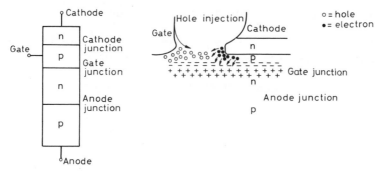

Figure 5.7 'Firing' of a controlled rectifier

loses all control once the main current is established. The rectifier can return to its non-conducting state only when the main current density is reduced below the critical value necessary for avalanche multiplication, and held at this low value long enough for charge recombination to occur. Simultaneously, the gate current must be zero. It is not practical to switch off large currents with a conventional mechanical switch; a method of switching off when d.c. supplies are involved is shown in Chapter 6. Fortunately the majority of power control applications involve energizing the load from the a.c. mains supply. Since the polarity of the mains voltage passes through zero, and reverses, during every cycle, then it is necessary only to remove the gate current to effect a switching-off of the main current.

It is customary in some quarters to analyse the action of an SCR by means of a two-transistor (pnp and npn) model in which the emitters represent the anode and cathode of the SCR. Such an analysis is false because it implies that the gate current must flow over the entire cathode or emitter junction, which it does not. Also the main load current must be carried by base junctions, which seems odd. Further, we might have to assume that the carrier migration time from the gate lead across the entire face of the junction determines the turn-on time; fortunately this is also incorrect.

The triac

The majority of cases concerning the control of electric power involve the use of alternating current. Since a thyristor conducts only when the anode is positive with respect to the cathode, the current in a.c. circuits must be zero for every other half cycle. Full power could never be achieved, which would be a disadvantage when operating motors or incandescent lamp dimmers. The disadvantage could be overcome in one of two ways. For small powers the use of a full-wave diode rectifier (see Chapter 6) would enable us to use control power through the whole of the supply cycle; alternatively, two thyristors could be used 'back-to-back', one conducting for half a cycle and the other conducting during the subsequent half cycle. Inevitably this led to the fabrication of the triac, a device which is in effect two thyristors 'back-to-back' within a single crystal structure.

The triac is thus a three-electrode semiconductor switch that can be triggered into the conduction mode, via a gate electrode, with any given polarity of the mains voltage applied to its terminals. The latter are known as main terminal 1 and main terminal 2. Figure 5.8 shows the construction schematically, together with the

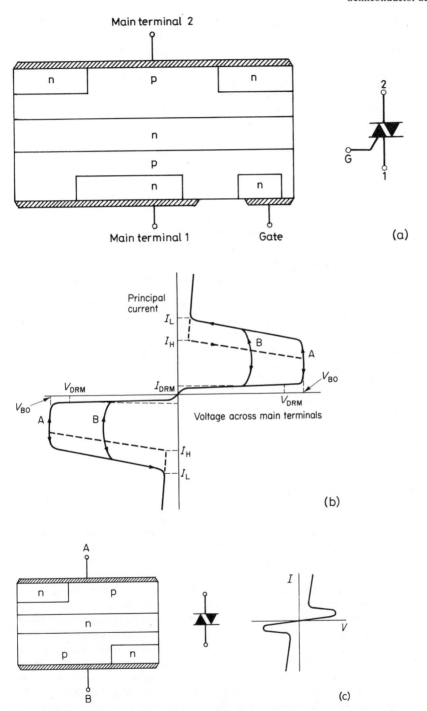

Figure 5.8 Multijunction devices: (a) Basic structure and circuit symbol; (b) principal voltage/current characteristics of the triac; (c) schematic, circuit symbol and re-entrant characteristic of a diac

principal voltage/current characteristics. With zero gate current the device breaks down at a voltage V_{BO}. Clearly, a device for any given application must be chosen with a V_{BO} value that exceeds the peak value of the repetitive applied voltage, V_{DRM}; if V_{DRM} exceeds V_{BO} curve A is followed. Curve B is the V/I trace when gate signals are applied to the triac. Once the latching current I_L has been exceeded the gate electrode loses control. The turn-on time is typically about 1 μs. To turn the device off, the current must fall below a minimum holding current, I_H. Since $I_L > I_H$ a hysteresis effect is evident. The usual way of reducing the principal current to zero is to lower the terminal voltage to zero. The turn-off times for triacs is about 50 μs.

The diac

Triggering of the triac is usually performed by including a diac in the gate lead. A diac is a glass encapsulated device with a structure similar to a miniature triac, but without the gate electrode (Figure 5.8). When terminal A is positive with respect to terminal B, two of the junctions of the sandwich illustrated to the right of the diagram are forward-biased. The third junction is reverse-biased and prevents conduction until the magnitude of the applied voltage is sufficiently high to initiate avalanche breakdown. Once this occurs the voltage across the device falls and conduction then takes place. Because of the symmetry of the device, reverse applied polarity brings about a similar condition in that part of the sandwich illustrated to the left of the diagram. The I/V characteristic curve for both applied polarities is thus as shown in the figure.

Practical applications of thyristors, diacs and triacs are given in Chapter 6 dealing with diode applications.

The unijunction transistor

This type of transistor consists of an n-type silicon bar with ohmic contacts (called base one, b_1, and base two, b_2) at opposite ends. (An ohmic contact is one that conducts current easily, irrespective of the direction of flow of the current.) A single rectifying contact is made between the two base contacts, usually closer to b_2 than b_1. This rectifying contact is called the emitter. During normal operation b_2 is held positive with respect to b_1 via a suitable load resistor (Figure 5.9). Provided the emitter junction is reversed-biased, there is no emitter current and the n-type silicon bar acts as a simple potential divider, the voltage at the emitter contact being a fraction, β, of the voltage from b_2 to b_1. To maintain this condition the emitter junction must remain reverse biased; this is ensured by keeping the emitter voltage, V_E, less than βV, where V is the voltage from b_2 to b_1. When V_E is allowed to exceed βV the junction becomes forward-biased and current flows in the emitter-to-b_1 region. The emitter injects holes into this region, and as they move down the bar there is an increase in the number of electrons in the lower part of the bar. The result is a substantial decrease in the bar resistance from the emitter to b_1. This decrease in emitter-to-b_1 resistance brings about a fall of emitter voltage. A condition arises whereby increases of emitter current are accompanied by decreases of emitter voltage, resulting in a negative resistance characteristic.

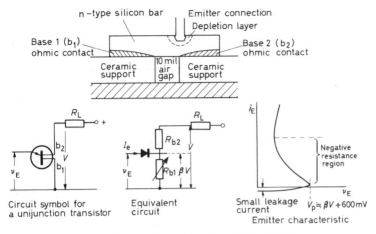

Figure 5.9 The unijunction transistor of the General Electric Company

Unijunction transistors are low in cost and are characterized by a stable triggering voltage, which is a fixed fraction of the voltage between b_2 and b_1. They have negative resistance characteristics which are uniform among units of the same type and which are stable with temperature and throughout the unit's life. Apart from its use in oscillators and time-delay circuits, the major use of a unijunction transistor is as a pulse generator to fire thyristors. This application is dealt with later.

Solar cells

The silicon photocell is a photovoltaic device that converts light directly into electrical energy. The selenium photocell makes this direct conversion, but the efficiency of a selenium cell is too low to allow it to be used as a solar battery. The silicon cell has an efficiency approaching 14 per cent in its present state of development, which is about 25 times greater than that of a selenium cell. (Efficiency in this context is the amount of electrical energy available from the device divided by the total radiant solar energy falling on the cell.)

Silicon cells are made by melting purified intrinsic silicon in quartz containers and adding minute traces of a Group V element, such as arsenic or phosphorus. The n-type silicon that results solidifies and is cut into slices. These slices after grinding and lapping are then passed into a diffusion chamber and boron is diffused into the n-type crystal from boron trichloride vapour. A pn junction results. The p- and n-type surfaces are then plated and terminal wires added.

It will be recalled that a barrier layer, in which very few charge carriers exist, is created between the p- and n-sides of the crystal, forming a pn junction. When discussing the pn junction as a rectifier we saw that the application of a reverse bias voltage increases the potential hill and prevents large numbers of electrons from flowing. The small leakage current that does result is attributed to the production of electron-hole pairs in the barrier layer, the energy coming from thermal sources. In the solar cell there is, of course, no reverse bias voltage, but nevertheless a potential hill exists across the junction. The incidence of radiant energy from the sun creates electron-hole pairs by rupturing the covalent bonds between atoms in the barrier layer. The holes are swept to the p-side and the electrons are swept to the n-side (Figure 5.10). If an external circuit exists, electrons flow round from the n-side to the p-side, dissipating energy in any load that is present. The source of the energy is the incident radiation which consists of photons of energy E, where $E = h\nu$, ν being the frequency of the radiation and h Planck's constant. The travelling of electrons to a negatively charged region may seem strange, but it must be remembered that the incidence of photons on the atoms in the barrier layer reduces the potential hill. The Fermi levels associated with junctions in the circuit are disturbed, giving a resultant e.m.f. of the polarity shown in Figure 5.10. In rather loose terms, there is an attempt to restore the potential hill to its former value.

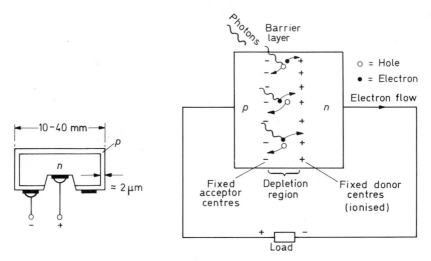

Figure 5.10 Principles of operation of a solar cell. The incidence of photons creates electron-hole pairs, the charge carriers are swept out by the field across the barrier layer. An electric current around the external circuit is then obtained

Solar cells are the source of power for energizing the transmitting and other electronic equipment in unmanned satellites. Wherever a source of light is available, solar cells can be used to energize low-powered transistor equipment instead of the more conventional batteries. They are also used in general photovoltaic work.

Transistors

The usefulness of electronics in modern technology would be extremely limited if we were restricted to those devices described in previous pages. The absence of an amplifying device would be a serious drawback. Fortunately, since 1948 the range of semiconductor devices has been extended to include transistors which, when combined with passive components in suitably designed circuits, have the ability to amplify small electric signals. Thus, for example, the tiny voltages and power available from a gramophone pick-up can be made to control very much larger powers in loudspeaker load circuits. When used as low-power switches in logic circuits, transistors can switch much larger powers in load circuits; digital computers then become a practical possibility. It is difficult to find any other recent invention that surpasses in importance the transistor; this device has revolutionized the whole of technology in the last few decades. The effect on the lives of those who live in technologically advanced countries is enormous.

The term 'transistor' arises from a combination of the italicized portions of the words *trans*fer and re*sistor*, since the device is made from semiconductor resistor material and a transfer of power is involved. Transistors are available in two varieties. Those that depend upon the presence of two pn junctions within a single crystal are called junction, or bipolar, transistors. The other variety consists of field-effect transistors, in which the magnitude of a load current is controlled by an electric field. This field is a created by applying a suitable potential to a control electrode known as a gate.

Field-effect transistors

Although it has been theoretically possible for many decades to make field-effect transistors, it is only comparatively recently that the significant advances in semiconductor technology have made possible the manufacture or reliable units in large numbers.

Basically there are two types of field-effect transistor, namely, the reverse-biased pn junction type and the insulated gate device.

The reverse-biased junction FET (JFET)

This type of device was first proposed by Shockley, who called it a unipolar FET because only one type of charge carrier is involved in the amplification process.

Figure 5.11 shows the cross-section of a JFET in idealized schematic form. Ohmic (i.e. non-rectifying) contacts are made at each end of the bar of n-type silicon; the majority carriers are therefore electrons in this case and we may for practical purposes ignore the presence of a small number of holes. The electron flow constitutes the current through the device and hence into the load circuit. The electron flow is from one end contact, called the source, to the other contact called the drain. The drain is maintained at a positive potential with respect to the source. For the Shockley device the n-bar has two p-type regions diffused into the body as shown in Figure 5.11. (An alternative arrangement is to have a cylindrical bar with ohmic contacts, constituting the source and drain, at the ends; p-type material is then formed by diffusion on the surface of the bar, but avoiding the two ends. We thus have an n-type core surrounded by a p-type region.) Metallized contact is then made with the p-type material; the contact is known as the gate, which constitutes the control electrode. The pn junction formed in the semiconductor material is reverse-biased. This is achieved by making the gate negative with respect to the source. Since the pn junction at the gate–channel interface is reverse-biased, a depletion layer extends into the body of the device thus reducing the channel width. Changes in channel width, brought about by varying the gate–source voltage, result in changes of resistance between source and drain, and hence changes in current through the device.

The shape of the depletion layer is not uniform throughout the length of the channel. For a given voltage between drain and source v_{DS}, the potential increases

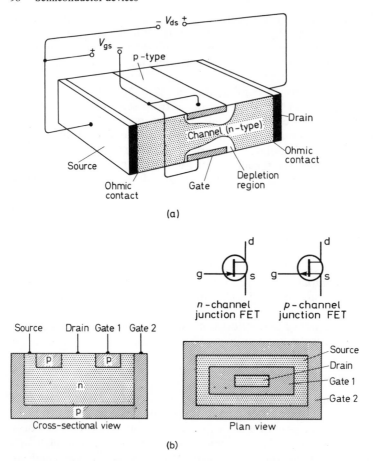

Figure 5.11 (a) The Shockley-type field-effect device; (b) a modern form of construction; for many applications gate 1 and gate 2 are connected in the external circuit

from source to drain along the channel; the reverse-bias voltage must therefore also increase for a given value of gate voltage at points approaching the drain. Consequently the depletion layer thickness increases towards the drain end of the channel.

Figure 5.12 shows a typical set of what are called output characteristic curves. This 'family' of curves shows the relationship between drain current, i_D, as a function of drain–source voltage, v_{DS}, for various fixed values of gate–source voltage, v_{DS}. These curves are of importance when considering the device as an amplifying agency as discussed in a later chapter.

When obtaining the curves, all voltages are relative to the source potential. For zero gate voltage, small increases in drain voltage result in a rapid initial rise of drain current. (Even for zero gate voltage the pn junction at points along the channel is reverse-biased because of the rising potential gradient along the channel from source to drain.) The graph of i_D against v_{DS} shows a rapid linear rise. In this region of operation the device acts as a resistor whose resistance is a function of the

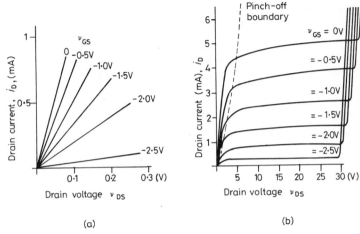

Figure 5.12 Drain characteristics for a JFET: (a) shows the position for low values of v_{DS}. The device acts as a resistor at these low values whose value is determined by v_{GS}; (b) the main output characteristics

resistivity of the n-type semiconductor material, v_{GS}, and the length and width of the channel. At low drain voltages the depletion layer does not extend into the channel to any great extent. Once v_{DS} exceeds only one or two volts, however, the channel width is reduced because the depletion layer penetration becomes significant. The drain current does not then rise so quickly with increasing values of v_{DS}. Soon the value of v_{DS} is high enough for the depletion layers at the drain end of the channel almost to meet; the channel is then said to be in the 'pinch-off' condition. It might be expected that the channel conductance falls to zero as the depletion layers from each side meet, thus cutting off the drain current. This does not happen, however, owing to the screening effect of the carriers concentrated at the centre of the channel. We find that once the pinch-off condition has been reached the drain current is almost independent of v_{DS}; the characteristic curve therefore flattens. The channel at this stage loses its ohmic properties and exhibits saturation characteristics.

If now the gate voltage is held at a fixed negative voltage, a similar mechanism operates, but the saturation current is lower because the whole effective channel is narrower. We see therefore that, for any fixed drain voltage above pinch-off conditions, the 'saturation' current is determined by the gate voltage.

The outstanding advantage of this type of transistor is the high input impedance that results from the reverse-biased pn junction. Input resistances as high as $10^8\,\Omega$ or $10^9\,\Omega$ are observed; the gate electrode is thus effectively isolated from the channel of the transistor. This means that the driving circuit or transducer has to supply negligible current and is therefore not disturbed by having an FET connected to it. We are assuming, of course, that v_{DS} is not high enough to initiate avalanche breakdown. At high values of v_{DS} breakdown does occur and the current rises rapidly, as shown in Figure 5.12(b).

FETs with p-channels are available. The mechanism is similar to that described above, but holes constitute the main channel current, and the polarity of the applied voltages must be reversed.

The metal-oxide-semiconductor transistor (MOST)

In an attempt to increase further the input resistance of field-effect devices, we have returned to an early idea whereby the gate electrode is electrically insulated from the conducting channel; in modern units the insulating layer between the gate and channel consists of an extremely thin deposit of silicon dioxide. The construction and mode of operation is thus significantly different from the reverse-biased junction types.

MOSTs are available in four varieties. Two of these are known as enhancement types and two are depletion types. We shall consider first the enhancement types.

Figure 5.13 shows in diagrammatic form the cross-section of an n-channel enhancement mode MOST. A lightly doped p-type silicon substrate has diffused into it two heavily doped n-regions (shown as n+). An insulating layer of silicon dioxide some 10^{-4} mm thick is then thermally grown over the surface. The heavily doped n-type regions are then exposed by etching away the silicon dioxide with hydrofluoric acid. Metal contacts are then deposited so as to make suitable ohmic contacts with the n-type regions, one of which acts as the source and the other the drain. Simultaneously, an aluminium layer is deposited on the surface of the oxide layer that remains between the n-type regions; this layer acts as the gate electrode.

Let us now consider an explanation of the shape of the drain characteristic curves showing the relationship between i_D and v_{DS} for given values of gate-to-source voltage, v_{GS}. With zero gate voltage, and zero drain voltage, $i_D = 0$. If now the drain voltage is increased from zero to some positive value, the pn junction at the drain becomes increasingly reverse-biased and no current can flow from source to drain. Consider now the case when v_{DS} is zero and the gate voltage is increased from zero to some positive value. Since the gate is now positive with respect to the source, an electric field is established between the gate and the p-type substrate. (It is usual to connect the substrate to the source either internally or by making an external electrical connection.) Initially, on increasing the gate voltage, the majority carriers (holes) are repelled and the region just below the oxide/substrate interface under the gate becomes depleted of charge carriers and behaves as though it were intrinsic. At this stage, even though the drain were made positive, the drain current would remain zero. Once a certain gate threshold voltage, V_T, has been exceeded, further increases in gate voltage attract minority carriers (electrons) to the oxide/substrate interface. The region under the gate then acts as though it were n-type because of the increasing electron density. Since increasing gate voltages result in increasing electron density, the channel, previously containing few charge carriers because of the light doping, is enhanced. Also the channel acting as n-type material connects the n-type drain and n-type source resistively; current can then flow if v_{DS} is made positive. For a given value of v_{DS}, increasing the gate voltage increases the conductivity of the channel, and thus increases the drain current, i_D.

The profile of the channel can be understood by realizing that, since both the gate and drain are made positive with respect to the source, the field at the source end must be greater than that at the drain end. The enhanced channel must therefore be thicker at the source end as compared with that at the drain end.

We are now in a position to explain the shape of the i_D/v_{DS} curve for gate voltages in excess of V_T. For any positive gate voltage above V_T, initial increases of v_{DS} from zero bring about linear increases in drain current; the channel is behaving resistively. As the drain voltage is increased, however, the channel at the drain end becomes thinner and consequently the rate of rise of i_D becomes less. Once v_{DS}

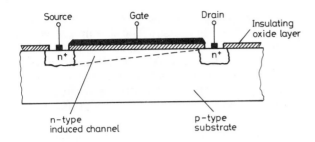

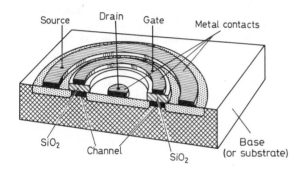

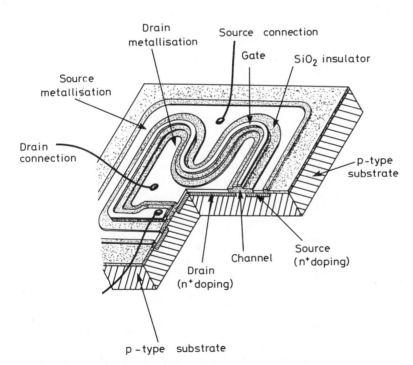

Figure 5.13 Diagrammatic representations of the structure of insulated gate MOSTs

becomes equal to v_{GS} the channel becomes very thin. It does not disappear because of the screening effect of the charge carriers constituting i_D. The channel, however, becomes 'pinched-off' and further increases in drain voltage do not bring about significant increases in drain current; the i_D/v_{DS} curve therefore exhibits a saturation effect. This almost constant drain current persists until the drain voltage is great enough to initiate avalanche breakdown, whereupon the drain current increases suddenly. Figure 5.14(a) shows the characteristics of an n-channel enhancement MOST.

By starting with an n-type substrate and making the gate negative with respect to the source a p-channel can be induced between source and drain. If a negative drain voltage is applied a drain current is established. The characteristics are similar to those of an n-channel MOST, but the polarities of the applied voltages must be reversed; the main channel current consists of holes (Figure 5.14(b)).

An enhancement mode MOST finds its main application in digital electronics because zero gate voltage cuts off the drain current. Such devices are suitable for the logic gates discussed in Part 3.

For some analogue electronic purposes it is desirable to avoid cutting off the drain current of a p-type substrate device by negative-going gate voltages (or

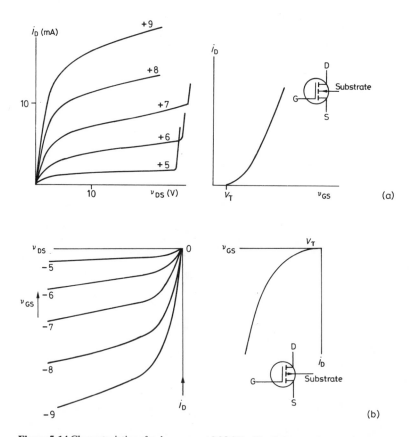

Figure 5.14 Characteristics of enhancement MOSTs. The i_D/v_{GS} curves are for an arbitrary fixed value of v_{DS}: (a) n-channel enhancement MOST; (b) p-channel enhancement MOST

positive-going gate voltages in an n-type substrate device). Under these conditions a depletion mode MOST can be used. The construction is similar to the enhancement modes except that the doping of the substrate is heavier and the drain, source and channel regions are all of the same conductive-type material; the drain and source are still heavily doped. The channel, as a result of the doping levels, has appreciable conductivity at zero gate voltage. In the case where n-type material is used the channel is n-type; the gate must therefore be driven negative in order to deplete the channel of charge carrier, and hence reduce the drain current. At zero gate voltage a considerable drain current can flow. The gate can be driven positive in which case the channel charge carrier density is further increased, and hence the drain current is greater. Figure 5.15(a) shows in diagrammatic form the cross-section of one type of construction together with characteristics and circuit symbols.

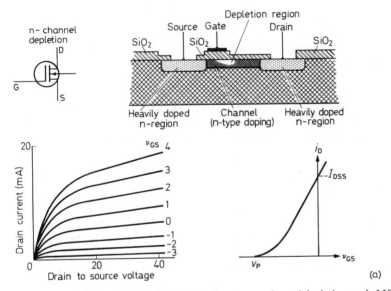

Figure 5.15 (a) Construction and characteristics for an n-channel depletion mode MOST. For a fixed drain voltage, i_{DSS} is the drain current that flows when $v_{GS} = 0$. (The circuit symbol for a p-channel MOST is similar to that shown, except that the arrow on the substrate lead points in the reverse direction)

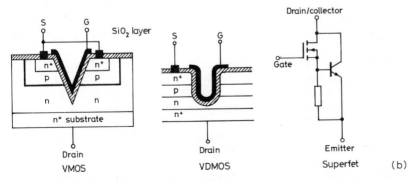

Figure 5.15 (b) Alternative geometries for MOSTs

The outstanding feature of MOSTs is their input resistance, which is of the order of 10^{14} or 10^{15} ohms. Unlike junction types, this input resistance is maintained even when gate bias voltages of the 'wrong' polarity are applied.

Alternative geometries

The type of FET discussed so far involves the movement of current in a channel that is parallel to the surface; since the current movement is across the device the construction is said to be lateral. For high-power applications, especially where high voltages are involved, the horizontal current path structure is unsuitable. Typical applications in the high-power field are in power control, mains control, the automotive industry and for high-power switching in computer applications. Lateral devices that are able to handle high currents are uneconomical to produce since they require comparatively high chip areas and production yields are low. Difficulties with registration of the photo-masked areas lead to limitations of gain–bandwidth product and may also result in high 'on' resistance switches.

One alternative geometry to the lateral arrangement that overcomes some of the problems is the vertical (VMOS) structure of the planar V-groove double-diffused MOSFET developed by the Japanese Electrotechnical Laboratory. The VMOS structure involves the growing of a thin SiO_2 layer into a V-groove that has previously been produced by precision etching in an n^+, n, p, n^+ crystal grown using standard epitaxial techniques. The main constructional features are shown in Figure 5.15(b). The metallic insulated gate and the source connections are at the top. The drain connection is at the bottom, and hence no extra chip area is required, as would be the case if it were at the top as in a lateral device. The transistor is operated as an enhancement MOST. Since two vertical channels are created on each side of the V-groove, the current carrying capacity is double that of a corresponding lateral MOST. The geometry is such as to produce short channels, thus minimizing charge carrier recombination. This in turn reduces delay time and hence faster switching speeds are possible at comparatively high channel currents. The width-to-length geometry is designed for low 'on' resistance, thus further enhancing the switching properties of such a device.

Load currents as high as 10 A can be switched with this form of construction. Since the control electrode is an insulated gate, it is easy to drive a VMOST from CMOS and TTL logic circuits (see Part 3 for definitions of CMOS and TTL). Further increases in load current can be achieved by suitable geometry modifications. The HEXFET arrangement, for example, has a top window hexagon and multiple vertical load current paths. Operating voltages of 100 V or more are possible with load currents up to 30 A, and 'on'-resistances of about 50 mΩ. Constant research will produce devices with performance figures superior to those quoted above.

The sharp V-groove unfortunately produces high fields at the tip, which results in a breakdown of the gate oxide layer at the tip. A further disadvantage is that long-term ion migration of aluminium gate material into the oxide layer leads to variations of gate threshold voltage. These difficulties may be largely overcome by using a U-groove instead of the V-shape.

In an attempt to produce improved switching properties various hybrid structures have been invented in which both MOS and bipolar constructions are combined to produce 'superfets' and insulated gate transistors (IGTs). We thus can obtain the best of both worlds in that the advantages of bipolar devices can be combined with those of FETs. Thus for switching we can make use of the low 'on' resistance and high 'off' resistance of bipolar devices, while we have the high input impedance, and fast turn-on and turn-off times of the FET.

Characteristic equations

The relationship between the drain current and the gate voltage for a JFET, operated in the saturation region with a fixed v_{DS} maybe expressed as

$$i_D = I_{DSS}\left(1 - \frac{v_{GS}}{V_p}\right)^n$$

where n is close to 2 for modern constructions. I_{DSS} is that saturation value of drain current that flows when $v_{GS} = 0$ (Figure 5.16). I_{DSS} depends upon the width-to-length ratio of the channel, the majority carrier mobility, the gate capacitance and the pinch-off voltage V_p. It is also temperature-dependent, being inversely proportional to T^n where n is approximately 1.5. This means that increases in temperature reduce i_D, so, from a thermal point of view, the device is stable. This is in contrast to a bipolar transistor in which increases in temperature increase the load current. In the latter case, thermal runaway is possible since rises of temperature increase the load current which increases the temperature and so on, producing a thermally unstable situation unless precautions are taken in circuit design.

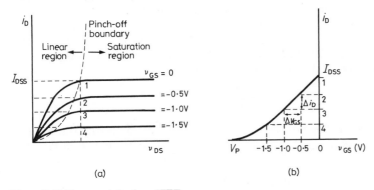

(a)   (b)

Figure 5.16 Characteristics for a JFET

V_p is the pinch-off, or threshold, voltage. From the given expression for i_D we see that that when $v_{GS} = V_p$, i_D is zero. In practice it is not so easy to determine V_p graphically. We could proceed along the lines suggested by Figure 5.16, in which particular values of i_D (e.g. 1,2,3,4) are observed for corresponding values of v_{GS}, and plotting them as shown in Figure 5.16(b).

Unfortunately the drain current does not cut off sharply and we must extrapolate the graph of i_D versus v_{GS} in order to find that value of v_{GS} that brings i_D down to zero. The difficulty arises because the square law expression for i_D does not hold at very low values of drain current. Since the square law expression does hold for values of i_D over quite a large range we could, however, convert the expression for i_D into its 'straight-line' form thus

$$\sqrt{i_D} = K\left(1 - \frac{v_{GS}}{V_p}\right) \text{ where } K = \sqrt{I_{DSS}}$$

$$\sqrt{i_D} = -\frac{K}{V_p} v_{GS} + K$$

Hence by plotting $\sqrt{i_D}$ against v_{GS} we obtain the straight-line graph of Figure 5.17, which is easy to extrapolate to give V_p.

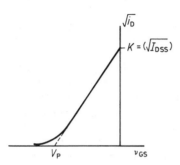

Figure 5.17 Graph of $\sqrt{i_D}$ versus v_{GS} makes it easier to estimate a value for V_p. The 'square law' for i_D does not hold for low values of i_D, hence the graph departs from its straight-line form when i_D is small

V_p is temperature – dependent, having a positive temperature coefficient (i.e. opposite to I_{DSS}). This is of little consequence at high values of i_D since the negative temperature coefficient of i_{DSS} predominates. At very low values of i_D, however (which can be encountered in the input stages of some operational integrated circuit amplifiers), the effects of the positive temperature coefficient of V_p predominate. Obviously there is an intermediate value of i_D in which the two opposite effects cancel; at this value of i_D we have a device that has a zero temperature coefficient, a fact that can be exploited in some circuit designs.

The transconductance of the device is defined by

$$g_m = \frac{\delta i_D}{\delta v_{GS}}$$

As we can see, this is the slope of the i_D/v_{GS} graph of Figure 5.16(b). Graphically, by observing the change in i_D that results from corresponding change in v_{GS}, we obtain

$$g_m = \Delta i_D / \Delta v_{GS}$$

The above equations and definitions apply also for depletion type MOSTs, since the drain current is a maximum for zero gate voltage, and reduces as the magnitude of the gate voltage is increased. The term 'pinch-off voltage' is not appropriate to enhancement mode devices because the drain current is zero for zero gate voltage. If, however, we define the threshold voltage V_T as that gate-source voltage at which drain current begins to flow, we may replace V_p in the above equations with V_T. The forms of the equations for all FETs are therefore similar.

Junction transistors

The junction transistor is a device in which two pn junctions are formed in close proximity within a single crystal. Since charge carriers of both polarities (holes and electrons) are involved in its mode of operation, a junction transistor is sometimes called a bipolar transistor. Early types were made of germanium, but silicon is now used because transistors made from silicon can operate at higher temperatures than their germanium counterparts. The technology of silicon has been advanced to such a state that large-scale mass production diffusion techniques are now well established. In order to achieve a satisfactory junction geometry during manufacture, the speed of diffusion of the impurities (arsenic, phosphorus and boron) must be considered. Technically it is easier to produce npn arrangements rather than pnp.

The term 'transistor' is commonly meant to imply the bipolar type. For all practical purposes transistors are manufactured by the planar epitaxial process; planar because the device consists of plane, or flat, layers of material grown upon (*epi*) a crystal arrangement (*taxis*). Figure 5.18 shows diagrammatically a typical planar epitaxial transistor. A low-resistivity single-crystal substrate acts as the collector region. (The collector of a bipolar transistor corresponds to the drain of an FET.) Electrical losses in the bulk of the collector region are then low owing to the small resistance involved. In order to be able to operate at high voltages (say up to 50 V or more) the pn junction formed between the collector region and the control electrode (known as the base) must be formed in high-resistivity material. The manufacturers therefore grow an extremely thin (10–20 µm) layer of almost pure silicon upon the heavily doped substrate; this is the epitaxial layer in which the transistor is fabricated. Thus is produced a device which has low collector losses yet

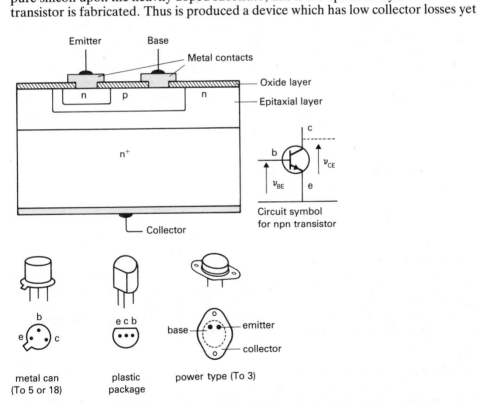

Figure 5.18 Bipolar transistor (npn). The circuit symbol for a pnp transistor is similar to that shown except that the arrow on the emitter line within the circle points towards the base

can operate reasonably high voltages. The base (corresponding to the gate of an FET, but not highly insulated from the body of the device) and the emitter regions are then produced by diffusing the necessary impurities into the epitaxial layers. The emitter is the electrode through which the main load current is injected in to the device; it corresponds to the source of an FET. The correct geometry is achieved by first growing a layer of silicon dioxide on the surface, and then etching away the required window by a photo-engraving process. Impurities are then allowed to diffuse downwards and underneath the oxide layer to form the base and emitter regions. The diffusion takes place in a furnace held at temperatures in excess of 1000°C. Suitable metallic contacts are then made to the regions and bonded to leads that connect with the external circuit. The whole device is then hermetically sealed into a suitable metal or plastic container. The metal containers for low-power devices are 'top-hat' shaped; two sizes are available, known as T05 and T018 cans. Large power transistors are housed in a container which can be bonded to a heat sink. When identifying lead configurations the leads are viewed as coming from the device and pointing towards the viewer; they must, therefore, be viewed from 'underneath' (as opposed to integrated circuits which must be viewed from the top for identification purposes).

Two kinds of junction transistors are available. In one type, *viz*, the pnp transistor, the sandwich is made of n-type material sandwiched between two p-type layers; in the npn transistor p-type mtaerial is sandwiched between two n-type layers.

Bipolar transistors may be used as amplifiers or as switches. In describing the amplifier action we will consider the npn variety, since these types are by far the commonest available. The description is also applicable to pnp types, but the applied voltage polarities must be reversed; also, for holes read electrons and *vice versa*.

For amplifier action the base–emitter junction must be forward-biased. This means that in the absence of a signal, the base must be maintained at a positive voltage (about 600 – 650 mV) with respect to the emitter. The ways in which this bias voltage can be applied are discussed in the chapter dealing with amplifiers. Readers will appreciate from the description of the diode action that maintaining this forward-bias voltage results in the flow of electrons, via the emitter lead, from the emitter across into the base region. In ordinary diodes the electrons recombine with holes in the p–side of the junction. The holes are injected at the ohmic connection with the p-material. The electron-hole movement constitutes the diode current. In transistors, however, the base region is lightly doped and so the chances of recombination are much reduced. Furthermore, the base region under the emitter is made extremely thin, and its width is less than the diffusion length of the free electrons (Figure 5.19). As a result, most of the free electrons injected into the base from the comparatively heavily doped emitter diffuse into the collector-base junction; this junction is strongly reverse-biased by applying a positive potential of several volts to the collector lead. Within the depletion layer at the collector–base junction the electrons come under the influence of a strong electric field which sweeps them into the collector region, and hence into the external load circuit. In a well-made transistor the collector current is almost as large as the emitter current. The small amount of recombination that takes place in the base region accounts in part for the small base current. As in normal diode action, some holes move across the base–emitter junction and combine with electrons in the emitter. Since the base–emitter voltage is also small (0.6 V) the power taken from the driving, or

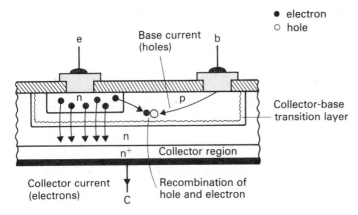

Figure 5.19 Transistor action when the e–b junction is forward-biased and the c–b junction is reverse-biased. (The collector current arrow shows the direction of the electrons, i.e. not conventional current flow)

signal, circuit is small. Hence it takes little power at the input circuit to make the emitter inject electrons into the base. These electrons, however, are accelerated, and acquire considerably more energy from the high-voltage supply powering the collector load circuit. (The acceleration is due to the strong field created by a voltage of, say, 10 V or more across the depletion layer.) Large voltages can therefore be developed in load resistors connected between the collector and the power supply rail. These large voltages are produced as a direct result of the application of small base–emitter voltages. The transistor thus acts as an amplifying device. (We are here assuming that the signal is applied to the transistor base–emitter terminals in such a way that the emitter is common to both input and output circuits – the so-called common-emitter mode. Other modes of operation are possible, as described below, but most common amplifying arrangement is the common-emitter mode.)

Transistor characteristics

Various attempts have been made to quantity the diffusion processes that give rise to transistor action. A mathematical model which is still frequently used is due to J.J. Ebers and J.L. Moll, and described in their paper 'Large Signal Behaviour of Junction Transistors' published in 1954. Their formulae relating the currents and voltages in a junction transistor are of the form

$$i_E = k_1 \left[\exp \left(\frac{ev_{BE}}{kT} \right) - 1 \right] + k_2 \left[\exp \left(\frac{ev_{CB}}{kT} \right) - 1 \right]$$

$$i_C = k_3 \left[\exp \left(\frac{ev_{BE}}{kT} \right) - 1 \right] + k_4 \left[\exp \left(\frac{ev_{CB}}{kT} \right) - 1 \right]$$

k_1 to k_4 are equation parameters related to reverse saturation currents; e,k and T are defined in the discussion on diodes. v_{BE} is the voltage between the base and the emitter, and in normal transistor operation takes a small positive value since the base–emitter junction must be forward-biased. v_{CB} is the collector–base voltage and is large (say 5 or more volts), and negative because the collector–base junction must be reverse-biased; thus the second term in each equation may often be ignored in normal transistor action, and hence

$$i_E = k_1 \left[\exp \left(\frac{ev_{BE}}{kT} \right) - 1 \right] \text{ and } i_C = k_3 \left[\exp \left(\frac{ev_{BE}}{kT} \right) - 1 \right]$$

At a junction temperature of 300 K (which is not unreasonable for a transistor operating correctly) e/kT is about $38.6\,V^{-1}$ i.e. $kT/e = 0.026\,V$, hence i_C can be expressed as

$$i_C = k_3 \left[\exp \left(\frac{v_{BE}}{0.026} \right) - 1 \right]$$

We see that the collector current is determined by the base–emitter voltage. Under normal operating conditions v_{BE} is much larger than 0.026 (i.e. 26 mV) and hence the 1 in the expression may be ignored. By taking logarithms on each side, $0.026\,(\log_e i_C - \log_e k_3) = v_{BE}$. If we take i_{C1} and i_{C2} to represent the collector currents at the corresponding base–emitter voltages v_{BE1} and v_{BE2}. Then

$$v_{BE2} - v_{BE1} = 0.026 \log (i_{C2}/i_{C1})$$

For a doubling of the collector current ($i_{C2}/i_{C1} = 2$), the necessary change in v_{BE} is $0.026 \cdot \log_e 2$, i.e. only about 18 mV. For a tenfold increase in collector current the base–emitter voltage needs to change only by about 60 mV.

Since the collector current is determined by the base–emitter voltage it may seem desirable to plot a set of output characteristic curves showing the way in which the collector current varies with collector–emitter voltage for various values of input voltage, v_{BE}. Such curves would not be very useful for circuit design purposes because it is unsatisfactory to use circuits that are designed to provide specific values of v_{BE}. The reasons are given in the next paragraph.

The base–emitter circuit includes a forward-biased pn junction. The base current expression is therefore of the form

$$i_B = I_0 \left(\exp \left(\frac{ev_{BE}}{kT} \right) - 1 \right)$$

I_0 is very temperature-dependent. The effect is that increases in temperature bring about net increases in the base current for fixed values of v_{BE}. It turns out that to compensate for this effect, v_{BE} must be reduced by about 2.1 mV per °C in order to return i_B to its lower-temperature value. For example, a 10°C rise in junction temperature would need a reduction of 21 mV in v_{BE}; without this reduction the collector current would more than double. Since the temperature variation is unknown, and not under the control of the circuit designer, we see that the design of transistor circuits to produce stable operation is not easy. The consequences of this are discussed in Chapter 7. Figure 5.20 shows the input characterstics for an npn silicon transistor used for the amplification of small signals. This information will be required when we discuss those circuits that are suitable for operation over a reasonably wide temperature range.

The equations that have been mentioned above are only approximate, but they do serve as a starting point for analytical work. To obtain an accurate representation of the performance of a given transistor, actual currents and corresponding voltages are measured, and the results plotted as graphs known as

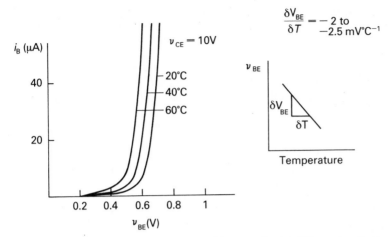

Figure 5.20 Input characteristic curves for a silicon transistor at different junction temperatures

characteristic curves. Many variables are involved; they are v_{BE}, v_{CE}, v_{CB}, i_C, i_B and i_E. Of the many possible two-dimensional graphs available, the input and output current/voltage curves are the ones of greatest practical value. We have seen one such set in Figure 5.20, i.e. graphs plotting i_B against v_{BE} for different junction temperatures.

The common-base mode

By choosing the graph of i_C as a function of v_{CB} for fixed values of i_E we obtain the output characteristics of the transistor used in the common-base mode. Figure 5.21 shows such a typical set. In explaining these curves we remember that the base–emitter junction must be forward-biased, so the emitter is some 600 mV negative relative to the base for an npn transistor. Even with zero collector–base voltage the electrons from the emitter have enough energy to diffuse into the

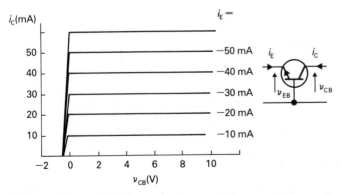

Figure 5.21 Characteristic curves for the common-base mode of operation. Note that in order to conform to the conventions associated with four-terminal networks the direction of i_E must be drawn as shown. Hence $i_C = -\alpha i_E$ (or $i_C = -\alpha i_E + I_{CBO}$ when reverse saturation current is included – see text). Note also that v_{EB} is negative

collector circuit and constitute a collector current. (The collector–emitter voltage is not zero but must also be about 600 mV.) The collector voltage (relative to the base) must be made slightly negative for the load current to be reduced to zero. For collector voltages greater than zero, all of the electrons that can be collected (i.e. the emitter current less the base current) find their way into the collector region; the collector current is therefore independent of collector–base voltage.

In addition to the emitter current that diffuses into the collector region, there is a reverse saturation current associated with the reverse-biased collector–base junction. This is the current that would flow between the collector and base if the emitter were open-circuit; this reverse saturation current, which includes the thermally generated electron-hole pairs in the depletion layer, is designated I_{CBO}, (the current between collector and base with the third electrode–emitter in this case – open-circuit). The total collector current is therefore given by

$$i_C = -\alpha i_E + I_{CBO}$$

α is slightly less than unity, because the collector current consists of the emitter current minus the base current. (In a good transistor α may be 0.99 to 0.999.) The minus sign arises as a result of the mathematical conventions associated with four-terminal networks. These are described in greater detail in the next chapter. The directions of i_E and i_C must be drawn as shown in Figure 5.21, hence $i_C = -\alpha i_E + I_{CBO}$. I_{CBO} is temperature-independent, but because its magnitude is so small (say 2×10^{-8}A) its effect can usually be neglected.

The common-emitter mode

Figure 5.22 shows the output characteristics for an npn transistor operated in the common-emitter mode in which i_C is plotted against v_{CE} for fixed values of base current. The base current is set by the base–emitter voltage, which must forward-bias the base–emitter junction by about 600 mV. For a given fixed base current, increases in collector–emitter voltage from zero to about 150 mV result in

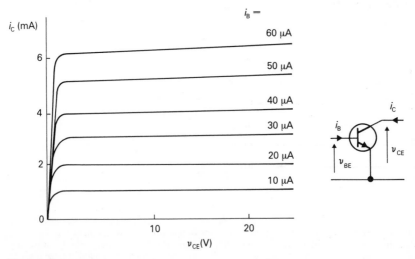

Figure 5.22 Common-emitter characteristics

a rapid rise in collector current. By the time that the collector voltage has risen to 150 mV all of the electrons that have diffused through the base are swept into the collector lead. Thereafter, further increases in collector voltage result in very little increase in collector current. It might be expected that i_C would be independent of v_{CE}. Measurements show that in this mode the collector current continues to rise slightly as the collector–emitter voltage is increases. This is because as v_{CE} rises, the reverse bias on the collector–base junction increases. The depletion layer there thickens and moves further into the base region. The effective base width therefore decreases. This improves transistor amplification so that although the base current is held constant, the collector current rises with increasing v_{CE}. This effect was described by Early and is often referred to as the Early Effect.

In several ways the common-emitter mode is a more useful configuration to use than the common-base mode. In the common-base mode the output current is less than the input (emitter) current; voltage amplifications is still possible because the electrons constituting the emitter current acquire considerable energy in their transition to the collector region. Current amplification, however, is less than unity. Also, since so much current is taken from the driving source, the effective input impedance is small (see Figure 5.23 for comparative details). With the common-emitter arrangement both voltage and current amplification are possible. Because little current is taken from the signal source, the input impedance is

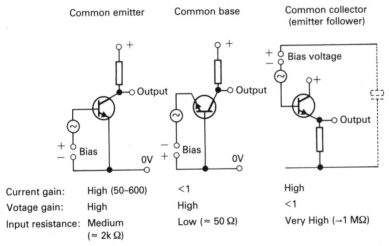

Figure 5.23 Comparison of three basic arrangements for bipolar transistors. (For practical circuits and descriptions see Chapter 3)

comparatively high. The common-emitter mode is therefore the one most usually used for amplification purposes. The output characteristics of Figure 5.22, when used in conjunction with the input characteristics of Figure 5.20, form a useful basis upon which the design of simple amplifiers may be made. In practice, however, designers do not normally use characteristics, but rely instead on design equations.

It is evident from Figure 5.22 that for a fixed value of collector–emitter voltage, an increase of base current brings about a corresponding increase in collector current. The ratio between these increases is known as the current gain of the

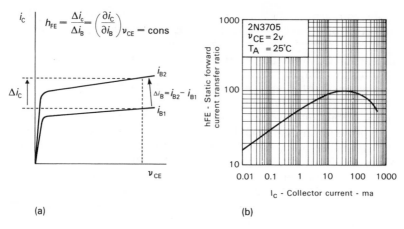

Figure 5.24 Estimation of h_{FE} and its variation with collector current

transistor and is given the symbol h_{FE} or β. Figure 5.24 shows how this parameter may be evaluated using graphical data. A more detailed description of the origin and meaning of the symbols used in connection with transistors is given in the section on transistor parameters and equivalent circuits in Chapter 7. For example, the difference between h_{FE} and h_{fe} is explained.

From Figure 5.24(a) it is seen that

$$h_{FE} = \left(\frac{\delta i_C}{\delta i_B} \right)_{v_{CE}} = \text{const.}$$

h_{FE} is not a true constant, but varies somewhat depending upon where on the set of characteristic curves it is estimated. h_{FE} also varies with the collector current, as illustrated in Figure 5.22(b).

The problem of the reverse saturation current associated with the reverse-biased collector–base junction is greater in the common-emitter mode than in the common-base mode. This is because, although the base current that enters the base lead can be kept constant, there is added to this current the hole component of the collector–base leakage current I_{CBO}. The electron component of I_{CBO} is swept into the collector region and is added to the main electron stream constituting the major part of the collector current. The hole component of I_{CBO} enters the base and moves across the base–emitter transition region. In doing so it alters the charge distributions and undergoes 'transistor amplification' as though it were 'normal' base current. The result is that the collector current can be expressed as

$$i_C = h_{FE}i_B + I_{CBO} + h_{FE}I_{CBO}$$

The second term is the direct electron contribution from the reverse saturation current, and the third term is the result of transistor amplification of I_{CBO}. Thus

$$i_C = h_{FE}i_B + I_{CBO}(1 + h_{FE})$$
$$= h_{FE}i_B + I_{CEO}$$

where I_{CEO} is that current that would flow between collector and emitter with the base lead open-circuit. We see that I_{CEO} is much larger than I_{CBO}; it is very temperature-dependent and is not under the control of the device user. Amplifier circuit designs must therefore incorporate features that minimize the effect of changes in I_{CEO}.

'Hidden' components

Although the description of transistor operation given above serves for many purposes, the circuit designer must always be aware that the transistor contains circuit elements not obvious from the simple circuit symbol. Charge storage associated with the collector–base and base–emitter junctions can be represented by appropriate capacitors. There are also stray capacitances between the three electrodes associated with the connections of the transistor to the leads. Additionally, there are equivalent additional resistances to be considered; these include not only the resistances associated with the bulk of the crystal material leading to the active region of the transistor, but also the small-signal equivalent resistances associated with each electrode. For example, by keeping the potentials at the base and collector constant we could determine the small-signal equivalent resistance associated with the emitter. Since

$$i_E = k_1 \left[\exp \left(\frac{e v_{BE}}{kT} \right) - 1 \right] - k_2 \left[\exp \left(\frac{e v_{CB}}{kT} \right) - 1 \right]$$

and assuming that the second term is small enough to be ignored

$$\frac{\delta i_E}{\delta v_{BE}} = i_E \cdot \frac{e}{kT}$$

Thus

$$\frac{\delta v_{BE}}{\delta i_E} = \frac{0.026}{i_E}$$

at 300°C. This represents an intrinsic emitter resistanc, r_e, in series with the emitter lead, the value of which is $26/i_E$ when i_E is in milliamps. For example, if $i_E = 1\,\text{mA}$ then $r_e = 26$ ohms. r_e is in fact the dynamic, or a.c. impedance of the

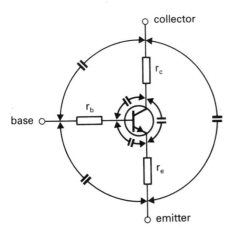

Figure 5.25 'Hidden' parasitic components

forward-biased base–emitter pn junction. We will find this information of value when considering amplifying circuits that use the transistor in the common-emitter and common-collector (i.e. emitter–follower) modes as explained in the next chapter.

We see therefore that the transistor could be represented by a circuit symbol of the kind shown in Figure 5.25. It would, of course, be inconvenient to use such a symbol, and we therefore use the standard symbol, but bear in mind that the additional components may have to be taken into account. Simplifications are possible if operation is at low frequencies. The capacitances involved are quite small and their effect may usually be neglected at frequencies up to about 5 or 10 MHz. We shall have more to say about the effect of r_e and r_b in Chapter 7.

Chapter 6

Diode applications

Diodes may be regarded as switches in most of the circuits in which they are incorporated.

When the diode is reverse-biased practically no current flows, and hence the switch is open. When the applied voltage is such that the diode is forward-biased then a large current can flow, and the voltage across the device is only about 600 mV for a silicon pn junction. In this condition the diode acts as a closed switch. Since an ideal switch has zero voltage across it in the closed position, the diode is seen not to be perfect. Usually, however, the 600 mV drop can be ignored. In circumstances where this is not the case (e.g. in biasing circuits or the voltage clamp circuits discussed below) then it is easy enough to take the 600 mV into account.

Figure 6.1 gives some example of the circuits for the reader to study. The reader is invited to determine the voltages V_1 to V_6 before reading the answers given in the

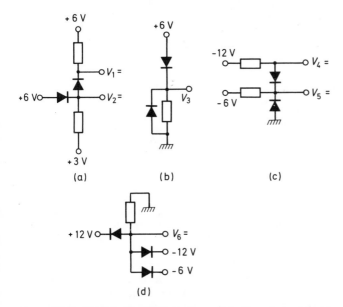

Figure 6.1 In the above circuits the voltages V_1 to V_6 are to be determined, taking into account the 0.6 V drop across any conducting diode

(Ans. $V_1 = 6$ V; $V_2 = 5.4$ V; $V_4 = -12$ V; $V_5 = -0.6$ V; $V_6 = -11.4$ V)

caption. These answers have been printed upside-down deliberately. Some thought must be exercised to obtain the answers, but the basic knowledge required is very small. From an examination of the circuit the reader needs to determine whether a given diode is biased in the forward or reverse direction. In the former case the diode is equivalent to an open switch; in the latter case the diode acts as a closed switch, but with 600 mV drop across it, the anode being positive with respect to the cathode.

The diode as a clamp and limiter

In electronic circuits the need may arise for the voltage at some point in the circuit to be restricted within specified limits. For example, the output of a servo amplifier driving a motor may be satisfactory during normal operation, but sometimes overloads within the amplifier cause the output to be excessively high, which may result in damage to the motor. By limiting the output, such damage can be avoided. Damage to transistors may occur when the input potentials to such devices are outside the normal working limits. In such circumstances it is highly desirable to incorporate a protecting circuit in the form of a clamping or limiting diode.

Figure 6.2 shows how diodes can be used as voltage limiters. The actions of the circuits can be understood if it is remembered that a forward-biased diode acts as a closed switch, whereas a reversed-biased diode acts as an open switch.

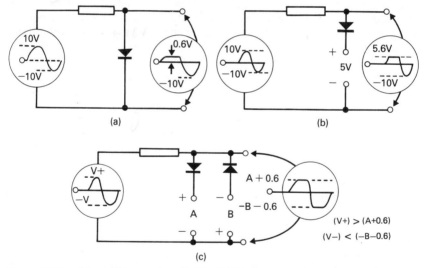

Figure 6.2 Use of the doide as a voltage limiter: (a) Diode prevents the output voltage from going positive by more than 0.6 V; (b) Output voltage is prevented from going more positive than the bias voltage (e.g. 5 V) plus the diode voltage (0.6 V); (c) Output voltage always lies between A + 0.6 and − B − 0.6 volts. It is assumed that V is numerically greater than A and B

It is often necessary in electronic apparatus to switch inductive loads. A relay coil controlled by a transistor switch is a very common example. Figure 6.3 shows how a diode may be used to protect the transistor from damage due to voltage overload; high back-e.m.f.s are produced when the inductive coil of the relay is suddenly switched off. When the transistor switch is on, the diode is reverse-biased by the

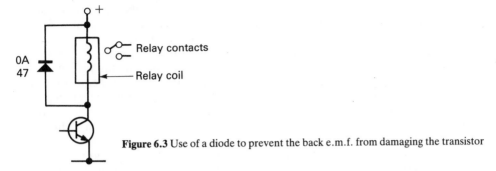

Figure 6.3 Use of a diode to prevent the back e.m.f. from damaging the transistor

voltage drop across the resistive component of the coil. On switch-off the diode is driven into forward-biased conduction by the back e.m.f. This effectively puts a closed switch across the relay coil thus preventing high voltages from being developed across the transistor. The energy stored in the magnetic field is then dissipated in the resistance of the coil.

The diode as a d.c. level adjuster

When a signal is transferred from one stage to another via a *CR* coupling network, the d.c. component is necessarily lost. This is because no steady voltages or currents can be transferred by a capacitor. In many instances, the loss of the d.c. component in a signal is desirable (as we shall see in the chapter on amplifiers).

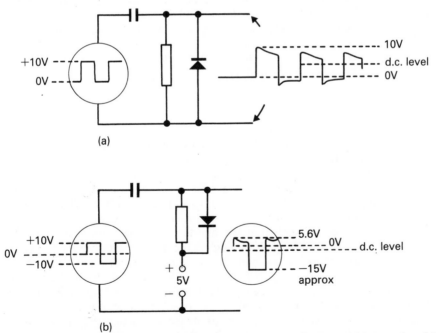

Figure 6.4 The use of diodes to restore or adjust the d.c. level of waveforms: (a) Restoration of d.c. level of waveform; (b) adjustment of d.c. level

However, there are occasions when it is necessary to restore the steady component to the waveform, or to introduce a d.c. component of some other desired magnitude. The circuits of Figure 6.4 show how this can be done.

In attempting to understand the operation of these circuits it should be remembered that the voltage across a capacitor is proportional to the stored charge; if the charge does not alter then the voltage across the capacitor terminals will not alter either. If, therefore, the potential of one terminal (relative to a reference potential) rises rapidly by a given amount, then the potential of the other terminal must also rise by the same amount if the charge held by the capacitor does not change.

Consider now the circuit of Figure 6.4(a). In the absence of the diode we have a straightforward coupling circuit, the time constant of which is assumed to be quite long compared with the period of the signal waveform. After an initial transient period the waveform settles to a position which is symmetrical about the zero voltage time (as shown in Figure 4.27). The d.c. level is 'lost' at the output.

When the diode is present, however, the output voltage rises initially to $+10$ V; the diode is reversed-biased and can therefore be ignored at this stage. Inevitably some charging of the capacitor takes place during the time that the input voltage is at $+10$ V. The output voltage therefore exhibits a droop in the top of the waveform. Immediately before the input waveform suddenly drops to zero the output voltage will be somewhat less than $+10$ V. The sudden drop in input voltage would, in the absence of a diode, make the output voltage fall by 10 V to some negative value because the fall period is too short to allow any change of charge on the capacitor. As soon as the output voltage reaches about -600 mV, however, the diode conducts and presents a very low resistance path to the capacitor, which rapidly discharges. The bottom of the waveform is therefore clamped to -600 mV. When next the input waveform jumps to $+10$ V the output will jump with it, and the cycle of events is then repeated. In this way the d.c. level of the signal is restored (or almost so, allowing for the diode voltage).

The d.c. level may be adjusted to some other level by the use of voltage bias circuits. Figure 6.4(b) shows one example.

Rectification

A most important application of diodes is their use in power supplies, where it is necessary to convert alternating voltages from the mains to steady voltages for feeding electronic circuits.

Figure 6.5 shows a circuit of a diode and a series resistor load R_L, the combination being supplied with an alternating sinusoidal voltage. This voltage may be obtained from an oscillator or the secondary winding of a transformer. When the applied voltage is such that the anode is positive with respect to the cathode, current flows in the circuit. If we neglect the resistance of the diode (which is non-linear, and therefore varies with the applied voltage) then the current waveform is sinusoidal during the first half-cycle. When the polarity of the voltage is reversed, the anode becomes negative with respect to the cathode and no conduction takes place. Figure 6.5 shows the position in graphical form. The voltage across the resistor R_L is proportional to the current, and is therefore a series of half-sine-waves. The alternating voltage has thus been converted to a series of unidirectional sinusoidal pulses, a process known as rectification.

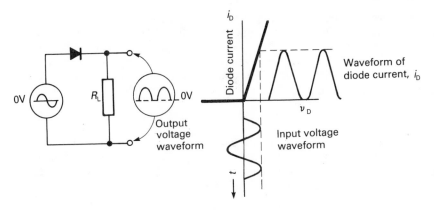

Figure 6.5 Basic principles of rectification. The diode voltage drop has been ignored

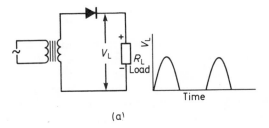

(a)

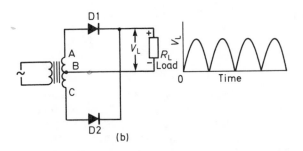

(b)

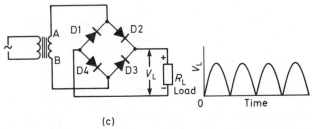

(c)

Figure 6.6 Rectifying circuits for power supplies: (a) Half-wave rectifier; (b) full-wave rectifier; (c) bridge rectifier

The waveforms have been superimposed on the voltage and current axes. The applied voltage varies sinusoidally; in one half-cycle the diode is forward-biased, and in the successive half-cycle the diode is reverse-biased. Only during the forward-biased period does conduction take place; the current then rises and falls with a half-sine-wave waveform. The voltage drop across the diode during conduction has been ignored.

Rectifying circuits

The half-wave rectifier is shown diagrammatically in Figure 6.6(a). During one half-cycle, the applied voltage has a given polarity, and during the succeeding half-cycle the polarity is reversed. Since the rectifier conducts for only one direction of applied voltage, current can flow only during half a period. It is for this reason that the arrangement of Figure 6.6(a) is known as a half-wave rectifier. When connected to an alternating voltage, the rectifier polarizes the resistor representing the load (i.e. the equipment being operated) in the way shown. On the diode a red dot, or a ring, indicates the cathode lead (i.e. the one going to the resistor).

Full-wave rectifier

By the addition of a second diode it is possible to have conduction in the load throughout the whole cycle. The arrangement, shown in Figure 6.6(b) is then known as a full-wave rectifier. If, at an instant of time, the polarity of A is positive with the respect to the centre-tap (B), then the polarity of C is negative to B. Diode D1 conducts since it is biased in the forward or conducting direction. The direction of the current is therefore from A through D1 and R_L and to the centre-tap. Half a period later C is positive relative to B, and D2 is now biased in the forward direction. The direction of the current is then from C, through D2, through R_L in the same direction as before, and back to the centre-tap. The voltage waveform across R_L is therefore as shown.

Bridge rectifier

The disadvantage in using the circuit of Figure 6.6(b) is that the transformer secondary must produce twice the voltage of that used in the half-wave rectifying circuit; this is because only half of the winding is used at any one time. This difficulty can be overcome by using four diodes in what is termed a bridge rectifier. The circuit arrangement is shown in Figure 6.6(c). When A is positive with respect to B, diodes D2 and D4 conduct. On a reversal of polarity between A and B, diodes D1 and D3 conduct. The resultant waveform across R_L is then as shown in the figure. The arrangement is in effect a full-wave rectifier, and is the popular choice for battery chargers. Since small efficient semiconductor power rectifiers are readily available at an economic cost, many manufacturers use the bridge rectifier circuit as the standard rectifying arrangement in their power packs. A more economical transformer can then be used.

Filter circuits for power supplies

It is evident that the rectifier circuits, as they stand, do not supply current at the steady or uniform voltage required by electronic apparatus. An examination of the waveforms shown in Figure 6.6 reveals a pulsating voltage. The rectified output must now be modified to level out these pulses and produce a supply voltage at the steady level required. Such modification is achieved by using filter circuits.

Filtering is accomplished by using capacitors and either inductors or resistors in the circuit. When used for this purpose, inductors are often referred to as 'chokes' since they choke off any variations of current and allow the easy conduction of only direct current.

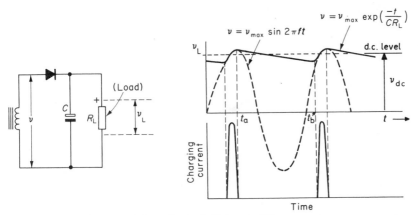

Figure 6.7 Approximate waveforms for the half-wave rectifier

The simplest filter arrangement, shown in Figure 6.7, consist merely of a capacitor in parallel with the load. The value of capacitance must be large in order to present as small a reactance as possible to the pulsating rectified output, and to store sufficient charge so that current may be maintained in the load during the period that the rectifier is not conducting. The reactance of the capacitor should be much less than the resistance of the load. For the kind of loads usually encountered, capacitors of 500–5000 μF are commonly used. The rectified pulses charge the capacitor to a voltage close to the peak value delivered by the rectifier. Because of the large value of C, the time constant CR_L is large compared with the periodic time (0.02 s) of the applied voltage. The voltage across R_L does not therefore fall sinusoidally, but decays exponentially according to the equation $v = V_{max} \exp(-t/CR_L)$. The fall of voltage may be reduced for a given load by increasing the value of C. There is a limit, however, to the value of capacitance used. It can be seen that the slower the rate of fall of voltage across R_L, the smaller is the time available to recharge the capacitor. The current pulse delivered by the rectifier must therefore have a greater peak value to deliver a given energy. All rectifiers have peak current ratings and these ratings can be exceeded if the value of C is too large, causing damage to the rectifier. For any given rectifier and associated circuit, the maximum value of C that can safely be used is specified by the manufacturer in his published data sheets.

The designer of rectifying circuits must observe an additional precaution. This precaution is concerned with the maximum peak inverse voltage that the rectifier can tolerate. During the time that the rectifier is not conducting, we see from Figure 6.7 that the peak or maximum inverse voltage applied to the diode is the sum of the voltage across the capacitor and that across the transformer secondary. The peak inverse voltage (p.i.v.) is thus approximately twice the peak voltage across the transformer secondary. The manufacturer's published data give the maximum p.i.v. that may safely be applied.

The output from the rectifier when a single capacitor is used as in Figure 6.7 is not smooth, but can be considered as being a small ripple voltage superimposed upon a comparatively large steady voltage. It is the steady component, V_{DC}, which is the wanted component; the alternating ripple voltage is unwanted, and must be removed using methods described a little later in the text.

An expression for the steady component, V_{DC}, can be deduced in the following way:

V_{DC} is the mean or average level of the waveform, i.e.

$$V_{DC} = \int_{t_a}^{t_b} \frac{V_{max} \exp[-t/CR_L]\mathrm{d}t}{t_a - t_b}$$

Since, in practice, the charging time is short, $t_b - t_a \approx \tau$ (the period of mains sinusoidal waveform) where $\tau = 1/f$, f being the frequency of mains supply.

By performing the integration and making some simplifying assumptions it is found that

$$V_{DC} = V_{max} \left(1 - \frac{1}{2fCR_L}\right)$$

where V_{max} is the amplitude of the transformer secondary voltage. For full-wave rectification the factor 2 must be replaced by 4. In practice the expression is of limited value because the load (as we shall see shortly) is rarely purely resistive; also the large value of C necessitates the use of an electrolytic capacitor and hence its actual value may be different from its nominal value by as much as 100 per cent.

The residual ripple that remains must, in the majority of cases, be removed. To reduce the ripple to a satisfactory level it is necessary to use a selective filter that will discriminate against the ripple voltage, whilst not attenuating the d.c. component seriously. This is easier to achieve with full-wave, rather than half-wave, rectification. For half-wave rectification the unwanted component has a fundamental frequency of 50 Hz; in full-wave rectification the unwanted component has a fundamental frequency of 100 Hz because two pulses are delivered to the capacitor during a single cycle of the mains.

The type of filter usually used in transistorized equipment takes the form of a simple resistor-capacitor potential divider as shown in Figure 6.8. The divider is frequency-selective since the capacitor presents a very small reactance to alternating components, and a very large reactance to steady components. The leakage current in an electrolytic capacitor of reasonable quality is usually small

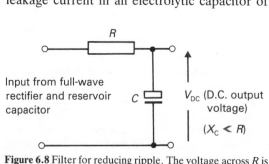

Input from full-wave
rectifier and reservoir
capacitor

R

C

V_{DC} (D.C. output
voltage)

$(X_C < R)$

Figure 6.8 Filter for reducing ripple. The voltage across R is usually allowed to be about 10–15 per cent of V

compared with the load current. To be effective the reactance, X_C, of the capacitor should be small compared with the resistance, R. Values of C of several thousand microfarads are usually employed. Since the value of C is rarely known with any accuracy a rough guide is to assume that the ripple is reduced by a factor of X_C/R. R is usually chosen so that with the known likely value of load current the voltage drop across R is about 10–15 per cent of the steady output voltage.

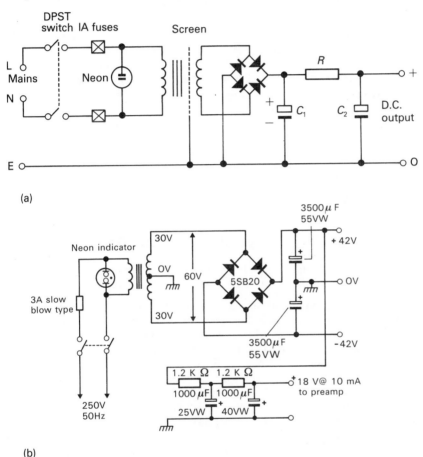

(a)

(b)

Figure 6.9 Two examples of power supply circuits: (a) Typical power supply for small currents; (b) a power supply suitable for an audio power amplifier

Figure 6.9 shows two typical simple power supplies. In (a) R is chosen as indicated above. C_1 is the largest value that the manufacturer of the bridge allows (say 5000 to 10 000 µF). Often it is convenient to make $C_1 = C_2$. The bridge type is not critical, but must obviously be chosen to handle the load current and peak value of the transformer secondary voltage. Generous safety margins should be allowed. The transformer secondary r.m.s. voltage should be chosen so that the peak value (amplitude) is slightly above the required steady output voltage plus the voltage drop across R. For electronic work it is a good idea to use a transformer that uses an electrostatic screen between the primary and secondary coils. This assists in preventing mains-borne noise from interfering with the electronic apparatus. Figure 6.9(b) shows a typical arrangement for a supply suitable for high-power audio amplifiers.

Stabilizers

One way of reducing ripple, and also maintaining the steady output voltage constant in the presence of variations of load current and supply voltage, is to use electronic stabilizers. The simplest of these employs a Zener diode.

The main characteristics of a Zener diode have already been discussed in Chapter 5. By applying sufficient reverse bias voltage, an avalanche condition is reached at which point there is a sudden and substantial increase in current. Thereafter, very small increases in reverse-bias voltage cause large increases in current. In effect large current variations are possible, but the voltage across the

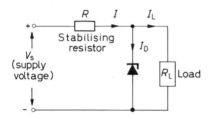

Figure 6.10 Stabilization using Zener diodes

device remains almost constant. Figure 6.10 shows the simple circuit arrangement. The component values may be estimated as follows:

(1) Determine the maximum and minimum values of load current, I_L. (Even if the minimum value of load current is considerable it may well be a good idea to assume zero value, since the load may well be inadvertently disconnected.)

(2) Determine the minimum value that the supply voltage is ever likely to be. In any case, for proper operation the supply voltage must always be at least 1 V (and preferably 2 or 3 V) higher than the avalanche voltage.

(3) Select a suitable Zener diode. Zener diodes are manufactured in a range of preferred values (e.g. 4.7 V, 6.8 V, 18 V, 39 V, 47 V, etc.). (It is possible to add two or more Zener diodes in series to make up an intermediate value. The current passing through the series must not exceed the maximum current rating for any one diode.) Three quantities need to be known about the device: the avalanche voltage, the wattage rating and the tolerance (10,5 and 1 per cent are typical tolerance figures). The wattage rating must be greater than $V_Z.I_Z$, where V_Z is the avalanche voltage and I_Z is the greatest current the Zener diode is called upon to conduct.

(4) Calculate the resistance value and wattage dissipation of the resistor R (Figure 6.10).

The stabilizing action is then as follows: the voltage across the load R_L will be reasonably constant and equal to the Zener voltage V_Z. We say 'reasonably constant' because some small variation of V_Z is noticeable when the current through the Zener diode varies. For a given fixed supply voltage, V_S, the voltage across the resistor R is $V_S - V_L$. If V_S is constant, the current through R must be constant. This means that the sum of the Zener current and the current through the load must be constant. If the load current through R_L increases, there must therefore be a corresponding fall of current through the Zener diode. Conversely, if the current through R_L falls, the current through the Zener diode must rise by a corresponding amount. We see, therefore, that if the current through R_L falls from its maximum amount to zero, the current through the Zener diode will rise by this amount; the diode must therefore be capable of conducting with safety the maximum load current.

Another important point to note is that when the load current rises to a maximum, the diode current must fall, but the latter must not be allowed to fall to zero. If this were to happen, voltage control would be lost. This can be seen by examining the Zener characteristic. As the Zener current falls, the knee of the reverse characteristic is eventually reached. At this point a current of 1 or 2 mA will be flowing. If we reduce the Zener current below this value the knee is passed and the origin is approached, i.e. the voltage across the diode falls from the Zener breakdown voltage to zero.

Consider now the position when the supply voltage, V_S, instead of being fixed, rises. This causes the supply current to rise. Almost all of the current rise is taken by the Zener diode. The Zener voltage is little affected, however, and the voltage across the load remains substantially constant. Because the supply current increases, the current through R increases. The voltage drop across R increases, thus absorbing most of the increase in supply voltage.

Stabilized supplies

When a higher degree of voltage stabilization is required than is possible by using a Zener diode, it is necessary to use more complicated electronic circuitry. Regulator circuits use Zener diodes as voltage references. The output voltage is monitored by

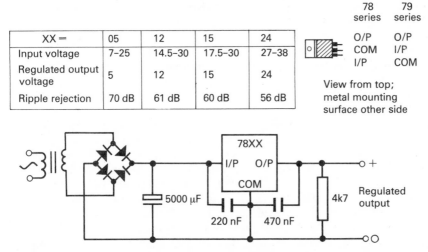

XX =	05	12	15	24
Input voltage	7–25	14.5–30	17.5–30	27–38
Regulated output voltage	5	12	15	24
Ripple rejection	70 dB	61 dB	60 dB	56 dB

	78 series	79 series
O/P	O/P
COM	I/P
I/P	COM

View from top; metal mounting surface other side

Figure 6.11 Regulated power supply for fixed output voltages of 5, 12, 15, or 24. For the 78 series the centre pin (COM) is connected to the metal mounting surface. The 220 nF and 470 nF capacitors (used to prevent oscillations) should be mounted close to the regulator IC. For negative output voltages use 79 XX where XX = 5, 12, 15, or 24. The bridge and 5000 µf capacitor need to be 'reversed' to give negative output voltages. Note that the COM and I/P leads of a 79 XX are interchanged; this means that the I/P lead is in the centre and is connected to the metal mounting surface

a d.c. amplifier and a comparison is made between the reference voltage and a suitable fraction of the output voltage. When differences exist, the circuitry automatically adjusts the output voltage to reduce the error to negligible proportions. The output voltage is then the required value.

Fortunately, the design of suitable regulators can be avoided by using one of many integrated circuit versions available. Figure 6.11 shows two types for load currents not exceeding 1 A. The presence of the 4.7 kΩ bleeder resistor should be noted. It is always a good idea to have a bleeder resistor to reduce the output voltage to zero on switch-off. Other types of regulator are available for higher currents. IC regulator circuits incorporate sophisticated features that include temperature-compensated circuitry and fairly elaborate arrangements to effect short-circuit protection. (See the later section on SCR applications for 'crowbar' protection.)

Switched mode power supplies

One of the disadvantages of the conventional regulated power supplies discussed so far is their comparatively low efficiency; this is especially so with supplies that incorporate voltage stabilizing circuitry. The power dissipated in the series pass transistor (acting as it does as a series variable resistor) is a significant portion of the input power from the mains. Rarely will the efficiency of a conventional power supply reach 45 or 50 per cent. A switched power supply on the other hand will deliver an output power that is as high as 75 per cent of the power supplied by the mains. Suppose, therefore, that an output of 12 V at 1 A is required from a supply. The output power is therefore 12 W and hence a conventional power supply working at 50 per cent efficiency will need 24 W from the mains. A switched power supply working at 75 per cent efficiency requires only 16 W from the mains. The power dissipated within the power pack as wasted heat is only 4 W for a switched supply, whereas the dissipation in a conventional supply is 300 per cent greater at 12 W. Conventional supplies are, therefore, necessarily bulky, heavy and have comparatively large heat sinks. The stray 100 Hz magnetic fields can also lead to some difficult 'hum' problems when designing compact equipment.

Principles of operation of switching regulators

Figure 6.12 shows in block schematic form the basic circuit for one form of switching regulator. It will be seen that the series transistor of a conventional stabilized power supply has been replaced by an electronic switch. It is here that the large saving in power is made since an ideal switch does not dissipate any energy. When the switch is open no current is passing, hence the power dissipation is zero; when an ideal switch is closed no voltage exists across it, so once again no power can be consumed. In practice, modern switching transistors approach the ideal sufficiently for little power to be dissipated in the device.

Merely replacing the series resistor of a conventional power supply with a high-speed transistor switch is not sufficient because of the severe discontinuities in the supply current to the output terminals. We must use some means of keeping the current flowing into the load during the periods that the electronic switch is off. It is for this reason that the components marked L, D and C in Figure 6.12 are used, representing a choke, high-speed diode and capacitor respectively. To see how

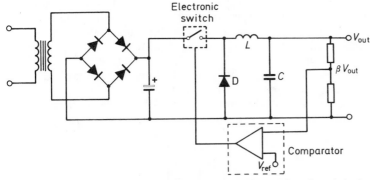

Figure 6.12 Schematic diagram showing the principle of operation of a switched mode power supply

these components effect the desired result in this circuit we must recall their basic properties from the discussions in earlier chapters.

The choke is the most important of the three; its properties largely determine the performance of the regulator. It should be remembered that inductors oppose any change in current in a circuit; they are also energy-storing devices, the stored energy being $Li^2/2$, where i is the current flowing. Figure 6.13 shows that part of the circuit of particular relevance to this explanation. When the switch, S, is closed the diode, D, is reverse-biased and therefore acts as an open switch. Since V_{out} is a good deal less than V_{in}, a high proportion of the supply voltage appears initially across the inductor.

The inductor has electrical 'flywheel' properties, however, and the current through it cannot rise immediately to its full value, but increases exponentially. Part of the current charges the capacitor and the remainder passes into the load. As soon as βV_{out} reaches a value just exceeding the reference voltage fed to the comparator, the output from the comparator changes rapidly and turns off the transistor switch, S. The flywheel effect attempts to maintain the current, the energy to do so coming from that stored in the magnetic field. The back e.m.f.

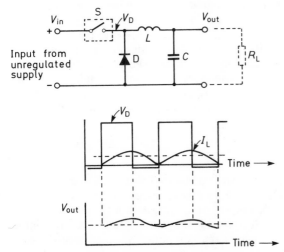

Figure 6.13 Basic operation of regulating circuitry

produced by the collapsing field forward-biases the diode D, and the maintained current continues to charge the capacitor and feed the load. When the inductor current falls below that required by the load, the capacitor makes up for the shortfall. In consequence the stored charge in the capacitor is reduced, and the voltage across this component (i.e. V_{out}) falls. Once the voltage falls below the required value the comparator output changes and switch, S, is restored to its closed position. The cycle of events is then repeated.

An alternative arrangement that eliminates the use of the conventional 50 Hz mains transformer is shown in Figure 6.14. The mains voltage is rectified directly by

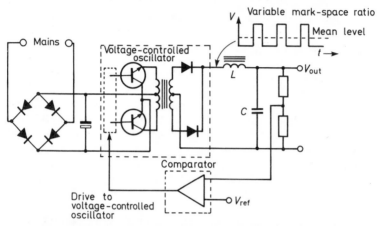

Figure 6.14 One form of switching supply that uses pulse width modulation

a bridge rectifier, and a simple capacitor gives a crude smoothing of the full-wave rectified waveform. The output voltage is then used to drive a high-frequency (25–50 kHz) inverter. The output from the inverter is then full-wave rectified. The two-diode arrangement, rather than a four-diode bridge circuit, is preferred because of the higher efficiency achieved (i.e. the losses associated with the two extra diodes in the bridge are eliminated). It will be seen that no switch is used immediately before the flywheel LC circuit. This is because the principle of switching is different from the former case. In the circuit of Figure 6.12 an almost constant mark-space ratio (of usually about 6:1) is used. In this circuit (Figure 6.14) the mark-space ratio is a function of the output voltage. The variable mark-space ratio is achieved in the feedback circuit to the inverter. When V_{out} falls below the design value the feedback circuit adjusts the mark-space ratio to increase V_{out} to its former value. The switching mechanism is therefore more complex than that previously described.

Polarity reversal

One of the advantages of using the diode, inductor-capacitor-flywheel arrangement is the ease with which the polarity of the supply voltage can be reversed. Figure 6.15 shows how this can be done. The principle of operation should be easy enough to follow from the figure. The energy stored in the magnetic field is made to charge the capacitor with the polarity shown when the components are connected as in Figure 6.15.

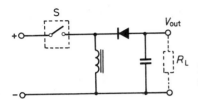

Figure 6.15 By rearranging the flywheel components as shown, a reversal of supply voltage polarity can be achieved

The use of switched mode circuits for power supplies is not without its drawbacks. Compared with conventional power supplies, the switched mode types have higher efficiency, and are small in size and weight, but they are complex units and are therefore more difficult to design. Fortunately for the majority of users, manufacturers have produced suitable ICs for the purpose. Readers who wish to involve themselves with the building of switched mode supplies could usefully consult the Texas Instruments Application Report 'Designing switching voltage Regulators with TL497', and Fairchild's Application Note on their µA 78540 IC.

The output ripple from these types of units is higher than that obtained from conventional regulated supplies. Their response to load variations is also slower than that of conventional supplies because of the complex feedback and switching circuitry.

High-voltage generation

When extremely high tension (EHT) voltages are required, say for cathode-ray tubes, photomultipliers, X-ray tubes, electron and ion guns then it is very expensive to produce the special transformers that would be necessary for use in circuits of the type discussed earlier. A much cheaper solution is to use a stack of diodes together with capacitors in what is known as a Cockroft–Walton voltage multiplier. The principle of operation of this type of high-voltage generator may be understood from a consideration of Figure 6.16. With end A of the transformer secondary positive with respect to end B, capacitor C_1 charges via diode D_1, to almost the peak secondary voltage with the polarity shown. When the polarity of the secondary voltage reverses, a voltage of almost twice the peak secondary voltage makes D_2 conduct and the charge on C_1 is then shared with C_2. Eventually, after several cycles of operation, C_2 charges to twice the peak transformer voltage. By

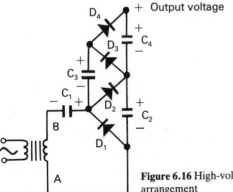

Figure 6.16 High-voltage generator using the Cockroft–Walton arrangement

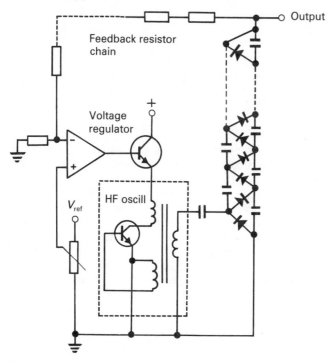

Figure 6.17 High-voltage Cockroft–Walton generator that uses a high-frequency oscillator

adding D_3, D_4, and C_3, C_4 we can obtain a voltage output of four times the peak value of the transformer secondary voltage. When A is positive with respect to B, diodes D_1 and D_3 conduct and transfer charge from C_2 to C_3. With reversed secondary voltage polarity D_2 and D_4 conduct and charge is transferred from C_3 to C_4. The method can be extended by adding further diodes and capacitors so as to produce any multiplication factor. Thus charge is 'pumped up' the column of capacitors. Although the final output voltage is high, each capacitor and diode need be rated for only twice the value of the peak transformer secondary voltage.

In many cases a high-frequency oscillator is used to drive a physically small and light radio-frequency type transformer as shown in Figure 6.17.

Power control circuits using SCRs and triacs

We will consider first the operating conditions and firing circuits for SCRs. Every manufacturer of SCRs publishes, or will supply, a complete set of ratings for his devices. The potential user must ensure that these ratings are not exceeded. It is usual to computer the circuit design so that reliable operation can be ensured for all rectifiers in a compatible series. (All rectifiers in a given series are rated for the same load current, but individual members of the series differ in respect of the maximum peak inverse voltages they can tolerate, and in their gate firing requirements.) Factors to bear in mind are variations in supply voltages, component tolerances, maximum forward and reverse voltages, maximum currents to be passed by the gate and main electrodes, and the operating temperatures.

A controlled rectifier that is not in a state of forward conduction has at least one pn junction reverse-biased irrespective of the polarity of the supply voltage. When the anode is negative with respect to the cathode, the anode and cathode junctions are reverse-biased, and the leakage current is very small since a silicon device is involved. If the reverse voltage is high enough, however, an avalanche condition obtains, as in normal junction diodes, leading to unwanted and uncontrolled current conduction. The characteristic in the forward direction (Figure 6.18) with

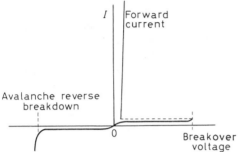

Figure 6.18 Characteristic of a controlled rectifier with zero gate signal; i.e. gate open-circuit

the gate open-circuited, differs from that in the reverse direction because only one junction, namely the control junction, is reverse-biased. Increases in rectifier forward voltage mean increases in reverse voltage for the control junction. If the peak applied forward voltage is large enough an avalanche condition is induced in the control junction. Once the avalanche current is established the rectifier suddenly commences to conduct just as though a current pulse had been injected into the gate electrode. The peak forward and reverse voltages for any given rectifier must not therefore be exceeded. From a knowledge of the maximum peak forward and reverse supply voltages likely to be encountered a rectifier with adequate ratings must be chosen; this insures against unwanted or accidental conduction.

Silicon rectifiers can usually be operated in the temperature range −50°C to +125°C. To prevent excessive temperature rise of the junction material, it is often necessary to mount the rectifier on a metal plate or moulding, called a heat sink. By ensuring good thermal contact with the heat sink, heat is conducted away from the rectifier, and the junctions remain within the specified temperature range. The size of the heat sink can be calculated from the published data. The sink may be no more than a small metal plate of 100 cm² area; on the other hand, heat sinks for use with devices handling load currents up to 100 A need large surface areas, and generally incorporate several cooling fins. (See the section on power amplifiers in the next chapter for an explanation of the calculations involved.)

It is in the design of the gating circuits that difficulties may be encountered. Controlled rectifiers are liable to damage if the maximum safe gate power is exceeded. In particular, the application of large negative voltages to the gate easily leads to failure of the rectifier. The tolerance limits of voltage, current and temperature are neatly summarized in what is called a 'firing diagram' or a 'set of gate control characteristics'. The gate characteristic is, of course, the normal junction characteristic between the gate and cathode leads (Figure 6.19(a)). For any given series of controlled rectifiers, the position of the characteristic is determined by the temperature and the manufacturing spreads for different rectifiers in the group. Figure 6.19(b) shows an upper and lower limit for the position of all possible characteristics for SCRs within a given series and operated within the allowed temperature range. Between the two limits, and near to the origin, there is a shaded area within which firing is possible, but not certain. The shaded area, shown enlarged in Figure 6.19(c) is bounded by the limiting characteristics and by upper voltage and current levels. The upper voltage level is determined by the low-temperature end of the scale. Even at the lowest temperature all rectifiers in the series will fire if the applied gate voltage exceeds this upper voltage level. The upper current level is set by the highest operating temperature likely to be encountered. For increasing rectifier temperatures, smaller gate currents are required to guarantee firing. A lower voltage level is fixed by the high-temperature end of the scale. Below this level of applied gate voltage no rectifier can fire. To guarantee safe, reliable firing it must be arranged that the firing point lies within the limiting characteristic curves and outside the shaded area. The remaining limit is set by the maximum safe gate power that can be dissipated. The position of the power parabola depends upon whether the gate power is dissipated continuously by the application of a steady voltage to the gate,

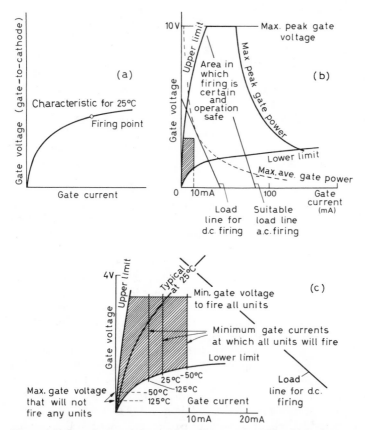

Figure 6.19 Gate characteristics and firing diagrams for SCRs: (a) Typical gate-cathode characteristics obtained by applying a positive voltage to the gate. If a positive voltage is applied between anode and cathode and the gate voltage increased, the gate current will increase along the curve. At a certain point the device will fire; (b) example of the limits of the positions of possible characteristics for a certain range of SCRs; (c) the shaded area in (b) enlarged to show detail

or discontinuously, as when the gate is fed by sinusoidal or pulse signals. The position of the firing point can be made to lie in the area of certain safe firing by choosing a suitable supply voltage and source resistance for the gate. From the diagram we can see that about 10 mA is a suitable gate firing current, a figure that is satisfactory for many commercial rectifiers. From the SCRs available at the present any particular example may need 10–15 mA of gate current to fire; the corresponding gate voltage lies in the range 3–10 V.

A basic circuit for d.c. firing is shown in Figure 6.20(a). For any particular SCR the appropriate design data must be consulted, but, as an example, the curves shown in Figure 6.19(b) will be used.

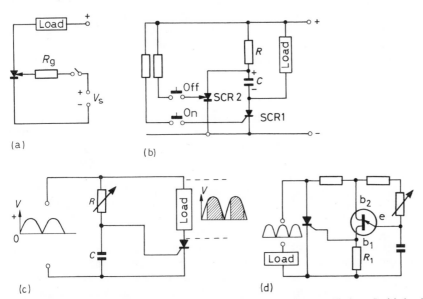

(a)

(b)

(c)

(d)

Figure 6.20 Basic firing circuits for d.c. supplies, including full-wave rectified a.c. In (c) the shaded area represents the time power is being supplied to the load; the actual area, which is variable, is a measure of the power supplied

A satisfactory d.c. load line is shown. (The definition and construction of load lines is discussed early in the following chapter.) The load line intersects the voltage ordinate at about 6 V, and the current abscissa at about 40 mA. A series resistor, R_g, is therefore required, of 6 V/40 mA = 150 Ω. We see from Figure 6.19(c) that the load line intersects a typical diode characteristic at the point (4 V, 10 mA) hence a typical gate dissipation is 40 mW.

Using the circuit of Figure 6.20(b) it can be seen how a controlled rectifier and small switches can be used to switch load currents of several tens of amperes with d.c. supplies. Switching off such a heavy load current cannot conveniently be achieved merely by breaking the supply line to the load. This would imply the use of a specially constructed switch that would be heavy and physically large. In Figure 6.20(b) a second SCR is used to stop the current. By pressing the 'off' switch, SCR2 fires and discharges C. For a short period a reverse voltage is applied to SCR1. This rectifier is consequently extinguished, and reverts to the blocked or non-conducting state. SCR2 extinguishes itself after discharging C because R is made sufficiently large to prevent sufficient current from flowing to keep SCR2 in the conducting state. For all controlled rectifiers there is a threshold current below which the rectifier no longer conducts. This is because the barrier layer at the control junction re-forms at very low currents. The maintaining, or holding, current is usually in the range 10–50 mA depending upon the rectifier type. When SCRs are used to control heavy currents in the way outlined above they are often referred to as 'contactless switches'.

When the main supply voltage consists of a series of rectified sine waves from the unsmoothed output of a half- or full-wave rectifier, it is possible to control the effective power into the load by delaying the firing. Such a firing delay circuit for d.c. (unsmoothed) is shown in Figure 6.20(c). As the supply voltage increases, the voltage at the gate does not rise immediately since it takes some time for the capacitor to charge to a sufficient voltage to fire the SCR. There is a phase delay between the supply the gate voltages determined by the values of C and R. Because of the non-linearity of the resistance of the gate junction, it is not possible to apply simple charging theory to calculate the time delay. The time delay is varied by varying R.

In a practical circuit, it is not safe to allow R to approach zero because the current into the gate will be excessive when the supply voltage rises. For example, if the maximum rated gate current is 30 mA and the supply voltage reaches a peak of 300 V then R must be at least 10 kΩ. Such a large resistance would prevent firing when the supply voltage is at some lower value because insufficient gate current is then available. To avoid this difficulty, and to give a wider range of delay times, a unijunction transistor may be used to fire the SCR. The unijunction transistor fires the rectifier by delivering to the gate short steep current pulses. The use of pulses is the most satisfactory and reliable method of firing a controlled rectifier. The basic circuit is shown in Figure 6.20(d). As the supply voltage rises, the voltage to the emitter rises, but with a phase delay determined by R and C. For any desired firing point in the half-cycle it is possible to adjust R to give the right phase displacement and voltage across C so that at the desired time the emitter voltage is β times the voltage across the transistor (i.e. the emitter voltage is then equal to the stand-off voltage). The emitter junction becomes forward-biased and a large current is established in the emitter–base-1 region thus discharging C. As a result, a voltage pulse appears across R_1 and fires the SCR. From the circuit it can be seen that when pulses are absent the gate–cathode voltage is never large, since the resistor connected to b_1 is small (commonly about 50 Ω).

When the supply voltage is alternating, circuitry must be used to protect the gate junction. A controlled rectifier is easily damaged if a large negative voltage is applied to the gate while the cathode is positive. The simplest way to protect the gate is to connect a diode in the gate lead, as shown in Figure 6.21. Delay can still

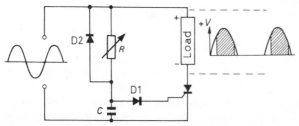

Figure 6.21 Basic delayed firing circuit for alternating supply voltages. D1 protects gate from excessive reverse voltage when cathode is positive. D2 discharges C during non-conducting half-cycle

be effected with an RC arrangement, but a second diode across R is necessary to dissipate the charge received during the half-cycle that the anode is negative. C is thus reset in readiness for the next firing half-cycle.

We see from the shaded areas that no power can be consumed in those half-cycles in which conduction is impossible. The circuit of Figure 6.21 is therefore limited in the amount of control it can exercise, since there is no possibility of obtaining anything approaching full power.

Practical thyristor circuits

For reasons given later in this section, SCRs are not so popular now that triacs are available. However, some useful circuits are given below that need only an SCR as the power controlling device.

Figure 6.22 shows a battery charging circuit that incorporates automatic overload protection. By an appropriate setting of R a fraction of the desired final battery voltage is available at the slider terminal. For all battery voltages that make the slider potential less than the Zener voltage, the Zener does not conduct and SCR2 is not fired. SCR1 can therefore be fired, the gate being supplied with current via R_2 and D_1.

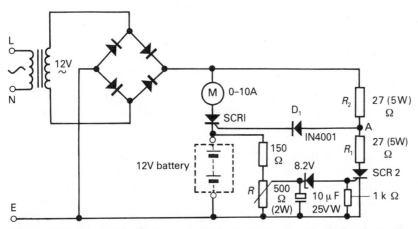

Figure 6.22 A battery charger with automatic overload protection. The semiconductor types are not critical, but must be adequately rated. SCR1 may well be a 16 A type, e.g. BTY87 100R. While SCR2 could well be a BTX18-100 (a 1 A thyristor). The bridge could be a 12.5 A type, e.g. PM7A2Q, while the Zener diode could be a BZY 88–8.2 V

As overcharging commences, the voltage of a lead acid 12 V battery under these conditions rises to about 13.5 V. The slider potential also rises and should be adjusted to ensure that the Zener diode conducts, thus initiating firing of SCR2. This thyristor then conducts during each successive half-cycle pulse from the bridge rectifier. The current through R_2 is then large enough to cause a substantial voltage drop across this resistor, and the potential at A never rises to a sufficiently high value to initiate that firing of the main charging thyristor, SCR1. If subsequently the battery voltage falls, the Zener diode no longer conducts and hence SCR2 cannot fire; the voltage at A is then high enough to allow SCR1 to fire and charging of the battery recommences.

A circuit which uses an SCR as an overload protection device is shown in Figure 6.23. Short-circuit protection is an attractive feature in any form of power supply, especially when it is automatic. With thermionic valves it is possible to use ordinary fuses as protective devices because valves can tolerate overloads for a period long enough to allow the fuse wire to melt. Circuits that use transistors and other semiconductor devices cannot be adequately protected with ordinary fuses, since such devices are damaged by overloads of such short duration that the fuse wire does not have time to melt. An effective method of overcoming the difficulty is to use semiconductor devices in a circuit that acts as a fast fuse. The arrangement shown in Figure 6.23 is often used as an overvoltage protection in regulated power supplies. Should some fault develop in the regulator which allows the unregulated high voltage to appear at the output terminals, then normally this would have a catastrophic effect on the equipment, especially if the latter incorporates digital circuitry that has a poor tolerance to high supply voltages. Used in this way the overvoltage protection circuitry is often known as a 'crowbar'.

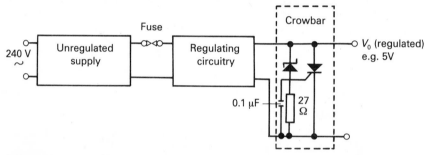

Figure 6.23 Crowbar overvoltage protection. Zener voltage should be slightly high than V_o (e.g. if $V_o = 5\,V$ then the Zener voltage should be 5.6 V. The crowbar then operates if V_o exceeds 5.6 + diode drop in the SCR, i.e. about 6.2 V)

In Figure 6.23 a Zener diode is used that has a breakdown voltage slightly in excess of the normal output voltage. Should a fault develop in the regulator, the output voltage can increase by only a small amount before the Zener diode conducts, and a firing pulse is delivered to the thyristor. The output terminals of the power supply are then effectively shorted swiftly and the equipment thus protected. To prevent damage to the unregulated supply, it is a good idea to incorporate a conventional fuse as shown.

Protection circuits in which the thyristor is isolated from the controlling circuit are possible now that optically isolated thyristor integrated circuits are available. A gallium arsenide infra-red emitting diode is combined with a light-activated thyristor within a standard 6-pin DIL plastic package. It is usual to fabricate the thyristor so as to be able to switch 240 V a.c. supplies. Since the light-emitting diode is not electrically connected to the thyristor, the mains circuit is safely isolated from the control circuit. The package is ideal as an interface between a computer and any mains-operated equipment that is to be computer controlled. Figure 6.24 shows how a 5 V logic circuit can be made to switch a mains-operated lamp. (See also below the circuits based on optically isolated triacs.)

Thyristors can be used to control the speed of universal a.c./d.c. motors of the type used in electrical drills and sewing machines. The circuit of Figure 6.25 gives a reasonable speed control with good torque characteristics owing to the feedback in

577 – 853

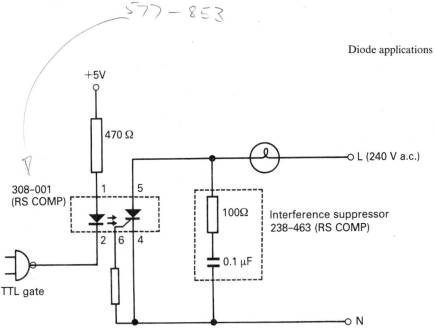

Figure 6.24 Lamp driver with TTL gate control. The interference suppressor is sold as a package. If separate components are used the capacitor must be suitable for mains operation (i.e. to BS 613 standard)

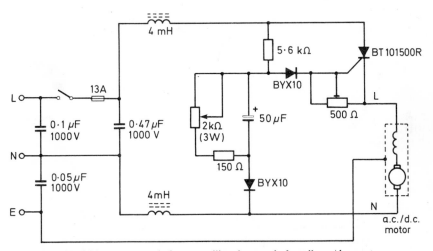

Figure 6.25 Variable power supply for controlling the speed of small a.c./d.c. motors

the form of the back e.m.f. of the motor. The circuit is therefore suitable when the motor is required to deliver power under varying load conditions. Heavy suppression of the radio-frequency interference, typical of these motors, is incorporated. (The more modest interference suppressor of Figure 6.24 could be used if it is essential to reduce the component count.) The 500 Ω preset should be adjusted to give satisfactory performance at slow speeds under lightly loaded conditions.

Triac circuits are to be preferred in the majority of cases when the control of mains power is involved. A single thyristor can only achieve 'half-wave' power control unless a bridge rectifier or another thyristor is added to the circuit. The

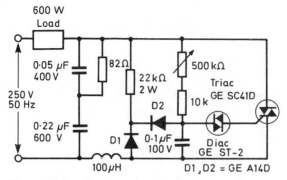

Figure 6.26 Complete incandescent lamp dimmer with r.f. filter

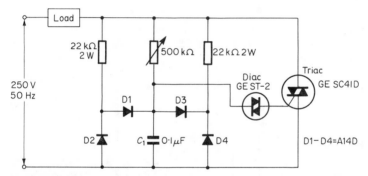

Figure 6.27 Wide range hysteresis-free phase control. Loads of up to 600 W may be accommodated with the GE SC41D triac; for greater loads a larger triac will be required. The circuit shown is capable of firing a wide range of triacs

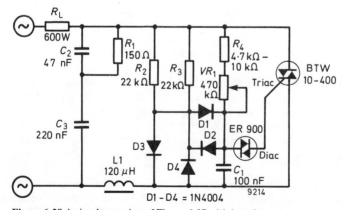

Figure 6.28 A simpler version of Figure 6.27 with interference suppression added

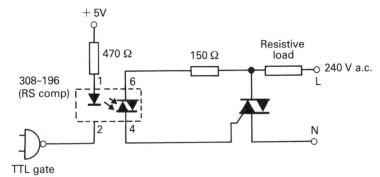

Figure 6.29 The use of opto-isolator triac module to fire a triac that can handle large currents

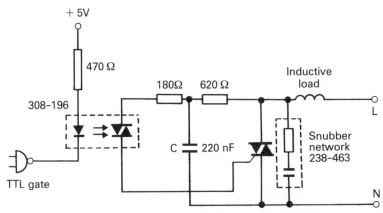

Figure 6.30 Circuit that drives an inductive load from a logic circuit. The triac type should be chosen to handle the maximum load current. The power factor is 0.75. (Type numbers refer to RS Component types)

increased cost and complexity of the circuit make the choice of thyristors unsuitable. Examples of circuits involving triacs are shown in Figures 6.26 to 6.30. It should be noted that circuits intended for incandescent lamp loads cannot be used with fluorescent lamps without modifications.

Exercises

1. Figure 6.31 shows various diagrams of circuits containing diodes and resistors. The voltage levels, relative to zero volts, are shown at several points in the circuits. State the voltage levels at those points marked V_1, V_2, etc.
2. Figure 6.32 shows a four-terminal network together with the waveform of the input voltage, v_i. Sketch the waveform of the output v_o, indicating the important voltage levels on the sketch.
3. Repeat exercise 2 for Figure 6.33.
4. Design a dimmer circuit, based on Figure 6.20(d), to control the level of illumination from a 250 V a.c. incandescent lamp.

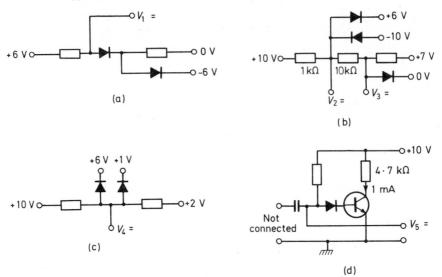

Figure 6.31 Circuits for exercise 1

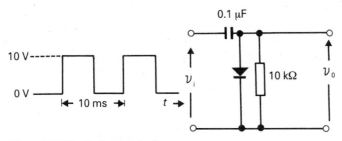

Figure 6.32 Circuits for exercise 2

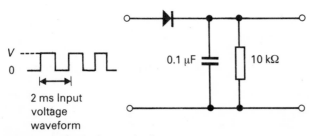

Figure 6.33 Circuits for exercise 3

5. What is meant by the terms 'p-type silicon' and 'n-type silicon'? Describe in qualitative terms the rectifying action of a pn junction. Draw the circuit of a bridge rectifier and smoothing capacitor. Show that the steady output voltage is given by

$$V_{out} = \sqrt{2}V_{\text{r.m.s.}} \left(1 - \frac{1}{4fCR_L}\right)$$

where Vr.m.s. is the secondary voltage from the transformer, f the supply frequency, R_L the load resistance, and C the smoothing capacitance.

6. Design a simple power supply capable of delivering 30 mA at 20 V. It may be assumed that the mains supply is constant at 250 Vr.m.s., 50 Hz, and that stabilization is not required.

7. Design a simple voltage stabilizer that could be added to a power pack you have already designed and built. The power pack delivers a nominal voltage of 20 V and the apparatus to be energized requires a steady voltage of 15 V at currents ranging from 30 mA to 150 mA.

Chapter 7

Transistor amplifiers

This subject involves several different symbols for currents and voltages. To avoid confusion it may be useful to repeat the statements made in the section on circuit analyses dealing with passive component networks. Instantaneous values are symbolized by lower case letters; where subscripts are used they will be in capital letters for absolute total values and lower case letters for alternating component values.

Average, r.m.s., steady or peak values of voltage or current are represented by capital (i.e. upper case) letters. As before, lower case subscripts represent only the alternating quantities whereas capital letter subscripts represent the total or absolute values. Double capital letter subscripts represent the power or battery supply voltages or currents.

In the case of resistances and impedances, capital letters are used to represent the appropriate quantities of actual physical components. Lower case letters represent the differential or incremental value of the quantity involved, e.g. $r = \delta V / \delta I = dv/di = v/i$. Figure 7.1 may help to clarify the situation.

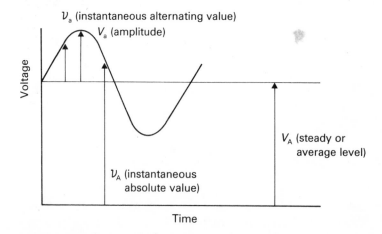

Figure 7.1 Conventions used to symbolize voltages (or currents). V_a may also be used to represent the r.m.s. value of the alternating component only

Transducers are used in laboratories and industrial locations to convert the quantities under observation into suitable electrical signals. Rarely is sufficient output available to actuate an indicating device directly. For example, electronic circuitry cannot of itself amplify the air pressure vibrations which we experience as sound. In sound reproduction we must first convert the pressure variations into electrical form by means of a microphone. Subsequently, the electrical signals that represent the sound variations are used to make records in the form of discs or magnetic tapes. To recover the information, pick-ups or tape-recorder heads are used to produce corresponding electrical signals from the groove modulations on discs or the variations of magnetization on a tape.

The magnitude of the voltages produced by high-quality microphones is often only a few tens or hundreds of microvolts; high-quality pick-ups and tape-recorder heads rarely have outputs exceeding a few millivolts. Such tiny voltages must be amplified considerably before the programme signals can be fed into loudspeakers to produce realistic sound levels. The electric output from photocells, strain gauges, glass electrodes, thermistors, moving-coil accelerometers, etc., all need to be increased, i.e. amplified, before the signal can be suitably recorded or indicated. Some intermediate electronic apparatus must therefore be provided, between the transducer and final indicator or load, to effect the necessary amplification. The design of a suitable amplifier depends upon the transducer and the load, upon the required power output, and on the frequency and nature of the signal.

Amplifiers may be classified according to the function they perform. If frequency is the classifying criterion then amplifiers are designated z.f., a.f. or l.f., r.f., v.h.f. or u.h.f. The z.f. (zero frequency) types are used to amplify steady voltages or currents. The term d.c. amplifier is often used for this type. Some ambiguity as to the meaning of d.c. arises, but most workers accept the term 'direct coupled'. We shall use the term 'd.c. amplifier' to mean an amplifier that can amplify steady voltages and signals whose frequency range extends down to zero frequency. All of the amplifiers for zero or very low frequency work are directly coupled between their stages.

Audio frequency (a.f.) amplifiers are used to amplify signals in the audible range, i.e. 20 Hz to 20 kHz (kilohertz). However, for faithful amplification of the signal waveform the amplifier must have a suitable response in a range of frequencies extending from about 10 Hz to 100 kHz. This is because non-sinusoidal periodic waves of audible frequency have harmonics extending beyond the upper limit of the audio range. The term l.f. (low-frequency) amplifier is to be preferred when audio work is not involved.

Radio-frequency (r.f.), very high frequency (v.h.f.) and ultra-high frequency (u.h.f.) amplifiers are used when the signal frequency is much higher than the audio range. The r.f. signals extend over a range that includes the familiar long-, medium- and short-wave communications bands, say from 200 kHz to 30 MHz (megahertz, i.e. millions of cycles per second). The v.h.f. range, used for television and frequency modulated (FM) radio transmissions, goes up to about 200 MHz while the u.h.f. band – used for colour television, for example – extends over many hundreds of megahertz.

The dividing lines between the different bands are not clearly defined. The techniques used in the design and construction of amplifiers vary enormously depending upon the frequency range to be handled.

A second classification may be made by considering the indicator or other apparatus to be controlled by the amplifier. When the latter is designed to give an undistorted voltage output, it is known as a voltage amplifier. Such an amplifier may be used to energize the deflector plates of an oscilloscope cathode-ray tube. On the other hand, servo and other electric motors, loudspeakers, potentiometric recorders, and so on, all require considerable power for their operation. An amplifier designed to supply such power is known as a power amplifier.

A third classification depends upon the position of the bias point. The design of amplifiers is discussed later in the chapter, but at this stage it may be said that the position of the bias point is determined by the prevailing circumstances. When a voltage amplifier is being designed, low distortion is of paramount importance and efficiency is rarely considered. To achieve a low distortion the transistor is biased in such a way that variations of collector current have the same waveform as the variations of signal current. Collector current is present throughout the whole of the signal cycle, and the input current swings are restricted to small values to ensure minimum distortion. The amplifier is then said to be operating in the Class A mode. The perfect Class A amplifier is an ideal voltage amplifier. The efficiency of such an amplifier is not high; in fact, the efficiency in terms of the output power to total power taken from the supply rarely exceeds 20 per cent. However, power output is of little concern in Class A voltage amplifiers; the aim is to produce an output voltage with the minimum possible distortion.

In a power amplifier, efficiency is an important consideration. Increases in efficiency are achieved by fixing the bias point so that the transistor is at, or very near to, the cut-off point (i.e. there is no forward-biasing of the base–emitter junction). Collector current is then present for only half the signal cycle, and under these circumstances the amplifier is said to be operating in the Class B mode. The efficiency of a Class B amplifier may lie in the range 50 to 78 per cent. Later it is shown that two transistors must be operated in what is termed a 'push-pull' arrangement in order to restore the complete waveform.

For loads consisting of tuned circuits the efficiency may be raised to over 90 per cent by biasing the transistor to beyond the cut-off point. The transistor is then operating as a Class C amplifier. The collector current is delivered to the tuned load in the form of pulses. The tuned load must have a high Q-value in order that the output is sinusoidal; there is thus a 'flywheel' effect. (In mechanics a flywheel of high inertia can be kept rotating at a constant angular velocity even though the supply energy is delivered in impulses.) This type of amplifier will rarely be used in laboratories except by those who work with r.f. heaters. Class C amplifiers are not, therefore, discussed in this book.

It is possible to generate signal power by altering the mark-space ratio of a square waveform in accordance with the signal. High efficiencies are achieved because the transistor absorbs little power. Apart form the transition between the 'on' and 'off' states, the transistor is either cut-off or is saturated. In both cases the product of the voltage across the device and the current through it is zero or very small; i.e. the power dissipated is low. A power amplifier using square waves is said to be operating in the Class D mode.

Complete amplifiers consist of several sections or stages. Each stage contributes to the overall amplification, and is a circuit arrangement of one or two transistors combined with appropriate passive components. The stages are connected in a chain, or cascade, the output from one stage being connected to the input terminals of the next stage in the chain. The initial stages are designed as voltage amplifiers, the function of which is to produce the desired level of amplification while simultaneously introducing the minimum possible distortion of the signal information. The final stages are designed to deliver sufficient power to operate the load; low distortion and high efficiency are major considerations in the design.

Apart from the amplification and power output requirements, two other factors need to be taken into account, viz. the input impedance and the output impedance of the complete amplifier. The design specifications of these two impedances are chosen so that the output devices and the transducers producing the input signals are operating under optimum conditions.

FET amplifiers

The main constructional features of FETs have already been discussed in the chapter dealing with semiconductor devices. The junction types pass maximum drain current when the gate–source voltage is zero. (It is not possible to have normal operation for gate voltages that are positive with an n-channel device – or negative with a p-channel device – since the gate–body pn junction would no longer be reverse biased.) Depletion mode insulated gate types also pass large currents for zero gate voltages. It is therefore possible to design amplifying circuits that are suitable for both types. As we shall see later, amplification of small signals with enhancement mode devices require different circuit arrangements

Figure 7.2(a) shows the output characteristics for an n-channel FET from which we can obtain information to plot the drain current/gate voltage curve of Figure 7.2(b). From these graphs we can estimate the important parameters associated with FETs; they are symbolized by g_m, the mutual conductance (known also as the transconductance), r_{ds}, the dynamic drain source resistance, and μ, the amplification factor.

$$r_{ds} = \left(\frac{\partial v_{DS}}{\partial i_D}\right) \; v_{GS} = \text{const.} \qquad g_m = \left(\frac{\partial i_D}{\partial v_{GS}}\right) \; v_{DS} = \text{const.}$$

r_{ds} can be obtained from Figure 7.2(a) and is the reciprocal of the slope of the i_d/v_{DS} curve at any selected operating point. In the saturation region r_{ds} is almost constant anywhere within the normal operating area; r_{ds} for junction devices is usually in the range 25 kΩ to 1 MΩ. (It should be noted that r_{ds} is the dynamic or a.c. resistance and is, of course, different from the value obtained for the quotient v_{DS}/i_D at any point. This latter value is termed the d.c. resistance and is of no importance in the context of amplifier performance.)

From Figure 7.2(b) we see that g_m is the slope of the i_D/v_{GS} curve; for modern devices g_m is usually in the range 1 to 8 mS (mS = millisiemen, i.e. milliamps per volt).

For the purpose of establishing simple equivalent circuits for the device it is useful to define the amplification factor, μ. A given change in the gate voltage has a much greater influence on the drain current than has an identical change in drain voltage. If the drain current is reduced by a very small amount by making the gate more negative (for an n-channel device) then the drain current can be restored to its former value by increasing the drain voltage by a much larger amount. The amplification factor is defined by

$$\mu = -\left(\frac{\partial v_{DS}}{\partial v_{GS}}\right) \; i_D = \text{const.}$$

(The minus sign is required to make μ a positive factor since $\delta v_{DS}/\delta v_{GS}$ is negative.)

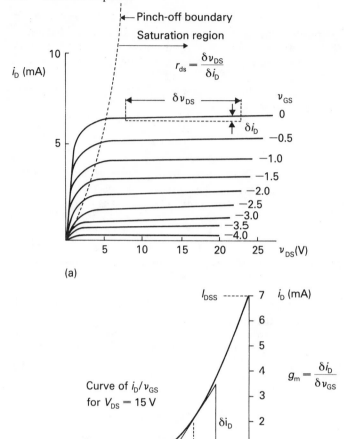

(a)

(b)

Figure 7.2 Characteristics for an n-channel FET. Depletion mode MOSTs have a similar form:
(a) Output characteristics; (b) transfer characteristics. It can be seen that the value obtained for g_m in
(b) depends upon the point chosen to take the tangent (i.e. on i_D)

It can be shown that

$$\frac{\partial y}{\partial x} = -\frac{\partial z}{\partial x} \Big/ \frac{\partial z}{\partial y}$$

hence

$$\frac{\partial v_{DS}}{\partial v_{GS}} = \frac{\partial i_D}{\partial v_{GS}} \Big/ \frac{\partial i_D}{\partial v_{DS}}$$

Since $\partial i_D / \partial v = 1/r_{ds}$, then

$$\mu = r_{ds} \cdot g_m$$

Two simple equivalent circuits can be devised that use these parameters. By regarding the FET as a transconductance device we use the Norton equivalent of Figure 7.3(a); alternatively the Thévenin equivalent of Figure 7.3(b) may be used. It must, however, be remembered that the values of g_m and r_{ds} vary with varying drain current, and hence the equivalent circuits are of limited value.

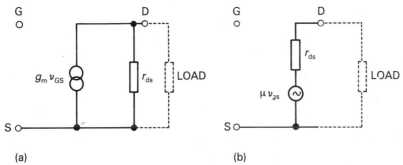

(a)   **(b)**

Figure 7.3 Equivalent circuits for a JFET: (a) shows the Norton form, and (b) the Thévenin equivalent. The input impedance is so great that the gate can be considered as being isolated from the source

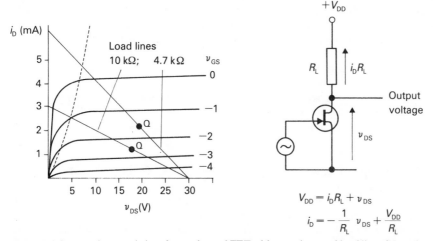

Figure 7.4 Output characteristics of an n-channel FET with superimposed load line. (Note the unsatisfactory circuit arrangement as an amplifier – see text.) The load lines are drawn for values of $R_L = 10\,\text{k}\Omega$ and $4.7\,\text{k}\Omega$; $V_{DD} = 30\,\text{V}$

The design of single-stage transistor amplifiers may be achieved by using graphical techniques, especially when the non-linearities of the device must be considered. Once the principles of operation are understood we can then see the circumstances under which certain restrictions may be imposed to obtain linear operation (or almost so). We may then, under these restrictions, use equivalent circuits to aid the analysis of performance.

Figure 7.4 shows an attempt to achieve voltage amplification. Variations of gate voltage give rise to variations of drain current. If a resistor is connected into the drain circuit as shown then drain current variations result in variations of voltage across the load resistor, R_L. It can be seen that the sum of the drain voltage, v_{DS}, and the voltage across R_L ($i_D R_L$) must equal the supply voltage V_{DD}. Hence

$$V_{DD} = v_{DS} + i_D R_L$$

$$\therefore i_D = -\frac{1}{R_L} v_{DS} + V_{DD}$$

This is a linear relationship of the form $y = mx + c$ where y is equivalent to i_D and x to v_{DS}. The graph representing the equation may be superimposed upon the i_D/v_{DS} characteristics; in these circumstances it is known as a load line. The load line equation must be satisfied by the 'amplifier' performance, but so must the equations represented by the output characteristic curves. This can be achieved only at the intersections of the load line with the characteristics (of which only a few are shown). Thus for any value of gate voltage, supply voltage and load resistance there can be only one value of drain current and one corresponding value of drain voltage. Also it should be noted that we must operate only in the saturation region of the characteristics.

Consider now the position where a signal is applied to the gate. Under quiescent conditions (quiescent = quiet or inactive) the signal voltage is zero and therefore $v_{GS} = 0$. We see immediately that we are then in a forbidden zone if a load resistor of $10\,k\Omega$ were used. The zone is forbidden because of the severe non-linearity of the characteristics in this region. By choosing a lower-valued load resistor (e.g. $4.7\,k\Omega$) we see that the intersection of the load line and the characteristic for $v_{GS} = 0$ lies within the normal operating area, but then another difficulty arises. In an n-channel JFET, positive-going signal voltages would forward-bias the gate-channel pn junction and hence the device would not function correctly. (For p-channel devices the polarities of the applied voltages are reversed.)

The way out of the difficulty is to superimpose the signal voltage upon some steady voltage, so that excursions of the signal voltage never forward-bias the gate junction of the JFET. The signal voltage is thus offset, or biased, by an appropriate amount known as the bias voltage. Under quiescent conditions v_{GS} = bias voltage. The intersection of the output characteristic for this value of v_{GS} and the load line is called the operating point, and is often given the symbol Q. From Figure 7.4 it can be seen that it is advisable to choose the position of Q so as to be at, or near, the centre of that portion of the load line that is within the normal operating area, i.e. between the pinch-off boundary and the v_{DS} axis. The signal voltages can then make their maximum excursions in both positive and negative directions without upsetting the normal operating conditions.

FET biasing circuits

Because of the similarity between the characteristics of JFETs and depletion mode MOSTs, bias circuits designed for JFETs are also satisfactory for depletion MOSTs. Examples using n-channel JFETs will be used, but the reader will appreciate that the same principles may be used with other types, paying suitable regard to the necessary changes in the polarities of the applied voltages.

In order to obtain a satisfactory performance from the amplifier, bias circuits must be devised that not only yield the proper bias voltage, but also take into account the stability of the position of the operating point Q. The characteristics of an FET depend upon temperature; also there are variations of parameter values within a batch of transistors which are nominally identical. It must always be borne in mind that the graphical data supplied by manufacturers are only statistical averages.

A bias circuit suitable for JFETs and depletion mode devices is shown in Figure 7.5. The function of R_1 is to maintain the d.c. level of the gate at zero potential. The necessary negative gate bias voltage is obtained by maintaining the source at a positive potential relative to the zero voltage reference line. This is achieved by inserting a resistor, R_2, between the source terminal and the zero voltage line. Under quiescent conditions the passage of the quiescent current through R_2 maintains the source at the necessary positive voltage; the gate is thus negatively biased with respect to the source.

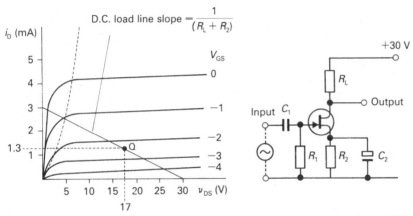

Figure 7.5 Single-stage, small-signal, common-source amplifier based on an n-channel JFET

In choosing suitable operating conditions and circuit component values, we must know the available supply voltage and the likely amplitude of the signal. As an example we will assume that a supply voltage, V_{DD}, of 30 V is available. Then

$$V_{DD} = i_D R_L + v_{DS} + i_D R_2 = v_{DS} + i_D(R_L + R_2)$$

The d.c. load is thus $R_L + R_2$; the corresponding load line is known as the d.c. load line. If the signal voltage is such that v_{GS} is driven almost to zero volts then in order to avoid operation within the very non-linear area to the left of the pinch-off boundary a load line corresponding to 4.7 kΩ must be used (see Figure 7.4). Such a large driving voltage will produce considerable distortion in the output waveform because of the non-linearity of the characteristics. It would be advisable to restrict the input drive to small values for which the non-linearity of the characteristics is not so noticeable; we could then use a load line corresponding to 10 kΩ. This has the advantage of using a smaller quiescent current and thus reduces the power dissipation (and hence temperature) within the device. The operating point should be chosen to be about in the middle of that portion of the load line between the pinch-off boundary and the v_{DS} axis. This indicates a bias voltage of −2.5 V. At this point the quiescent current i_D is about 1.3 mA and the quiescent drain–source voltage is, say, 17 V. Since the voltage across R_2 must be 2.5 V then $R_2 = 2.5\,\text{V}/1.3\,\text{mA} = 1.9\,\text{k}\Omega$. The nearest preferred value of 1.8 kΩ would have to be used; the change in position of the operating point would be negligible. Since the voltage across R_2 and the FET is 19.5 V the voltage across R_L must be 10.5 V hence $R_L = 10.5\,\text{V}/1.3\,\text{mA} = 8\,\text{k}\Omega$. Here an 8.2 k$\Omega$ would be used in practice. Half-watt resistors for R_2 and R_L would be more than adequate.

Consider now the position where temperature changes occur, or where a change of transistor with different characteristics (but of the same nominal type) is made. Any changes that increase the drain current increase the voltage across R_2; the bias voltage therefore increases. Increases in bias voltage, all other factors remaining constant, result in decreases in drain current. The net result is that the actual increase in drain current is only a small fraction of what it would have been had the bias voltage remained constant. A negative feedback system is therefore created which stabilizes the position of the operating point.

In the presence of signal voltages a similar mechanism operates. If the signal voltage rises, for example, the drain current rises, and the voltage across R_2 must also rise. As a result not all of the signal voltage, v_s. appears across the gate–source terminals of the FET. The voltage gain of the amplifier is thus less than it would otherwise have been. This negative feedback effect at signal frequencies can be removed by connecting a large-valued capacitor across R_2. The capacitor provides a very low impedance path between the source and the zero-voltage line at signal frequencies, thus allowing all of the signal voltage to be developed across the gate–source terminals. The maximum gain at signal frequencies is thus obtained. For the slow variations of temperature, however, the reactance of C_2 is too large to be effective at these very low frequencies; the full stabilizing effect of R_2 on the bias conditions can thus be realized.

In order to obtain the maximum possible gain, estimates of suitable values of C_2 can be made by ensuring that $\omega C_2 \gg g_m$. This condition may be justified in the following way. The output voltage is given by

$$v_o = i_D R_L$$

$$\therefore \frac{dv_o}{dv_{GS}} = \frac{di_D}{dv_{GS}} R_L = g_m R_L = \frac{v_o}{v_{gs}}$$

If the whole of the signal voltage, v_s, appears across the gate–source terminals, as will be the case if $\omega C \gg g_m$, then $v_s = v_{gs}$ and hence the voltage gain is $g_m R_L$. At high frequencies this is not difficult to achieve. If, however, C_2 is not made large enough, then at the lower frequencies the impedance of the R_2, C_2 combination becomes significant and introduces some negative feedback. If Z is the impedance of R_2 in parallel with C_2 then

$$v_s = i_d Z + v_{gs} \ .$$

thus

$$v_s/v_{gs} = g_m Z + 1$$

Under these circumstances the gain is given by

$$\frac{v_o}{v_s} = \frac{v_o}{v_{gs}} \cdot \frac{v_{gs}}{v_s} = g_m R_L/(1 + g_m Z)$$

hence for maximum gain $g_m Z$ must be as small as possible. Since $g_m Z = g_m R_2/(1 + j\omega C_2 R_2)$ it can be seen that ωC_2 must be made much larger than g_m. Usually, making ωC_2 greater than $10\,g_m$ will suffice. If, for example, near maximum gain was required at frequencies as low as about 16 Hz ($\omega = 100$), then, using an FET with a g_m of, say, 2 mS, C_2 should be greater than 200 μF. Values of 100 to 500 μF for C_2 are usually satisfactory in this circuit.

In order to avoid upsetting the bias conditions the input signal cannot be connected directly to the gate unless the d.c. level of the signal is zero. Since this cannot always be guaranteed it is advisable to incorporate d.c. isolation of the signal by inserting C_1 into the input line.

The capacitor, C_1, must not only be an open-circuit for d.c. but must ideally be a short-circuit for alternating signals. The transfer of the signal voltage via a CR coupling network has already been extensively discussed earlier in the book. If the signals are known to have a sinusoidal waveform then we can calculate the time constant C_1R_1 from a knowledge of the turning (or break) frequency of the coupling network. The output from the network is no more than 3 dB down at a frequency of $1/(2\pi C_1R_1)$. In our example, when the lowest frequency to be handled is 40 Hz then $C_1R_1 = 1/(2\pi 40) = 0.004$ s. The input resistance of a JFET is of the order of $10^8\,\Omega$, hence the input impedance of the amplifier is largely determined by the impedance of the C_1R_1 network. At frequencies above 40 Hz the reactance of C_1 can be taken as being negligibly small and so R_1 effectively determines the input resistance; this can be made almost any value in the range up to about 10 MΩ and is chosen to suit the driving signal source. Values in excess of 10 MΩ are undesirable since the input leakage current to the reverse-biased gate, although very small ($\approx 10^{-8}$ A), is not negligible if very large values of R_1 are used. For example, if R_1 were 50 MΩ then the resulting uncontrolled voltage across R_1 due to the leakage current would be 0.5 V; this would upset the bias conditions sufficiently to spoil the performance of the amplifier.

Since, presumably, the object of using an FET is to construct an amplifier that has a high input impedance, R_1 will be quite large; as an example we will take R_1 to be 1 MΩ, in which case $C_1 = 0.004 \times 10^{-6}$F, say practically 4.7 nF. This will ensure that, for sinusoidal signal voltages, the response will not be more than 3 dB down at 40 Hz. It is not often, however, that the signal voltage is sinusoidal. In cases where the waveform is unknown we should make the input time constant, C_1R_1, at least five times the period of the lowest frequency signal component that must be amplified. In our example, let us say again that 40 Hz is the lowest frequency of concern then the period is 25 ms. With $R_1 = 1$ MΩ, $C_1 = 125 \times 10^{-2} \times 10^{-6} = 125$ nF. With this value even square waveforms would be handled without undue distortion.

The application of input signals with a square waveform is a severe test for any amplifier; deficiencies in transient response and frequency response are soon revealed. It is worth looking into this aspect in further detail.

Square-wave response in amplifiers

The term 'frequency response' necessarily implies the use of sine-wave signals. In plotting the frequency response of an amplifier we obtain a graph that gives us information about the output from the amplifier at any suitable single frequency we wish to choose. Of all periodic waves only the sine wave has a single frequency associated with it; other periodic waves, e.g. square waves, are synthesized from sets of sine waves of differing frequency. amplitude and relative phase. An amplifier with a given frequency response may therefore process satisfactorily sine-wave signals that have any frequency within the 3 dB response points, i,e. within the range f_{tl} and f_{tu}, the lower and upper turning frequencies, respectively. The amplifier would not, however, process satisfactorily square-wave signals that have repetition rates that lie in the same frequency range. For those signals that have repetition rates close to f_{tl}, a dropping or sag of the top to the waveform will be evident (a mirror image of this sag is produced with the bottom of the waveform). This is because the low-frequency response of the amplifier is inadequate. At the other end of the frequency range square-wave signals that have repetition frequencies near to the upper turning frequency would experience a deterioration of the waveform's leading edge. This deterioration is the result of a poor high-frequency response. In amplifiers of limited bandwidth both sag and leading-edge deterioration may be present simultaneously.

Sag

This quantity is shown diagrammatically in Figure 7.6(d). It is usual to express sag as a percentage of the initial voltage V, i.e. sag, S, is given by

$$S = \frac{\text{Voltage fall}}{V} \times 100\%$$

The low-frequency response in a.c. amplifiers is determined largely by the coupling capacitors. In the simple amplifier already discussed only one coupling capacitor is used, and hence the low-frequency response is as is shown in Figure 7.6(c). We have in effect a high-pass filter where C is the coupling capacitor and R represents the input resistance of the amplifier. On applying a square wave of amplitude V to the input terminals, the voltage applied to the base of the amplifier will immediately rise to V. The top of the voltage waveform cannot be maintained flat because of the charging of C. (In so far as C charges – and discharges – the coupling

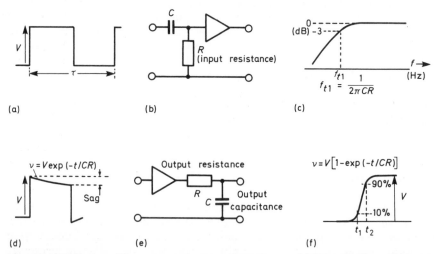

Figure 7.6 Effect of amplifier components on transient response: (a) Input waveform; (b) input equivalent network of amplifier; (c) frequency response; (d) output waveform; (e) equivalent output network; (f) Rise time

capacitor is behaving in an undesirable way, as explained in Chapter 4.) The instantaneous voltage, v, of the top of the waveform is given by

$$v = V \exp(-t/CR)$$

After half a period of the input waveform ($\tau/2$) the value of v has fallen to

$$v = V \exp(-\tau/2CR)$$

The sag therefore is given by

$$S = \frac{V - V \exp(-\tau/2CR)}{V} \times 100\%$$

If f is the repetition rate of the square wave then $\tau = 1/f$; *also* $CR = 1/2\pi f_{tl}$ (f_{tl} being the lower turning frequency) hence

$$S = [1 - \exp(-\pi f_{tl}/f)] \times 100\%$$

We can see from this expression that the lower turning frequency must be much smaller than the repetition rate of the square wave if excessive sags are to be avoided. Suppose, for example, that f were only a little more than $3f_{tl}$, say $f = \pi f_{tl}$, then $S = (1 - e^{-1}) \times 100 = 63$ per cent. This enormous sag could not be tolerated in any respectable low-frequency amplifier designed to handle square waves.

When the sag is required to be less than, say, 5 per cent then the time constant must be long compared with the periodic time of the input signal. Then $f_{tl} \ll f$ and an approximate expression for S can be obtained by invoking the exponential theorem ($e^{-x} \approx 1 - x$ when x is small). Under these circumstances

$$S = \frac{100\pi f_{tl}}{f} \%$$

Hence for a 5 per cent sag $f_{tl} \approx f/60$. Where the repetition rate of the square wave is about 60 per second the lower 3 dB frequency needs to be only about 1 Hz if excessive distortion of the waveform is to be avoided. We see therefore that the estimate of 4.7 nF for the input capacitor of the simple amplifier would make the latter suitable for only sine-wave operation at low frequencies. Where low-frequency square waves are involved, the input time constant needs to be 5 or 10 times the period of the square wave having the lowest repetition rate.

Rise time

The rise time of an amplifier is the time taken for the output to rise from 10 per cent to 90 per cent of its final value when the input voltage suddenly rises from zero to some steady value. Theoretically we assume that the jump in input voltage occurs in zero time. The output voltage cannot jump to its final value in zero time, however, because it takes a finite time to charge the stray capacitances involved in the circuit.

Let us assume that for our simple amplifier the output resistance is R and the stray capacitance C. The output end of the amplifier may, for this purpose, be regarded as a low-pass filter as shown in Figure 7.6(e). The voltage across C when the leading edge of a square wave of amplitude V is applied to the input network is given by

$$v = V[1 - \exp(-t/CR)]$$

$v = 10$ per cent of V at time t_1, and 90 per cent of V at time t_2, therefore $(t_2 - t_1)$ is the rise time. Hence

$$0.1V = V[1 - \exp(-t_1/CR)]$$
$$\text{and } 0.9V = V[1 - \exp(-t_2/CR)]$$

Therefore $\exp(-t_1/CR) = 0.9$ and $\exp(-t_2/CR) = 0.1$

Therefore $t_1/CR = 0.1$ and $t_2/CR = 2.3$

Hence $t_2 - t_1 = 2.2CR$

Since the upper turning frequency, $f_{tu} = 1/2\pi CR$, the rise time, t_r, is given by

$$t_r = t_2 - t_1 = \frac{2.2}{2\pi f_{tu}} = \frac{0.35}{f_{tu}}$$

We can therefore begin to see a quantitative connection between the steady-state frequency response and the transient response.

With more complicated amplifiers the expressions are not so easy to derive. Nevertheless an empirical law based on many observations gives the rise time as

$$t_r = \frac{0.35 \text{ to } 0.45}{f_{tu}}$$

Since f_{tu}, especially in video and high-frequency amplifiers, is so much larger than f_{tl}, the upper turning frequency in effect gives the bandwidth of the amplifier, B. Hence $t_r = (0.35 \text{ to } 0.45)/B$.

Estimation of amplifier gain

To estimate the gain of the amplifier shown in Figure 7.5 from graphical data the dynamic or a.c. load line is required. From a d.c. point of view the load for the transistor consists of $R_L + R_2$, and the corresponding load line is the one required for the design of the bias circuits. From a signal or a.c. point of view, R_2 is short-circuited by C_2 and hence no alternating signal can be developed across it. The only alternating voltage available as an output is that developed across R_L. This output is fed to some following circuit that itself must have an input impedance Z. The effective a.c. load for the amplifier is thus R_L in parallel with Z. If, however, the following stage has a high input impedance (as it will be if it is also an FET amplifier stage) Z will be very large compared with R_L. For simplicity we will assume this to be the case in our example, and hence our a.c. load will be taken to be R_L. (In practice, the input impedance of the following stage must always be considered, and may, if low enough, have to be combined in parallel with R_L.)

Figure 7.7 shows the position with the amplifier of Figure 7.5. A change in input voltage from -2 to -3 is 1 V; this is the peak-to-peak input voltage with a signal having an amplitude of 0.5 V and centred on the bias voltage of -2.5 V. This results in a change of drain voltage form 14 V to 19.5 V output voltage. Hence the gain is 5.5.

The gain can be calculated analytically by considering the equivalent circuit for the amplifier as shown in Figure 7.8. From the drain characteristics of Figure 7.7 it can be seen that r_{ds} is about $100 \text{ k}\Omega$ and g_m about 0.7 mS. Since Z is assumed to be large compared with R_L we see that nearly all of the alternating component of the

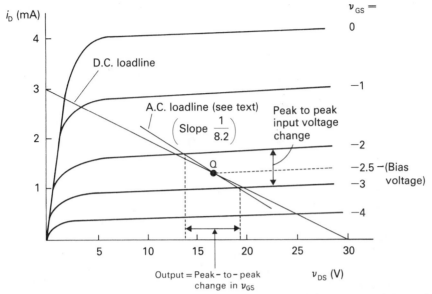

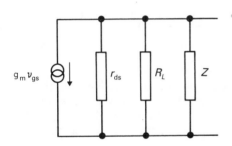

Figure 7.7 Graphical estimation of the gain of the amplifier shown in Figure 7.5, from the a.c. load line

Figure 7.8 Equivalent circuit for the amplifier of Figure 7.5. Since $R_L \ll r_{ds}$ and it is assumed that $Z \gg R_L$ then the gain is given by $A = -g_m R_L$

drain current ($g_m v_{gs}$) passes through R_L. The output voltage is thus $g_m v_{gs} R_L$. Since v_{gs} is the input signal voltage, the gain is seen to be $g_m R_L$. In our example the gain is given by $0.7 \times 8.2 = 5.7$, which is in reasonable agreement with the estimate of 5.5 from the graphical data.

We must not forget that a phase reversal of 180° takes place in the amplifier. When it is essential to take this into account, the gain must be expressed as $|g_m R_L| \angle 180°$ or $-g_m R_L$.

Enhancement mode devices require a modified bias circuit. The resistor in the emitter lead must be retained in order to generate the necessary bias voltage. D.C. negative feedback then ensures stability of the operating point under changing temperature conditions, or when the actual transistor used has characteristics which are different form the statistical averages used in circuit design. These factors have been described above.

The reader will recall that the characteristics of an n-channel enhancement FET are similar to those shown in Figure 7.9. It will be appreciated that the bias voltage must be of the same polarity as the supply voltage (i.e. positive for an n-channel device). Since the voltage drop across the emitter resistor causes the voltage at the emitter terminal to be positive with respect to the zero voltage reference line, then

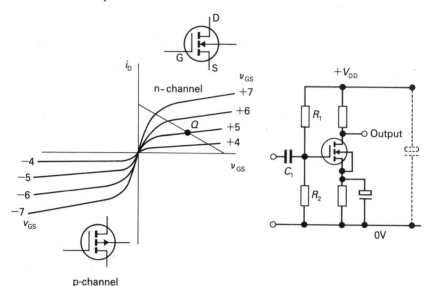

Figure 7.9 Characteristics of enhancement type FET with the biasing arrangement of an amplifier based on an n-channel enhancement MOST

we can no longer have the gate voltage with as quiescent voltage of zero; the gate must be biased to a positive voltage relative to the source. This is achieved by using a potential divider across the supply lines as shown in Figure 7.9. The values of the resistances R_1 and R_2 are chosen to maintain the quiescent voltage of the gate, relative to the zero voltage line, at V_{BIAS} plus the voltage across the emitter resistor. In order that the signal source does not upset the d.c. biasing conditions, a capacitor, C_1, is interposed as before. The input impedance is then C_1 in series with R where $R = R_1R_2/(R_1 + R_2)$. The resistors R_1 and R_2 are in parallel so far as the signal is concerned because the positive and negative supply lines are at the same

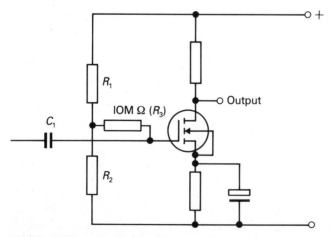

Figure 7.10 Alternative biasing arrangement for an enhancement mode device that increases the input impedance. The circuit is also suitable for JFETs and depletion mode MOSTs

signal potential; they are effectively shorted together for signal purposes by the power supply. In Figure 7.9 this is indicated by the large-value capacitor of the smoothing circuitry of a mains power supply. We can assume that the insulated gate impedance is so high compared with R that it does not effect the impedance.

When we wish to have an input impedance greater than the parallel combination of R_1 and R_2 the alternative arrangement in Figure 7.10 may be used. Because the input impedance of the gate is anything up to $10^{15}\Omega$ the current through the $10\,M\Omega$ resistor, R_3, can be taken as zero; hence the gate potential is the same as that at the junction of R_1 and R_2. The input driving circuit, however, 'sees' a capacitor C_1 in series with R_3 and the parallel combination of R_1 and R_2; the input impedance is thus very high. By choosing appropriate values for R_1, R_2 and R_3, the circuit of Figure 7.10 is suitable for n-channel JFETs and depletion mode devices as well as enhancement MOSTs.

From the example given above we see that the voltage gain of an amplifier based of a JFET is not large. This is typical of all FETs, and gains of 5 to about 10 are about all that can be expected. For this reason FETs are seldom used if voltage amplification is the only aim. FET amplifier circuits are used when the primary aim is to capitalize on the very high input impedances that can be achieved. When large voltage amplification is important, bipolar transistors are used, since gains approaching 400 are possible in single-transistor amplifier stages.

Bipolar transistor amplifiers

A simple single-stage, common--emitter amplifier can easily be designed using graphical techniques similar to those used for FET amplifiers. The problem involves the choice of the correct position of the operating point within a restricted operation region, and, having made the choice, to stabilize the position of the operation point against changes of temperature, transistor sample, and supply voltages. Unlike FETs, the controlling electrode of a bipolar transistor (i.e. the base) must draw current for satisfactory transistor action. We must ensure that the base–emitter junction is forward-biased by about 600 mV, but for bipolar transistors it is customary to regard the device as being current operated, and to fix the bias conditions so that under quiescent conditions a steady bias current is drawn. The collector current is, of course, determined by the base–emitter voltage (remembering that $i_C = k_3[\exp(ev_{BE}/kT) - 1]$, but the design of amplifying circuits in which a bias voltage is fixed is unsatisfactory for the reasons given in Chapter 5. As stated there, the temperature dependence of v_{BE} and k_3 (a reverse-saturation current) make it virtually impossible to construct a practical circuit to yield a fixed bias value for v_{BE}. Graphical designs of transistor amplifiers may therefore be made on a set of output characteristics which relate collector current to collector–emitter voltage for fixed values of base current. The safe operating area is shown with a dotted perimeter on the characteristics in Figure 7.11. The limitations are (a) that the collector current must not exceed the manufacturer's specified level, (b) collector–emitter voltages must exceed about 0.3 V so as to avoid the 'knees' of the characteristics (i.e. operation must be in the 'constant' current region), (c) the maximum power dissipation must not be exceeded, and (d) the base current must always exceed zero; this means in effect that the base–emitter junction must always be forward-biased.

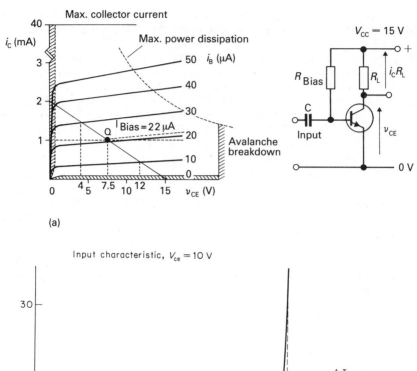

(a)

(b)

Figure 7.11 (a) Output characteristics for a bipolar transistor, with circuit diagram for a prototype simple amplifier. The load line is drawn for a load of 7.5 kΩ; (b) input characteristic of a silicon transistor

Within this region load lines can be constructed for various load resistors in the same way as is applicable to FETs. Figure 7.11 shows a typical set of characteristics together with the circuit diagram of a simple prototype amplifier. Assuming the supply voltage is 15 V then maximum output voltage swings can be achieved if the operating point position is chosen that gives a collector voltage of half the supply voltage, i.e. 7.5 V. If we choose a reasonable quiescent current (say 1 mA) then the value of R_L is 7.5 kΩ and the value of the bias current needed is 22 μA. The base–emitter voltage must be forward-biased by 0.6 V, hence $R_{BIAS} = (15 - 0.6)$V/22 μA = 655 kΩ; the nearest preferred value of 680 kΩ would be used, the resulting drop in bias current being negligibly small.

Remembering that the positive and zero voltage supply lines may be considered as being joined so far as signal voltages are concerned, the input impedance of the amplifier is C in series with R, where R is the parallel combination of R_{BIAS} and the dynamic resistance of the base–emitter junction. The latter can be determined from the slope of the input graph of i_B against v_{BE} at the quiescent operating point (i.e. for $i_B = 22$ μA in the above example). For small-signal amplifying transistors the input resistance is often about 1–5 kΩ. Since R_{BIAS} is so much greater than this, then effectively the input impedance of the amplifier is C in series with about 2 kΩ; we see that this is very much smaller than the figures obtained for FET amplifiers. On the other hand the voltage gains for bipolar transistor amplifiers are very much greater than those of FET amplifiers. Gains of up to 300 or 400 are possible. In the example of the simple amplifier of Figure 7.11, we see that a peak-to-peak variation of base current of 20 μA brings about a peak-a-peak variation of collector voltage of 8 V; this is the peak-to-peak output voltage. By examining the input characteristic for the base–emitter junction a swing of 10 μA either side of the bias current (i.e. 20μA peak to peak) would be brought about by a peak-to-peak change of the input voltage, v_{BE}, of typically 25 mV. The gain in this case is 8 V/25 mV = 320, which is much larger than that obtained with a comparable FET amplifier.

Temperature effects and bias stabilization

Two significant factors affect the stability of the position of the operating point on the load line. They are (a) the reverse leakage current of the collector–base junction, and (b) the variation with temperature of the base–emitter voltage v_{BE}, necessary to produce a given collector current. We have already seen (Chapter 5) that one approximate expression for the collector current is given by

$$i_C = h_{FE}i_B + I_{CBO}(1 + h_{FE})$$

The second term is significant in devices with h_{FE} values of, say, 100–900.

Bias circuits

Two techniques are used to maintain the operating point in a relatively stable position: they are (a) stabilization techniques, and (b) compensation techniques. Stabilization techniques involve designing circuits that use d.c. feedback to maintain the position of the operating point. The technique is a linear one and is used in Class A voltage amplifiers. Power amplifiers, particularly those using Class B arrangements, use compensation techniques because stabilization circuits consume too much power. Compensation involves the use of non-linear temperature or voltage-sensitive elements to provide a compensating voltage or current. The undesirable effects of temperature on the performance of the transistor circuits are thus considerably reduced.

Figure 7.12 shows the basic circuit arrangement from which commonly used bias circuits involving stabilization techniques are derived. The circuits are typical of those used for small-signal, single-stage Class A amplifiers. The factors controlling the stability of these circuits can be assessed by solving the numerous equations that

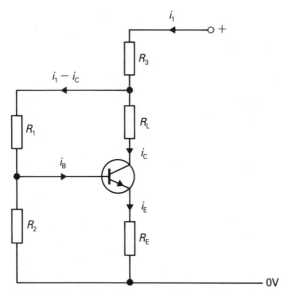

Figure 7.12 Basic circuit arrangement from which some commonly used bias circuits are derived

can be established by applying Kirchhoff's Laws to the circuit of Figure 7.12. Various stability factors have been defined, all of which show how the collector current varies with I_{CBO}, I_{CEO}, h_{FE} and v_{BE}.

In the case of a common-base amplifier in which the base is common to both the input and the output we have

$$i_C = \alpha i_E + I_{CBO}$$

A stability factor S_1 may be defined as

$$S_1 = \frac{\partial i_C}{\partial I_{CBO}}$$

in which I_{CBO}, the reverse saturation current of the base–collector junction, varies with temperature, and it is assumed that circuit conditions maintain a constant value of i_E and α. Then $S_1 = 1$. When S is small, as it is in this case, the implication is that the thermal stability of the circuit is very good; the higher the value of S, the poorer is the thermal stability.

Space limitations prevent analyses of all of the stability factors, but we will indicate the method of analysis by taking appropriate cases associated with Figure 7.12.

Consider the position in which R_3 and R_E are zero and R_2 is removed (i.e. infinitely large). The amplifier is then identical to the prototype amplifier of Figure 7.11. We can now obtain an expression for the stability factor that enables us to calculate the change in i_C that occurs as a result of a change in I_{CBO}, all other quantities (h_{FE}, v_{BE}) remaining constant.

$$S_1 = \frac{\partial i_C}{\partial I_{CBO}}$$

$$i_C = h_{FE} i_B + I_{CBO} (1 + h_{FE})$$

For constant base current and h_{FE}

$$S_1 = \frac{\partial i_C}{\partial I_{CBO}} = h_{FE}I_{CBO} \text{ (when } h_{FE} \gg 1)$$

For an h_{FE} of, say, 100, it can be seen that the S_1 is 100 times greater than that obtained for the common-base case; thermal stability of the prototype common-emitter amplifier is very poor compared with an equivalent common-base amplifier.

The thermal stability of the prototype amplifier can be greatly improved by including a resistor, R_E, in the emitter lead. The emitter potential is then $i_E R_E$ and, since $i_E \approx i_C$ this potential, near enough, is $i_Q R_E$ where i_Q is the quiescent collector current. The base voltage must be some 600 mV greater than the emitter voltage for correct transistor amplifying action. It is often arranged that this base voltage is obtained from a potential divider. One practical bias circuit is obtained by modifying the circuit of Figure 7.12, making $R_3 = 0$. To obtain an expression for the stability factor we can, for analysis purposes, take the Thévenin equivalent of R_1 and R_2 energized by V_{CC}, as shown in Figure 7.13(b). V is the voltage at the

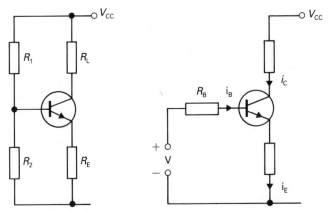

Figure 7.13 (a) Commonly used bias circuit to obtain stability of the operating point; (b) the equivalent circuit in which R_1 and R_2 have been replaced by their Thévenin equivalent

junction of R_1 and R_2 with the base lead disconnected, and R_B is the Thévenin equivalent resistance, equal to the parallel combination of R_1 and R_2. The voltage across R_B is given by

$$R_B i_B = V - (v_{BE} + i_E R_E) = V - (v_{BE} + i_C R_E + i_B R_E)$$

$$i_B(R_B + R_E) = V - v_{BE} - i_C R_E$$

Now

$$i_C = h_{FE}i_B + I_{CBO}(1 + h_{FE})$$

$$= h_{FE}\left[\frac{V - v_{BE} - i_C R_E}{R_B + R_E}\right] + I_{CBO}(1 + h_{FE})$$

$$\therefore i_C(R_B + R_E + R_E h_{FE}) = h_{FE}(V - v_{BE}) + I_{CBO}(1 + h_{FE})(R_B + R_E)$$

Hence

$$S_1 = \frac{\partial i_C}{\partial I_{CBO}} = \frac{(1 + h_{FE})(R_B + R_E)}{R_B + R_E + R_E h_{FE}}$$

We shall see that typical figures for component values may be $h_{FE} = 100$, $R_E = 2\,k\Omega$ and R_B about $10\,k\Omega$. This makes S_1 about $100\,(10 + 2)/10 + 200$, i.e. $S_1 = 5.7$. In the prototype amplifier of Figure 7.11 S_1 was 100, which meant that the change in i_C was about 100 times the change in I_{CBO}; in the common-emitter of Figure 7.13(a) the change in i_C is only about 5.7 times the change in I_{CBO}. There has thus been a substantial improvement in the thermal stability of the circuit, brought about by connecting a resistor in the emitter line. In the absence of an emitter resistor, an increase in temperature produces an increase of $h_{FE} I_{CBO}$ in the collector current. When the emitter resistor is present, however, any increase in collector current causes the emitter voltage to rise; for a fixed base voltage, v_{BE} must therefore fall. Falls in v_{BE} cause falls in collector current that largely offset the rises due to temperature increases.

Other stability factors can be defined as

$$S_2 = \frac{\partial i_C}{\partial h_{FE}} \text{ and } S_3 = \frac{\partial i_C}{\partial v_{BE}}$$

where changes in h_{FE} and v_{BE} are due to thermal variations. In all cases, analysis shows that increased thermal stability results from the inclusion of an emitter resistor.

From the expressions obtained it will be seen that thermal stability is improved by increasing the value of R_E. It must be remembered, however. that for a given quiescent current and supply voltage, increasing R_E causes a bigger fraction of the supply voltage to be dropped across it; less voltage is therefore available for the transistor and the load resistor, R_L. This means that the maximum permissible output voltage swing is reduced. In designing a circuit of the form of Figure 7.13(a) we usually allow 10–15 per cent of the supply voltage to be dropped across R_E. Knowing (say from graphical sources) a likely or desired value of quiescent current, we can calculate the emitter voltage and a value of R_L. The base voltage must be $600\,mV$ above the emitter voltage (for an npn silicon transistor). Knowing the base potential, relative to the zero voltage line, we can now estimate values for R_1 and R_2. In the above analysis of stability factors we assumed the base potential to be reasonably constant; variations of base bias current must therefore not disturb the potential at the base unduly. This can be achieved by allowing the bias potential divider chain to take about 10 times the bias current. R_1 and R_2 can then be estimated.

As an example, suppose we wished to build an amplifier to operate over a frequency range of 40 Hz to 500 kHz. A supply voltage of 18 V is available and the quiescent current is to be 1.0 mA. The voltage across R_E should be about 2.0 V, hence R_E is 2.0 kΩ. Probably the nearest preferred value available is 2.2 kΩ; in any case we have seen that it is an advantage to choose as high a value for R_E as is consistant with other design considerations. To allow a maximum output voltage swing before waveform clipping occurs, the quiescent voltage at the collector should 'sit' halfway between the supply rails. The voltage across R_L is then $V_{CC}/2$ which gives a value of R_L of 9 kΩ; in practice we would use 9.1 kΩ as the nearest preferred value. The value of v_{CE} under these circumstances is 6.7 V. The d.c. load line has a slope of $-1/(9.1 + 2.2) = -1/11.3(\text{mA V}^{-1})$, and passes through the points $(1\,\text{mA}, 6.7\,\text{V})$ and $(0, 18\,\text{V})$. Assuming that the transistor characteristics are as shown in Figure 7.11, these results give a value of bias current of $20\,\mu\text{A}$. The bias chain R_1, R_2 would then need to take $200\,\mu\text{A}$. Since the base voltage is $2.2 + 0.6 = 2.8\,\text{V}$, $R_2 = 2.8\,\text{V}/200\,\mu\text{A} = 14\,\text{k}\Omega$, the nearest preferred value to which is $15\,\text{k}\Omega$. The voltage across R_1 is $18 - 2.8 = 15.2\,\text{V}$, therefore the value of R_2 must be $15.2\,\text{V}/220\,\mu\text{A}$ (i.e. $200\,\mu\text{A} + 20\,\mu\text{A}$ of bias current) $= 68\,\text{k}\Omega$ for the nearest preferred value. It may seem that the base–emitter potential would then be slightly higher than required, but the accompanying rise in collector current would increase the voltage across R_E, hence v_BE would be readjusted to only a little more than $600\,\text{mV}$.

In order not to upset the d.c. bias conditions the signal source must be isolated from the base at zero frequency. An isolating capacitor, C_1, must therefore be connected as shown. The capacitance of C_1 must be large enough so that there is little or no opposition to the signal at frequencies within the desired pass band. For reasons explained below the emitter resistor must be by passed with a high-value capacitor. The emitter, for signal frequencies, is therefore connected to the zero voltage line. The signal source therefore 'sees' as an input impedance C_1 in series with R, where R is the parallel combination of R_1, R_2 and the base–emitter resistance. The latter will be about $2.5\,\text{k}\Omega$ and hence the effective value for R is about $2\,\text{k}\Omega$. Sine wave signals down to $40\,\text{Hz}$ would not be attenuated by more than $3\,\text{dB}$ if $C = 1/(2\,\pi\,40 \times 2500) = 1.6\,\mu\text{F}$ (arithmetically); practically, a value of $2.2\,\mu\text{F}$ may well be used. For waveforms other than sinusoidal, the worst case would be a square-wave signal with a repetition rate of 40. As explained earlier in the chapter, the value of C would need to be considerably increased, say to as much as $50\,\mu\text{F}$.

After discussing negative feedback we shall examine the consequences of leaving R_E unbypassed. In the meantime we note that for maximum voltage gain a bypass capacitor must be connected across R_E. It will be recalled from the chapter on semiconductor devices that the transistor itself has an effective emitter dynamic resistance of r_e equal to $(kT/(ei_\text{C}))$ which at about $300\,^\circ\text{C}$ works out to be $25/i_\text{C}$ ohms when i_C is in milliamps. The total effective emitter resistance is therefore $R_\text{E} + 25/i_\text{C}$, i.e. $2.5\,\text{k}\Omega + 25\,\Omega$ in our case. The bypass capacitor can be connected only across R_E, and therefore its capacitance must be large enough so as to present, at low signal frequencies, a reactance that is small compared with r_e. A value of

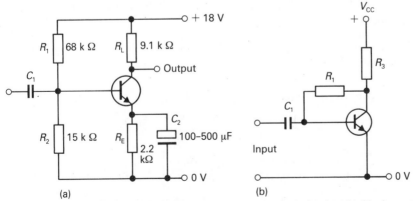

(a)   (b)

Figure 7.14 (a) Shows circuit component values for the example design described in the text; (b) shows an alternative biasing arrangement that uses feedback to stabilize the position of the operating point

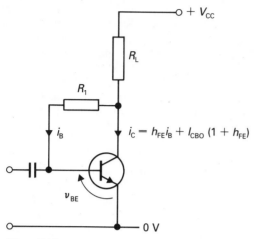

Figure 7.15

500 μF has a reactance of only 8 ohms or so at 40 kHz. In practice values of 100–500 μF may be used, since we can usually tolerate some loss of gain associated with the lower values of C_2.

Figure 7.14 shows the completed circuit diagram of the example amplifier, together with an alternative form of biasing that involves omitting R_2 and connecting the emitter directly to the zero voltage line. R_L of Figure 7.12 is made zero and R_3 becomes the new R_L. Analyses using the principles described above can be made for the alternative biasing arrangement.

Current mirror stabilization

We have, so far, considered the effect of R_E on the stability factor S_1 and seen how an improvement can be made on the performance of the circuit. This improvement is also noted for stability factors S_2 and S_3, thus minimizing the effects on i_C of variations of h_{FE} and v_{BE} as well as I_{CBO}. It is possible, however, to achieve stable amplifier operation in a different way.

Let us consider the circuit of Figure 7.14(b), reproduced for analysis purposes in Figure 7.15. The sum of the voltages across R_L and R_1 added to v_{BE} is equal to the supply voltage, hence

$$V_{CC} = R_L[i_B + h_{FE}i_B + I_{CBO}(1 + h_{FE})] + i_B R_1 + v_{BE}$$

$$\therefore i_B = \frac{V_{CC} - v_{BE} - R_L I_{CBO}(1 + h_{FE})}{R_1 + R_L(1 + h_{FE})}$$

$$\therefore i_C = h_{FE}\left[\frac{V_{CC} - v_{BE} - R_L I_{CBO}(1 + h_{FE})}{R_1 + R_L(1 + h_{FE})}\right] + I_{CBO}(1 + h_{FE})$$

$$i_C = \frac{h_{FE}(V_{CC} - v_{BE})}{R_1 + R_L(1 + h_{FE})} + I_{CBO}\left[\frac{(1 + h_{FE})(R_1 + R_L)}{R_1 + R_L(1 + h_{FE})}\right]$$

From this we see that

$$S_1 = \frac{\partial i_C}{\partial I_{CBO}} = \frac{(1 + h_{FE})(R_1 + R_L)}{R_1 + R_L(1 + h_{FE})}$$

$$S_2 = \frac{\partial i_C}{\partial h_{FE}} = \frac{(R_1 + R_L)(V_{CC} - v_{BE} + R_1 I_{CBO})}{[R_1 + R_L(1 + h_{FE})]^2}$$

$$S_3 = \frac{\partial i_C}{\partial v_{BE}} = \frac{h_{FE}}{R_1 + R_L(1 + h_{FE})}$$

In all cases, small values of R_1 result in small values of the stability factors. When $R_1 = 0$ the stability factors are at their minimum for this circuit; $S_1 = 1$, S_2 tends to zero, and S_3 is approximately equal to $1/R_L$. Thus the position of the operating point will be very stable, and this means that the collector current will be almost constant even in the presence of variations of I_{CBO}, h_{FE} and v_{BE}. When $R_1 = 0$ the circuit obviously cannot be used as an amplifier, nor could it be used as a switch in logic and computer applications, because the output terminal is then connected to the input terminal. Nevertheless, we can capitalize on the constant current properties of the circuit by using the arrangement for bias purposes in integrated circuits. The standard bias circuits so far discussed in connection with discrete component amplifiers, use several large-valued resistors. The circuits are therefore unsuitable for integrated circuit amplifiers because such resistors use large areas of the silicon chip compared with transistors; this would mean that fewer components could be fabricated on a given chip area, resulting in uneconomic manufacture. In addition, the resistor values need to be reasonably close to the design values, and this is difficult and expensive to realize in IC technology. Where emitter resistor stabilization is used it is necessary to use bypass capacitors of several tens or hundreds of microfarads; such capacitors cannot be fabricated on a silicon chip.

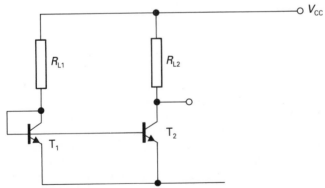

Figure 7.16 The Widlar stabilization circuit, also known as a current mirror. (The absence of the circle for the transistor symbols indicates that an integrated circuit is involved)

In integrated circuits the stabilization of the position of the operating points, and hence the collector currents, of transistor amplifiers can be achieved by using the principles involved in the circuit of Figure 7.16, where T_2 and R_{L2} form the transistor amplifier and T_1 and R_{L1} are the stabilizing components. It will be seen that the T_1, R_{L1} combination is the amplifier discussed above, in which the resistor connecting the collector and base is made equal to zero. The base of T_1 is directly connected to the base of T_2. The circuit, known as the Widlar stabilization circuit, acts in the following way. T_1 and T_2 are assumed to be identical, and so close together that they can be considered to be in the same thermal environment. These conditions are easy to satisfy with integrated circuit technology since the transistors are fabricated at the same time under identical conditions, and with identical geometry. If the temperature increases, thus causing a tendency for the collector current of T_1 to rise, there must be a similar tendency for the collector current of T_2 to rise by the same amount. T_1 is however stabilized against changes of collector current, and v_{BE} will adjust so as to keep the collector current of T_1 constant. Since the two bases are interconnected, the collector current of T_2 reduces as a result of the change in v_{BE}.

Since the collector current of T_1 is stable, the collector current of T_2 must also be stable. If $R_{L1} = R_{L2}$ the collector current of T_2 is constrained to be equal to that of T_1; for this reason the Widlar stabilization circuit is often known as a current mirror.

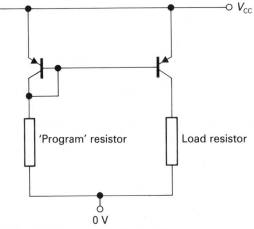

Figure 7.17 Current mirror biasing. The programming resistor has a value equal to $(V_{CC} - 0.6)/$(desired load current)

Fortunately, the input characteristics of a transistor are not markedly dependent on v_{CE} and so the equality of the two collector currents is well maintained even if R_{L1} differs from R_{L2} by a considerable amount. This allows us to 'programme' the current in T_2 by choosing an appropriate value for R_{L1}, and thus have a similar current flowing in the load R_{L2} as shown in Figure 7.17. We shall return to this circuit and its useful modification when we discuss integrated circuits in greater detail.

Transistor parameters and equivalent circuits

When transistors are incorporated into circuits, it is natural for many electronics engineers to analyse the behaviour of the circuits in order to obtain a better understanding of the design principles involved. For those not wishing to engage in complicated circuit analyses, it is still necessary to know something of the procedures involved so that they can read the literature and understand some of the techniques used.

The analysis of transistor circuits is made easier if the actual transistor can be replaced by an equivalent circuit. A large and bewildering number of parameters have been defined to describe transistor behaviour. Space does not permit a discussion of every parameter system that has been devised. Some aspects of one approach to the subject of transistor parameters will be discussed here. We should maintain an open mind about parameters and not hesitate to use a different system in those circumstances in which it could obviously be better to use the different approach. In many analyses, however, the system outlined below proves to be satisfactory.

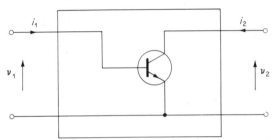

Figure 7.18 'Black-box' four-terminal network representation of a transistor

In so far as the transistor is a linear circuit element (i.e. working on the straight portions of the appropriate characteristics), it may be represented as a 'black box' with two input terminals and two output terminals (Figure 7.18). The internal workings of the box are of no concern when considering circuit analyses; the transistor's behaviour from an external point of view is all that needs to be known. Four variables are associated with the box, viz. v_1, i_1, v_2 and i_2, representing input voltage and current and output voltage and current, respectively. Because linear operation is assumed these signal voltages and currents are related by sets of simultaneous linear equations. Any two variables can be taken as known and the other two can be calculated from the equations. There are six ways in which two variables from the four can be selected. Not all selections produce useful results. The most fruitful are those about to be described.

Suppose the two currents involved, i_1 and i_2, are known. The voltages, v_1 and v_2, must be related to the currents. The relationships can be expressed by the following equations:

$$v_1 = z_{11}i_1 + z_{12}i_2 \tag{7.1}$$

$$v_2 = z_{21}i_1 + z_{22}i_2 \tag{7.2}$$

The z coefficients have numeral subscripts to show their position, i.e. z_1's are in the first line, the first being z_{11} and the second z_{12}. In the second line there are z_2's z_{21} and z_{22} in the first and second position, respectively. Expressed in matrix form the equations become:

$$\begin{pmatrix} v_1 \\ v_1 \end{pmatrix} = (Z)\begin{pmatrix} i_1 \\ i_2 \end{pmatrix} \quad \text{where } Z = \begin{bmatrix} z_{11} & z_{12} \\ z_{21} & z_{22} \end{bmatrix}$$

The z coefficients are called the z parameters of the transistor and are a suitable description of the transistor's external behaviour. The various terms of the defining equations represent the a.c. or signal variations only, and it is assumed that all supply voltages and bias currents are present. The latter play no part in circuit analysis, however, and can be ignored.

When the transistor is open-circuited to a.c. at its output, i_2 is zero and Equation (7.1) becomes $v_1 = z_{11}i_1$. z_{11} is therefore the input impedance, v_1/i_1, when the output is open-circuited. When the output is short-circuited to a.c., $v_2 = 0$, therefore $-z_{21}/z_{22} = i_2/i_1$. This ratio is the current gain. In the common-emitter mode $-z_{21}/z_{22} = h_{fe}$. By opening and shorting the input and output terminals to a.c. and using Equations (7.1) and (7.2), it can be seen that z_{11} is the input impedance with output o.c. (open-circuited); z_{21} is the forward transfer impedance with the output o.c., z_{22} is v_2/i_2 = output impedance with input o.c., and z_{12} is the reverse transfer impedance with the input o.c.

If $z_{12}i_1$ is added to both sides of Equation 7.2 and $z_{21}i_1$ is transposed, then Equations (7.1) and (7.2) may be written as:

$$v_1 = z_{11}i_1 + z_{12}i_2 \qquad (7.3)$$

$$v_2 + (z_{12} - z_{21})i_1 = z_{12}i_1 + z_{22}i_2 \qquad (7.4)$$

These equations show that the 'black box' may be replaced by either of the networks of Figure 7.19. The networks are known as the T-parameters, Figure

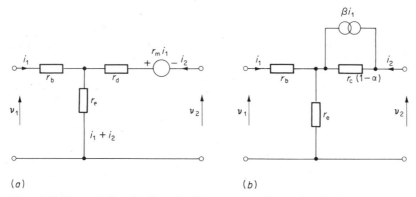

(a)   (b)

Figure 7.19 The equivalent circuits, using T-parameters, of a transistor in the common-emitter mode: (a) Voltage generator form; (b) current generator form

7.19(a) being the voltage generator form, and Figure 7.19(a) being the current generator form. From Figure 7.19(a), using Kirchhoff's Laws, it can be seen that:

$$v_1 = (r_b + r_e)i_2 \qquad (7.5)$$

$$v_2 + r_m i_1 = r_e i_1 + (r_e + r_d)i_2 \qquad (7.6)$$

By comparing Equations (7.3), (7.4), (7.5) and (7.6) and equating coefficients, it is easy to obtain a relationship between the T-parameters and the z-parameters.

The T-parameters are often used as a first approach to transistor equivalent circuits. The resistances r_e, r_b and r_d are, respectively, the resistances associated with the emitter, base and collector. Typical values for a small a.f. transistor are r_e = 25 Ω, r_b = 500 Ω and r_d = 20 KΩ. r_m, the mutual resistance, is 0.98 MΩ, the corresponding h_{fe} being 50. The accuracy of the figures is not high. It must be remembered that there are large tolerance spreads in transistors of the same nominal type from the same manufacturer.

From the foregoing, it can be seen how a whole system of parameters may be constructed. The principles for setting up the equations are the same in every case. Two of the four variables are selected and their dependence on the other two, using suitable coefficients, is stated. For example, for certain purposes it may not be found convenient to use the z-parameters. In high-frequency work the y parameters are preferred. They are defined from the equations:

$$i_1 = y_{11}v_1 + y_{12}v_2 \tag{7.7}$$

$$i_2 = y_{21}v_1 + y_{22}v_2 \tag{7.8}$$

They are called the y or admittance parameters because each one has the dimensions of an admittance, i.e. a current divided by a voltage.

Both the z- and y-parameters are difficult to measure in the laboratory; a set is therefore defined in which all of the parameters can be easily and accurately determined. Such a set is known as the h-parameters, h standing for hybrid. Many manufacturers now prefer to describe their transistors in terms of h- parameters. The defining equations are:

$$v_1 = h_{11}i_1 + h_{12}v_2 \tag{7.9}$$

$$i_2 = h_{21}i_1 + h_{22}v_2 \tag{7.10}$$

Examination shows that h_{11} has the dimensions of an impedance, h_{22} the dimensions of an admittance, and h_{12} and h_{21} are pure ratios. This is the reason for calling them hybrid parameters.

In the USA they prefer not to use the numeral subscripts. Instead they use h_i, h_r, h_f, h_o. A second letter subscript shows the mode of operation, b, e and c standing for common (or grounded) base, emitter and collector, respectively, Thus:

$$h_{ib} = h_{11}$$

$$h_{rb} = h_{12}$$

$$h_{fb} = h_{21}$$

$$h_{ob} = h_{22}$$

In the common-emitter mode numeral subscripts are primed thus:

$$h_{ie} = h'_{11}$$

$$h_{re} = h'_{12}$$

$$h_{fe} = h'_{21}$$

$$h_{oe} = h'_{22}$$

The letters i, r, f and o stand for input, reverse, forward and output respectively. h_{ie} is therefore the input resistance v_1/i_1 of a common-emitter transistor with the output short-circuited to a.c. (see Equation 7.9). Under the same circumstance $h_{fe}(= h'_{21}) = i_2/i_1$. This is the current gain in the common-emitter mode. h_{oe} is the output impedance with the input open-circuited ($i_1 = 0$). (When $i_1 = 0$ it is implied that the input current, I, is constant so that the variation or change in input current, i_1, is zero.) h_{re} is the reverse voltage feedback ratio v_1/v_2 when $i_1 = 0$.

Although there is a large, and for the newcomer bewildering, number of parameters, the latter are all logically derived from a simple basic pattern. Once this is appreciated much of the initial confusion is dispelled. A table for the interconversion of z, y and h parameters is given in Appendix 3.

In designing transistor amplifiers there are four quantities of particular interest: the current gain A_i, the voltage gain A_v, the input impedance Z_i, and the output impedance Z_o. These quantities may be obtained by using the equivalent circuits of Figure 7.19. Today, however, most designers are using the hybrid matrix parameters (i.e. h-parameters) as a starting point for their calculations. Manufacturers prefer to publish the performance details of their transistors in h-parameters form because, as previously mentioned, these parameters can all be evaluated in the laboratory with reasonable accuracy. If a designer finds that a given analysis would proceed with greater ease if he were to use, say, the y of z parameters, it is quite easy for him to use a parameter conversion table.

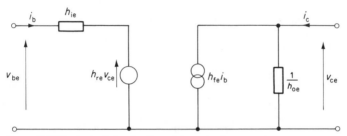

Figure 7.20 Equivalent circuit of transistor using h-parameters

The defining equations of the h-parameters suggest the transistor equivalent circuit of Figure 7.20. In attempting to find the quantities A_i, A_v, Z_i and Z_o, we replace the actual circuit with an equivalent one. Thus all components that do not affect the a.c. performance can be eliminated. For example, in Figure 7.14(a) the emitter is connected to the negative supply rail so far as signals are concerned, because the impedance of C_2, the bypass capacitor, is almost zero at the frequencies involved. Of course, from the biasing and d.c. point of view the emitter resistor and bypass capacitor are essential for the proper and stable working of the transistor; these components can, however, be ignored in circuit analysis.

Estimates of the values of the h-parameters can be made from graphical sources. Figure 7.21 shows how this can be done for h_{oe}, h_{fe} and h_{ie}. (When the estimate of current gain is made using static values for the ratio, i.e. $\Delta I_C / \Delta I_B$ the symbol h_{FE} is used. In practice the difference between h_{FE} and h_{fe}, the small-signal ratio, is small enough at low frequencies to be ignored.)

From Figure 7.20 it will be seen that the current $h_{fe}i_b$ must be shared between the load and $1/h_{oe}$. This means that the effective current gain into the load circuit is not quite equal to h_{fe}. Several authors use the symbols β or β_F to represent the effective current gain. From the graphs of Figure 7.21 we see that h_{oe} is the slope of the output characteristic at the operating point. For bipolar transistors this characteristic is almost parallel to the v_{CE} axis, hence h_{oe} is very small; $1/h_{oe}$ is consequently very large, and for the load impedances usually encountered in practice $(1/h_{oe}) \gg Z_L$. The consequence of this is that h_{fe} is almost the same as β. If the output characteristic were actually parallel to the v_{CE} axis then h_{fe} would equal β. In this book the symbol h_{fe} (or h_{FE}) has been used interchangeably with β.

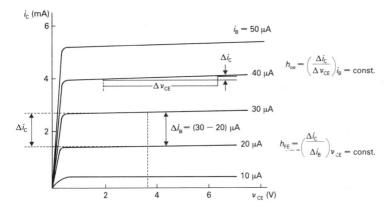

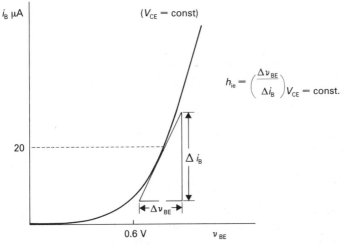

$$h_{oe} = \left(\frac{\Delta i_c}{\Delta v_{CE}}\right)_{i_B} = \text{const.}$$

$$h_{FE} = \left(\frac{\Delta i_c}{\Delta i_B}\right)_{v_{CE}} = \text{const.}$$

$$h_{ie} = \left(\frac{\Delta v_{BE}}{\Delta i_B}\right)_{V_{CE}} = \text{const.}$$

Figure 7.21 Estimation of h-parameters from graphical data

It is important to remember that h_{FE} is not known with any accuracy, and hence it is often pointless in developing complicated models and formulae unless there is a real advantage in doing so. Frequently, approximate formulae give all the vital information that is necessary. Unfortunately the developing of approximate formulae and 'rules of thumb' can be done only by understanding the fundamentals of the subject. We do, however, now have enough information to see how approximate expressions can be obtained. Take for example, the amplifier circuit of Figure 7.22(a), and let us assume that we need to know the voltage gain and input impedance for low-frequency signals. The equivalent circuit is also shown in Figure 7.22(a). Knowing that h_{ie} is about $2\,k\Omega$, that $(1/h_{oe}) \gg R_L$ and assuming that R_E is adequately bypassed for the signal frequencies of interest, the circuit can be reduced to that shown in Figure 7.22(b). From this we see that the voltage gain is $v_o/v_i = h_{fe}i_b R_L/(h_{ie}i_b) = h_{fe}R_L/h_{ie}$, and the input impedance is h_{ie}.

At first sight it would seem that the voltage gain is dependent on h_{fe}, but we see that this is not the case when we recall the discussions in Chapter 5. From the

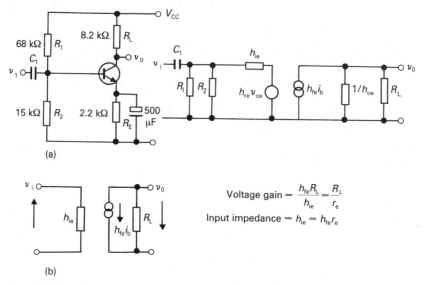

(a)

(b)

Voltage gain $= \dfrac{h_{fe}R_L}{h_{ie}} = \dfrac{R_L}{r_e}$

Input impedance $= h_{ie} = h_{fe}r_e$

Figure 7.22 Development of approximate formulae from equivalent circuits: (a) Circuit diagram of a single-stage l.f. amplifier together with equivalent circuit; (b) Simplified equivalent circuit. R_1 and R_2 are both much greater than h_{ie}; the reactance of C_1 is taken to be negligible; $1/h_{oe} \gg R_L$; h_{re} is very small. (See text for definition of r_e)

Ebers–Moll equations it was shown that an intrinsic emitter resistance, r_e, exists and that $r_e = 26/i_c$ ohms when i_c is in milliamps.

$$r_e = \frac{\partial v_{BE}}{\partial i_E} = \frac{v_{be}}{i_e} = \frac{v_{be}}{i_b(1 + h_{fe})} = \frac{h_{ie}}{(1 + h_{fe})}$$

hence $h_{ie} = r_e (1 + h_{fe})$, and the voltage gain is given by $A_v = h_{fe}R_L/h_{ie} = R_L/r_e$. (For most transistors $h_{fe} \gg 1$.) We see therefore that almost any type of transistor will give the same voltage gain. It is also easy to remember that the input impedance is equal to $(1 + h_{fe}) r_e$, so that even if the value of h_{ie} is not known (which is often the case) the input impedance can be estimated from a knowledge of the quiescent current and h_{fe}. (Again, since $h_{fe} \gg 1$, $h_{ie} = h_{fe} r_e$). Of all h-parameters the h_{fe} figure is the most likely to be known. We see therefore that for the amplifier of Figure 7.22 and assuming a quiescent current of 1 mA and an h_{fe} figure of 100, the input impedance will be about 2.6 kΩ and the voltage gain about 315.

It is of interest to note that if the amplifier is biased so that half of the supply voltage is dropped across R_L (which is usually the case to accommodate maximum drive voltages) then the maximum gain that the amplifier can produce is about $20V_{CC}$. Such a result can be derived as follows:

With half the supply voltage across R_L the quiescent current is given by $V_{CC}/2R_L$. When the emitter resistor is adequately bypassed then the gain of the amplifier is R_L/r_e. It has been shown that $r_e = 26/(i_E \text{ mA})$, hence for a given quiescent current the corresponding value of r_e is $26/(I_q \text{ mA})$ohms, i.e. $(0.026/I_q)$ kΩ. Therefore the maximum gain is given by

$$A_{max} = \frac{R_L}{r_e} = \frac{R_L I_q}{0.026} = \frac{V_{CC}}{2 \times 0.026} \approx 20 \, V_{CC}$$

Since $i_E = i_B (1 + h_{fe})$ and $r_e = 26/i_E$, then the input impedance may be expressed as $Z_{in} = r_e (1 + h_{fe}) = 26/(i_B \text{ mA})$.

The dependence of the input impedance on r_e can be considerably reduced by having only part of the emitter resistor bypassed. The input impedance is then given by

$$Z_{in} = (r_e + R)h_{fe}$$

where R is the unbypassed portion of the emitter resistor. In the extreme, the bypass capacitor may be omitted altogether; the input impedance is then $h_{fe} (r_e + R_E)$ which is approximately equal to $h_{fe} R_E$ since usually $R_E > r_e$. The price to be paid is the reduction in voltage gain, which now becomes

$$h_{fe} R_L/[h_{ie} + R_E (1 + h_{fe})]$$

$$\text{i.e. } h_{fe} R_L/[(r_e + R_E)(1 + h_{fe})]$$

Since $h_{fe} \gg 1$ and $r_e \gg R_E$ the voltage gain becomes R_L/R_E. Under these circumstances the gain of the amplifier of Figure 7.22 is only about 3.7 and the input impedance about 220 kΩ. These changes are examples of what happens when negative feedback of the type discussed here is applied. We shall shortly be discussing feedback in greater detail and noting the improvements in amplifier performance that result.

Multistage amplifiers.

When more gain is required than can be obtained from a single stage, two or more stages are coupled together. The simplest arrangement is to use RC coupling. Figure 7.23 shows a simple two-stage amplifier, each stage being similar to the

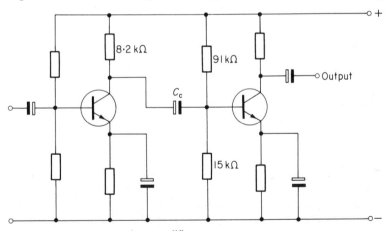

Figure 7.23 Two-stage transistor amplifier

circuit of Figure 7.14(a). Direct connection between the collector of the first transistor and the base of the second transistor is not possible without upsetting the bias conditions in the second stage. The d.c. conditions are maintained by interposing a coupling capacitor, C_c, which effectively is an open-circuit to steady potentials, but which allows signal voltages to be transferred.

The term 'gain' when used in connection with a transistor amplifier is ambiguous; the term should be qualified so that it is quite clear that 'power gain', 'current gain' or 'voltage gain' is intended. The current gain of the first stage of Figure 7.23 is not the h_{fe} figure for the transistor. When the first stage is replaced by an equivalent circuit, as in Figure 7.24, it can be seen that the current $h_{fe}i_b$ must be shared

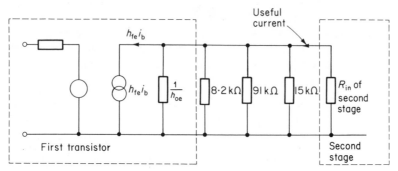

Figure 7.24 Equivalent circuit of Figure 7.23

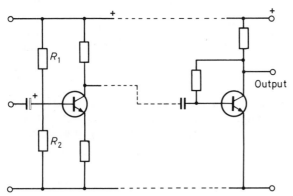

Figure 7.25 A two-stage amplifier in which the emitter bypass capacitor of the first stage has been omitted in order to increase the amplifier's input impedance

between five impedances, all of which are in parallel. These impedances are $1/h_{oe}$, the load resistor $8.2 \, k\Omega$, the bias chain of the second stage, $91 \, k\Omega$ and $15 \, k\Omega$, and finally the input resistance R_{in} of the second stage, which may be about $1.5 \, k\Omega$. A simple application of Ohm's Law shows the fraction of $h_{fe}i_b$ that can be counted as 'useful' current.

The amplifiers described above, and their design by graphical methods, provide a useful introduction to the principles of operation of transistors in amplifier circuits, but such amplifiers are rarely used in practice. Since high-gain operational integrated circuit amplifiers are now inexpensive and readily available, they are the obvious choice for amplifier tasks.

Feedback circuits

The subject of negative feedback has been briefly mentioned on several occasions in the previous text. We now need to examine the subject of feedback in more detail in order to understand the changes in circuit performance that result from its use.

Ordinary amplifiers of the type already discussed suffer from various forms of distortion, and their performance is altered by the ageing of components and variations of supply voltages. Straightforward amplifiers are not therefore accurate measuring devices. All modern high-performance amplifiers use negative feedback, and thus are accurate amplifying devices that can be used as the basis of reliable electronic measuring equipment.

Distortion

We will now discuss the kinds of distortion encountered in electronic circuits, and then proceed to show how negative feedback reduces the effect of such distortion.

The output of an amplifier is said to be distorted if a change of waveform occurs between the input and output terminals. The output waveform may contain frequency components not present in the original signal, or, where complex signals are involved, the phase relationship between the various components of the signal may be altered. The relative amplitudes if these components may also be altered.

The actual output of an amplifier is necessarily limited. Although the gain of an amplifier may be, for example, 1000, this does not imply that any magnitude of input voltage is amplified 1000 times. An input voltage of 100 mV (r.m.s.) does not produce an output voltage of 100 V in the types of amplifiers we are discussing. There is a linear relationship between output and input voltages over only a restricted operating range. The relationship between input and output voltages is known as the transfer characteristic and this characteristic is curved at the ends. The gain of the amplifier therefore varies with the instantaneous magnitude of the input signal, and non-linear distortion is said to be present. Curvature of the dynamic characteristics of the transistors contributes to non-linearity of the transfer characteristics (Figure 7.26). The application of a sinusoidal input voltage results in a periodic output waveform that is non-sinusoidal. Fourier analysis shows that spurious harmonics are present, the result being known as *harmonic distortion*.

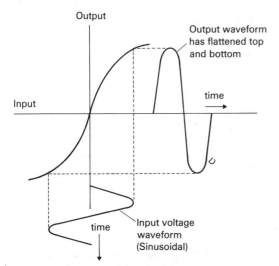

Figure 7.26 Curvature of the transfer characteristic leads to the introduction of harmonic distortion

Total harmonic distortion, *D*, is measured as the root of the sum of the squares of the r.m.s. voltages of the individual harmonics, divided by the r.m.s. value of the total signal, *V*, i.e.

$$D = \frac{\sqrt{(V_{H2}^2 + V_{H3}^2 + V_{H4}^2 \ldots V_{HN}^2)}}{V} \times 100\%$$

where V_{H2}, V_{H3}, etc, are the r.m.s. values of the harmonic components. V_{H2}, for example, is the second harmonic, i.e. that component which has a frequency that is twice that of the fundamental (i.e. lowest frequency) component.

Intermodulation distortion is a form of non-linear distortion whereby the amplification of a signal of one frequency is affected by the instantaneous amplitude of a simultaneously applied signal of relatively large amplitude and low frequency. Combination frequencies are produced which have values equal to the sum and difference of the frequencies of the two applied signals. The amplitude of the high-frequency signal is largest when the low-frequency signal is passing through its zero voltage values and least when the low-frequency signal is at its peak or trough. Since the amplitude of the high-frequency signal varies, amplitude modulation is said to be occurring.

Attenuation distortion is caused by the variation of the gain of an amplifier with frequency. If, for example, a complex waveform has a high-frequency harmonic, and the gain at that frequency is very low, clearly this harmonic must be almost absent in the output waveform.

Phase distortion is present when the relative phases of the harmonic components of a signal are not maintained. Such distortion is caused by the presence of reactive and resistive components in the circuit. In cathode-ray oscilloscope amplifiers, television video amplifiers and in radar circuits, phase distortion is highly undesirable. It is often said that phase distortion is unimportant in audio amplifiers as the ear is insensitive to moderate changes in phase. While it is true that the ear is insensitive in this respect, it is not true that demands on the audio amplifier can be relaxed. Spurious phase shifts adversely affect the movement of the speaker diaphragm, especially during transients. The quality of a sound depends, among other things, upon the attack and decay times. To obtain similar attack and decay times in the reproduced sound, phase distortion should be reduced to a minimum.

Transistor and circuit noise, and 50 Hz components ('hum') are usually classified as distortion when introduced by an amplifier into a signal otherwise free of them.

Most forms of distortion may be markedly reduced by using negative feedback. Other benefits are also obtained, but before discussing these we will first derive the general feedback formula.

Feedback is said to occur in amplifiers when part of the output signal is fed back and combined with the input signal. The relative phases of the input signal and the fraction fed back from the output are very important. When the output fraction is in phase with the input signal, reinforcement takes place and feedback is then said to be positive. This is usually an unstable arrangement that leads to uncontrolled oscillations. When we wish deliberately to produce oscillations, then the circuitry is designed so that positive feedback occurs at only the desired frequency of oscillation.

When part of the output is fed back in antiphase (i.e. 180° out-of-phase) the effective input voltage (or current) to the amplifier is the input signal minus the fraction fed back; the feedback is then said to be negative. In practice the relative phase of the fraction fed back is often not precisely 180°, but may be some other

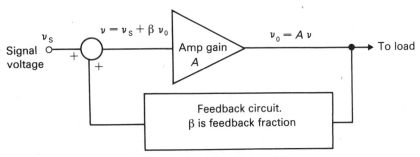

When $|v| < v_s$ Feedback is negative
When $|v| > v_s$ Feedback is positive

Figure 7.27 Block diagram representation of a feedback amplifier

value; we shall examine the consequences of this in due course. In the meantime we shall represent the gain of the amplifier by the letter A in bold type, where $A = A\angle\theta$, A being the magnitude of the gain and θ the phase angle between the output and input signals. The fraction of the output fed back and combined with the input is given the symbol β. The feedback path may also introduce a phase shift (of, say, α) and hence $\beta = \beta\angle\alpha$.

Figure 7.27 is one way of representing the overall general situation. The amplifier (e.g. of the straightforward type already discussed) is represented by the triangle and has a gain of A. A fraction of the output, β, is fed back and combined with the signal. We shall deduce the general feedback equation using voltages, but current symbols could equally well be used. So far as the amplifier is concerned it 'sees' an input voltage, v_i, which is the combination of the signal voltage, v_s, and the fraction of the output voltage fed back, βv_o, where v_o is the output voltage. Hence

$$v_o = Av_i = A\,(v_s + \beta v_o)$$
$$v_o(1 - \beta A) = Av_s$$

therefore

$$\frac{v_o}{v_s} = \frac{A}{1 - \beta A} = A'$$

This is the general feedback equation from which we may deduce the consequences of making the feedback positive or negative. A is known as the open-loop gain and is equal to v_o/v_i. A' is the closed-loop gain, so called because we have connected the input and output terminals via a feedback circuit, thus forming a closed loop; $A' = v_o/v_s$ and is the effective gain from the user's point of view. The product βA is known as the loop gain. This term arises because a voltage v_i at the input of the amplifier is amplified by a factor of A and then attenuated by a factor of β in the feedback path; thus around the loop v_i has been operated upon by the factor βA.

Negative feedback

When a fraction of the output is fed back in antiphase, stable operation as an amplifier usually results; we are then using negative feedback. A practical way of making the feedback negative is to use an amplifier that has a 180° phase difference between v_o and v_i. (The single-stage transistor amplifier is an example in which the output is 180° out of phase with the signal input at the base terminal.) $A = A\angle180°$ = $A(\cos 180 - j\sin 180) = -A$; the feedback network is designed to have zero phase shift, i.e. $\beta = \beta\angle0$, thus

$$A' = \frac{-A}{1 + \beta A}$$

for negative feedback.

The minus sign indicates that $A' = A'\angle180$. When the open-loop gain is very large (say 10^5 or 10^6) we can say that $A \to \infty$ and thus $\beta A \gg 1$ for those values of β usually encountered. Under these circumstances $A' = -1/\beta$ for practical purposes. The gain of the feedback amplifier is therefore independent of A provided the latter is large. Variations of supply voltages, ageing of components, and other causes of the variations in gain of the main amplifier are therefore relatively unimportant in a negative feedback amplifier. The gain with feedback depends only on β and this can be made very stable by choosing simple feedback circuits that use very stable circuit components.

Let us now see how the use of negative feedback improves amplifier performance. Initially we shall assume that there is no phase shift in the feedback path, and that the amplifier without feedback always has a phase shift of 180°; we are thus considering simple negative feedback.

Effect of negative feedback on gain stability

Let us suppose that the gain of an amplifier is -10^6 and that 1/100th of the output voltage is fed back in antiphase, i.e. $\beta = 10^{-2}$. The gain of the negative feedback amplifier is then:

$$A' = \frac{-10^6}{1 + 10^{-2}.10^6} \approx -100 \text{ (i.e. } 100\angle180)$$

If now a serious upset in the amplifier reduces the gain from -10^6 to -10^4, the gain of the negative feedback amplifier becomes:

$$A' = \frac{-10^4}{1 + 10^{-2}.10^4} \approx -100$$

which is the same as before. The gain of the feedback amplifier has not been altered by a large change in the gain of the main amplifier. This independence of gain results from the fact that the input to the main amplifier is the difference between the signal voltage and the voltage fed back. If the gain in the main amplifier falls, the difference voltage will increase slightly and so the output remains almost constant.

We can deduce the fractional change in the closed-loop gain, dA'/A', which results from a given fractional change in the open-loop gain, dA/A, as follows:

$$A' = -A(1 + \beta A)$$

therefore

$$\frac{dA'}{dA} = -\frac{1}{(1 + \beta A)^2}$$

hence, dividing both sides by A' gives

$$\frac{\mathrm{d}A'}{A'} = \frac{\mathrm{d}A}{A} \cdot \frac{1}{1 + \beta A}$$

So a percentage change in the open-loop gain is reduced by a factor of $(1 + \beta A)$ to give the corresponding change in the closed-loop gain.

Effect of negative feedback on the frequency response

The upper curve of Figure 7.28 represents the frequency response of an amplifier without feedback. It has a gain of $-A_1$ at frequency f_1, and a gain of $-A_2$ at f_2. A_2/A_1 is small, indicating a restricted bandwidth. If now negative feedback is applied, the gain at f_1 is $A'_1 = -A_1/(1 + \beta A_1)$ and at f_2 is $A'_2 = -A_2 (1 + \beta A_2)$. Therefore

$$\frac{A'_2}{A'_1} = \frac{-A_2}{(1 + \beta A_2)} \cdot \frac{(1 + \beta A_1)}{-A_1}$$

When $\beta A_1 \gg$ and $\beta A_2 \gg 1$, $A'_2/A'_1 \approx 1$. thus the gains at the two frequencies are approximately equal. This is shown in the lower curve of Figure 7.28. Negative feedback thus increases the bandwidth of the amplifier.

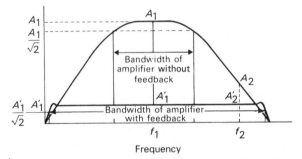

Figure 7.28 The effect of negative feedback on bandwidth. Note the effect on gain at different frequencies. Without feedback the amplifier has gains of A_1 and A_2 at f_1 and f_2 respectively. A_1 is much greater than A_2. With feedback, however, the gains at f_1 and f_2 are A'_1 and A'_2. These gains are equal

The ways in which negative feedback are realized practically are exemplified by many circuit diagrams given later in the book. We may connect a simple potential divider consisting of two high-grade resistors across the output terminals; an appropriate fraction of the output voltage is then available at the tapping point to be fed back to the input terminals. As we shall see, this arrangement is frequently used with IC amplifiers.

Negative feedback may be introduced into a single-stage amplifier of the type shown in Figure 7.22 merely by omitting the bypass capacitor across R_E. The base–emitter signal voltage is then the signal voltage, v_s, minus the a.c. voltage developed across R_E, viz, $i_C R_E$. The voltage fed back is seen to be $i_C R_E$. The output signal voltage is $-i_C R_L$, and hence $\beta = -i_C R_E/i_C R_L = -R_E/R_L$. If the open-loop gain of the stage is large (and with the bypass capacitor present A may be, say, 200 to 300) then in the absence of the bypass capacitor a good approximation for the gain is $-1/\beta$, i.e. $-R_L/R_E$.

Distortion in negative feedback amplifiers

One of the most troublesome imperfections in an amplifier is the presence of distortion in the output waveform. The types and causes of distortion have already been outlined earlier in the chapter. A substantial reduction in distortion can be achieved by the application of negative feedback. When negative feedback is applied to an amplifier we find that the overall performance is relatively independent of the characteristics of the amplifier.

To obtain a quantitative expression for the reduction, suppose that a signal, v, is applied to an amplifier without feedback. The resulting output would be $Av + D$, where D is the distortion component. If feedback is now applied and a signal v_s of sufficient magnitude to give the same output as before is used, the voltage at the input of the amplifier does not consist of the pure signal alone. If the distortion with feedback is d then $-\beta d$ is fed back along with the same fraction of the distortionless component of the output voltage ($-\beta d$ because the feedback is negative). Let us consider only the distortion component. So far as the amplifier itself is concerned it 'sees' a base input voltage of $-\beta d$. This is amplified and distortion is added so that distortion output voltage is $-A\beta d + D$. This is equal to the distortion component d, so we have

$$d = -A\beta d + D$$

Therefore

$$d(1 + \beta A) = D$$

i.e.

$$d = \frac{D}{1 + \beta A}$$

The distortion with feedback is therefore reduced by $(1 + \beta A)$ over what it would have been in the absence of feedback.

In making the comparison it is assumed that the outputs are the same with and without feedback. This is necessary because the voltage excursions in the output stage of the amplifier (where nearly all of the distortion is introduced) must be the same in both cases. As the amplifier with feedback has a lower gain than the same amplifier without feedback, the input signal to the feedback amplifier must be raised sufficiently to make the outputs equal. In doing this it is assumed that the driver stage supplying the input at the higher voltage does not contribute to the distortion by being itself over-driven.

The previous paragraphs indicate that negative feedback has the effect of straightening the effective dynamic characteristic. If, however, any stage of the amplifier is so overloaded that the transistors are driven beyond cut-off, or conversely are 'bottomed', then feedback is not able to reduce the resulting distortion. On the contrary, it leads to greater distortion because, having straightened the main part of the characteristic, the discontinuities at the overload points are more severe than they would be were feedback not applied. (A transistor is said to 'bottom' when it is driven into saturation, i.e. is passing the largest possible current. Under these conditions the collector voltage becomes almost zero and reaches its bottom value.)

Hum, when introduced into the amplifier from sources such as the mains transformer and supply lines, is reduced by negative feedback. This does not mean that hum originally in the signal is reduced, because so far as the amplifier is concerned the hum voltage itself constitutes a signal. With negative feedback the ratio of hum to the wanted signal remains the same. Care must therefore be taken that no hum is induced in the input section of the amplifier, which may not be within the feedback loop, e.g. the lead to the first transistor.

Distortion due to phase shifts that vary with frequency is also reduced provided that the open-loop gain is reasonably large over the range of frequencies of interest. Remembering that

$$A' = \frac{A}{1 - \beta A} \text{then } |A'|\angle\phi = \frac{|A|\angle\theta}{1 - |\beta|\angle\alpha.|A|\angle\theta}$$

(ϕ is the phase shift of the feedback amplifier, i.e. the closed-loop phase shift; θ is the open-loop phase shift, and α is the phase shift in the feedback path)

Suppose we achieve negative feedback by having a 180° phase shift in the open-loop amplifier and zero phase shift in the feedback path. At low frequencies the closed-loop gain is given by

$$|A'|\angle\phi = \frac{|A|[\cos 180° + j \sin 180°]}{1 - |\beta A|[\cos 180° + j \sin 180°]} = \frac{-|A|}{1 + |\beta A|}$$

When

$$\text{When } |\beta A|\gg| \text{ then } |A'|\angle\phi = -\frac{1}{\beta}, \text{i.e.} \left|\frac{1}{\beta}\right| \angle 180°.$$

The phase shift of the closed-loop gain is 180°. If, for example, $A = 10^4$ and $\beta = 10^{-2}$ at low frequencies, then $A' = 10^2\angle 180°$. If now we operate at a higher frequency, the open-loop gain may be reduced to, say, 10^3; the reduction is due to stray capacitances associated with the amplifier. Let us suppose that these capacitances cause a phase shift in the open-loop amplifier so that now $A = |A|\angle 135$. The feedback circuit is assumed to be unaffected by frequency changes and thus remains at $10^{-2}\angle 0$. The resulting closed-loop gain is then

$$|A'|\angle\phi = \frac{10^3[\cos 135 + j \sin 135]}{1 - 10^{-2}.10^3[\cos 135 + j \sin 135]}$$

If this is evaluated it will be found that $|A'|\angle\phi$ is approximately equal to $93\angle -176°$. The 45° phase change in the open-loop amplifier has been reduced to less than a 4° change in the closed-loop amplifier at the higher frequency of operation. Phase distortion has thus been considerably reduced.

Effect of negative feedback on input and output impedances

Feedback in transistor amplifiers may be achieved in four different ways. Transistor amplifiers may be designed as voltage amplifiers or current amplifiers; in each case the feedback may be derived from either the output voltage or the output current.

With voltage amplifiers, the feedback signal is a voltage and must be added to the input voltage in series. When the feedback voltage is derived from, and is a fraction of, the output voltage we are said to be using series-voltage feedback; alternatively the feedback voltage may be proportional to the output current, in which case it is current-derived, and is often referred to as series-current feedback. Figure 7.29(a) shows an example in which a fraction of the output voltage is fed back to the

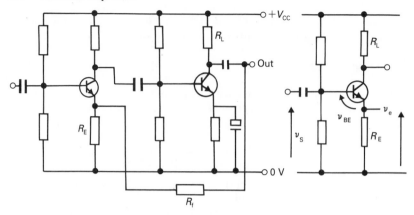

(a) Voltage-derived voltage feedback
(series–voltage feedback, but see text)

(b) Current-derived
voltage feedback
(series–current feedback)

Figure 7.29 Examples of voltage feedback. In the case of (b) the signal component v_e at the emitter is proportional to i_e (and $i_e \approx i_c$). v_e added to v_{be} equals v_s, i.e. $v_{be} = v_s - v_e$. Since the voltage fed back is added in series with the signal voltage, the circuits are known as series feedback amplifiers. In (a) some current-derived voltage feedback is also present in the first stage because the emitter resistor is not bypassed

emitter of the first stage; the voltage fed back is thus in series with v_{be}, the signal voltage. R_f and R_E form a potential divider, and thus the voltage fed back is $R_E/(R_E + R_f)$ of the output voltage, v_o, i.e. $\beta = R_E/(R_E + R_f)$. We thus have voltage-derived voltage feedback. Such feedback is also known as series-voltage feedback.

Figure 7.29(b) shows the case in which the voltage fed back is proportional to the current through R_E; this current can be taken as being the same as the load current. (The base current component of the emitter current can be neglected for transistors having reasonably high h_{FE} values). In this case the feedback voltage is current-derived, hence this is an example of series-current feedback. Although Figure 7.29(a) shows an example of series-voltage feedback it will be evident that the first stage is similar to the amplifier of Figure 7.29(b). We see therefore that both series-voltage and series-current feedback are employed in the same feedback amplifier. Multiple feedback paths are common in transistor amplifiers.

In the case of current amplifiers a feedback current is involved, which must be combined with the signal current in parallel; these amplifiers are therefore shunt types. The feedback current may be derived from the output voltage or from the output current giving rise to shunt-voltage and shunt-current feedback respectively. Examples are shown in Figure 7.30.

In general, it should be noted that voltage-derived negative feedback stabilizes the closed-loop voltage gain against changes in the open-loop gain. Current negative feedback attempts to stabilize the output current.

Irrespective of the type of negative feedback used, the benefits described earlier are all obtained. The input and output impedances, however, may be either increased or decreased depending upon the feedback arrangement used. The effect of feedback on these impedances can be appreciated by representing the actual amplifier by one of four equivalent circuits. It will be seen that two of these are

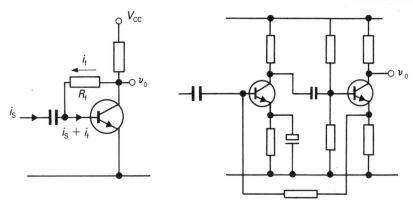

(a) Voltage-derived current
    feedback
(shunt–voltage feedback)

(b) Current-derived current
    feedback
(shunt–current feedback)

Figure 7.30 Examples of current feedback. In (a) the current through R_f is i_F and $i_F = i_{bias} + i_f$. As the input current i_s rises the collector current rises and v_o falls. Consequently i_F is reduced which means that i_f is negative (i.e. 180° out of phase with i_s). Since the feedback and signal currents are added in parallel, the circuits are known as shunt feedback amplifiers

Norton equivalent circuits, the other two being Thévenin equivalent circuits. Since a Norton equivalent can be transformed into the corresponding Thévenin equivalent it will be seen that all four methods of representing the actual amplifier are really equivalent. Figures 7.31 shows the four possible representations. The choice of a suitable equivalent depends upon the nature of the amplifier circuit. Amplifiers which use bipolar transistors are often conveniently analysed using Figure 7.31(c); this equivalent circuit corresponds to the h-parameter representation of a bipolar transistor in which $K_i = h_{fe}$, $Z_i = h_{ie}$ and $Z_o = 1/h_{oe}$. FETs may be represented by (a).

Ideal current amplifiers have zero input impedance, and the output current is not affected by load variations. The input circuit, like an ammeter, does not impede the input current; this is also true for a transresistance amplifier. Ideal voltage amplifiers have an infinite input impedance (like an ideal voltmeter) and hence there is no disturbance to the circuit to which it is connected; the output impedance of an ideal voltage amplifier is zero, hence the output voltage is not affected by load variations.

The ideal transconductance amplifier has an infinite input impedance and infinite output impedance, thus it senses the input voltage correctly, and its output current is unaffected by load variations. The ideal transresistance amplifier has zero input impedance and zero output impedance, and hence presents no impedance to the input current, and has an output voltage that is constant in spite of load variations.

The required input and output impedances of a given real amplifier are determined by the specifications of the system into which the amplifier is to be incorporated. These impedances can be realized by the application of specific forms of negative feedback. Space limitations prevent a full discussion of every form, but the method of analysis is illustrated by the following examples. We shall first analyse the equivalent circuits, and then see how the results can be applied to real circuits.

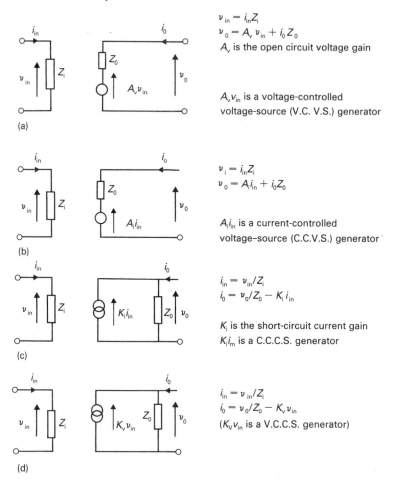

$v_{in} = i_{in}Z_i$
$v_0 = A_v v_{in} + i_0 Z_0$
A_v is the open circuit voltage gain

$A_v v_{in}$ is a voltage-controlled
voltage-source (V.C. V.S.) generator

(a)

$v_i = i_{in}Z_i$
$v_0 = A_i i_{in} + i_0 Z_0$

$A_i i_{in}$ is a current-controlled
voltage–source (C.C.V.S.) generator

(b)

$i_{in} = v_{in}/Z_i$
$i_0 = v_0/Z_0 - K_i i_{in}$

K_i is the short-circuit current gain
$K_i i_m$ is a C.C.C.S. generator

(c)

$i_{in} = v_{in}/Z_i$
$i_0 = v_0/Z_0 - K_v v_{in}$
($K_v v_{in}$ is a V.C.C.S. generator)

(d)

Figure 7.31 Equivalent circuits in which the four gain factors are defined: (a) Voltage amplifier equivalent circuit in which the generator is voltage controlled; (b) transresistance amplifier equivalent (A has dimensions of resistance); (c) current amplifier equivalent circuit; (d) transconductance amplifier. K_v has the dimensions of conductance and hence this circuit is referred to as a transconductance amplifier

In all the equivalent circuits the sense of the voltages and currents are shown by arrows for the general case of feedback. The expressions are therefore general too. With amplifiers, however, stable operation is achieved only when the feedback is negative. The general expressions must therefore be modified. For negative feedback βA must be negative, i.e. either A must involve a 180° phase reversal between the input and output signals, or β must involve a 180° phase shift in the feedback path, but not both simultaneously.

Voltage amplifiers in which voltage-derived feedback is used are usually analysed with the v.c.v.s. equivalent circuit (Figure 7.31(a)). The technique for finding an expression for the output impedance of an amplifier is to apply a voltage to the output terminals of the equivalent circuit, and then calculate the resultant current flowing into these terminals when the input voltage signal source is replaced its

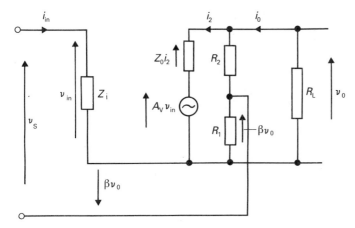

Figure 7.32 Equivalent circuit for a v.c.v.s. feedback amplifier

internal impedance. If the applied voltage and resulting current are represented by v_2 and i_2 respectively, then the output impedance with feedback is given by $Z_o' = v_2/i_2$. For example, for a four-terminal network described by an h-matrix, $v_2/i_2 = 1/h_{22}$ (= $1/h_{oe}$ for a bipolar transistor operated in the common-emitter mode). Applying this technique to Figure 7. 32, $v_s = 0$, v_o becomes v_2 and i_o becomes i_2. We may include the internal source impedance with Z_i. The current taken by R_L and the R_1, R_2 chain may be temporarily ignored; these components can be added in parallel at a later stage in the analysis. We then have

$$v_2 = A_v v_{in} + Z_o i_2$$

and since $v_{in} = \beta v_2$

$$v_2(1 - \beta A_v) = Z_o i_2$$

hence

$$\frac{v_2}{i_2} = Z'_o = \frac{Z_o}{1 - \beta A_v}$$

For negative feedback βA must be negative. Since in this case β is positive, we must arrange that in the actual amplifier a 180° phase reversal takes place between the input and output voltages. Thus for negative feedback.

$$Z_o' = \frac{Z_o}{1 + \beta A_v}$$

The effective output impedance with negative feedback is thus reduced. For large values of βA this reduction is considerable, and in many practical cases Z_o' may be considered as approaching zero. This output impedance of the complete circuit is Z_o' in parallel with R_L in parallel with $(R_1 + R_2)$.

The input impedance may be estimated by assuming that the output voltage is proportional to the input voltage, so $v_o = A v_{in}$. (A is a fraction of A_v, i.e. $A = K A_v$)

$$Z_i = \frac{v_{in}}{i_{in}} = \frac{v_s + \beta v_o}{i_{in}} = \frac{v_s + \beta A v_{in}}{i_{in}} = \frac{v_s}{i_{in}} + \beta A Z_i$$

Since v_s/i_{in} is the effective input impedance as 'seen' by the input driving circuit, when feedback is applied

$$Z_i' = Z_i (1 - \beta A)$$

For negative feedback $Z_i' = Z_i (1 + \beta A)$; the input impedance is therefore increased for this type of feedback circuit.

For current-derived feedback in voltage amplifiers (e.g. of the type shown in Figure 7.29(b)) then it will be found that both the input and output impedances are increased when negative feedback is applied.

Figure 7.30 shows circuit diagrams of current amplifiers in which the feedback current may be proportional to the output voltage, as in (a), or proportional to the output current, as in (b). They are both shunt types (being current amplifiers) one having shunt-voltage feedback and the other being an example of shunt-current feedback. Shunt-current amplifiers are often conveniently analysed with the aid of the equivalent circuit shown in Figure 7.31(c).

To analyse a shunt-voltage feedback system, transresistance amplifier equivalent circuits may be used. As an example, consider Figure 7.33 showing the circuit of a simple shunt-voltage amplifier together with its transresistance equivalent circuit. (Note here that A_i has the dimensions of an impedance and may be referred to as the transresistance factor.) An expression for the overall gain of the transresistor amplifier without feedback is defined as $A_R = v_o/i$. This gain has the dimensions of an impedance, unlike the straightforward ratios for voltage gain and current gain. From Figure 7.33(b) we see that v_o is proportional to $A_i i$ and hence $v_o = KA_i i$

$$\begin{aligned}
v_o &= A_R i \quad \text{where } A_R = KA_i \\
&= A_R(i_f + i_{im}) \\
&= A_R(\beta v_o + i_{im}) \text{ since } i_f \text{ is proportional} \\
&\qquad\qquad\qquad \text{to } v_o, \text{ i.e. } i_f = \beta v_o
\end{aligned}$$

Therefore

$$\frac{v_o}{i_{in}} = \frac{A_R}{1 - \beta A_R} = A_{Rf} \text{ (i.e. } A_R \text{ with feedback)}$$

For negative feedback

$$A_{Rf} = \frac{A_R}{1 + \beta A_R}$$

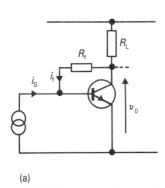

(a)

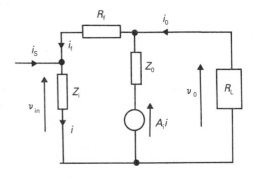

(b)

Figure 7.33 Simple shunt-voltage feedback amplifier together with its transresistance equivalent circuit

To find an expression for the input impedance when feedback is applied we shall assume that R_L is temporarily removed; we are thus finding the input impedance when the output terminals are open-circuit.

Since

$$v_o = \frac{A_R i_{in}}{1 - \beta A_R} \text{ and } v_{in} = (i_{in} + \beta v_o)Z_i = \left[i_{in} + \frac{\beta A_R i_{in}}{1 - \beta A_R} \right] Z_i$$

Then

$$Z_{if} = \frac{v_{in}}{i_{in}} = Z_i \left[1 + \frac{\beta A_R}{1 - \beta A_R} \right] = \frac{Z_i}{1 - \beta A_R}$$

For negative feedback A_R is negative and hence $Z_{if} = Z_i/(1 + \beta A_R)$ i.e. the input impedance is reduced when this type of negative feedback is applied.

To find the effect of the feedback on the output impedance we may proceed as follows. In the absence of feedback (i.e. with R_f removed)

$$v_o = A_i i_s + i_o Z_o$$

If now R_f is returned to the circuit, feedback is introduced and, assuming that the feedback current is small enough to leave i_o almost changed,

$$v_{of} = A_i(i_s + i_f) + i_o Z_o \quad (v_{of} \text{ is the output}$$
$$\text{voltage with feedback)}$$

Taking $i_f = \beta v_{of}$, then $v_{of}(1 - \beta A_i) = A_i i_s + i_o Z_o$

Therefore

$$v_{of} = \frac{A_i}{1 - \beta A_i} i_s + \frac{Z_o}{1 - \beta A_i} i_o$$

For negative feedback A_i must be negative, therefore comparing the two expressions for the respective output voltages, one without feedback and the other with negative feedback, it will be seen that with negative feedback the output impedance has been reduced by a factor $(1 + \beta A_i)$.

The foregoing analyses, to obtain expressions showing the effect of various forms of negative feedback on the input and output impedances of amplifiers, are not intended to be used for circuit design purposes, but are useful in gaining an insight to circuit performances. In many cases where circuit design is involved the designer has to use equivalent circuits and invoke standard circuit analytical techniques.

The effect of negative feedback on input and output impedances is summarized in Figure 7.34.

Type of negative feedback	Input impedance	Output impedance
Series–voltage	Increased	Decreased
Series–current	Increased	Increased
Shunt–voltage	Decreased	Decreased
Shunt–current	Decreased	Increased

Figure 7.34 Effect of various forms of negative feedback on input and output impedances

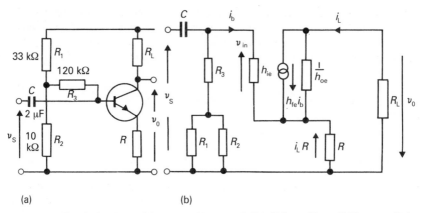

Figure 7.35 Circuits for the problems posed in example 7.1. $[1/h_{oe} \gg (R_L + R)]$ for a small-signal amplifier. (See also later section on bootstrapping)

Example 7.1. Figure 7.35(a) shows the circuit diagram of a simple single-stage amplifier. The transistor parameters are $h_{fe} = 100$ and $h_{ie} = 1.5\,\text{k}\Omega$. What value of unbypassed emitter resistor, R, is required to give a closed-loop trans-conductance of -10^{-3}S? With this value of R, calculate the input impedance.

A convenient equivalent circuit is shown in Figure 7.35(b). h_{oe} is not given and may be taken as being small enough to be ignored (i.e. the current through $1/h_{oe}$ is small enough to be neglected in comparison with i_L). Note that in this case the arrows showing the polarities of the voltages and directions of the currents take into account the phase reversal between the base and collector voltages, and therefore the equivalent circuit is valid for negative feedback.

$$i_b = \frac{v_{in}}{h_{ie}} = \frac{v_s - i_L R}{h_{ie}} = \frac{v_s - h_{fe}i_b R}{h_{ie}}$$

Therefore

$$i_b \left(1 + \frac{h_{fe}R}{h_{ie}}\right) = \frac{v_s}{h_{ie}}$$

The closed circuit transconductance is defined as $-i_L/v_s$ for negative feedback and this, in our example, must equal -10^{-3}S, therefore $i_L/v_s = 10^{-3}$. Taking $i_L = h_{fe}i_b$ then multiplying each side of the previous equation by h_{fe}

$$h_{fe}i_b = \frac{h_{fe}\,v_s}{h_{ie} + h_{fe}R} = i_L$$

Therefore

$$\frac{i_L}{v_s} = \frac{h_{fe}}{h_{ie} + h_{fe}R} = \frac{1}{r_e + R} = 10^{-3}$$

where $r_e = h_{ie}/h_{fe}$, i.e. the intrinsic emitter resistance. From the given data $r_e = 1500/100 = 15\,\Omega$, therefore $R = 985\,\Omega$.

To find the input resistance we note from discussions elsewhere in the book that the input resistance at the base of the transistor is approximately $h_{fe}(r_e + R) = 100\,\text{k}\Omega$. The input impedance is therefore C in series with R_{in} where $R_{in} = 100\,\text{k}\Omega$

in parallel with $(R_3 + R_1)$ in parallel with R_2. R_{in} is therefore approximately $100 \, k\Omega$ in parallel with $128 \, k\Omega$, i.e. about $56 \, k\Omega$. The input impedance is therefore $[56 \, k\Omega + 1/(j2\pi f. \, 2 \times 10^{-6})] = (56 - j80/f) \, k\Omega$, where f is the frequency of the signal voltage. The response would therefore be $3 \, dB$ down when $80/f = 56$, i.e. $f = 1.4 \, Hz$.

The reader should consult the later section on bootstrapping to see how the circuit of the amplifier for this exercise can be easily converted into a practical high input impedance amplifier.

The emitter follower

An important case of a voltage amplifier that uses voltage-derived voltage feedback is the common-collector circuit, more usually known as the emitter-follower because the emitter output signal voltage is in phase with and almost equal to the input signal voltage applied to the base. The output voltage is taken from the emitter lead and the collector is connected directly to the supply line as shown in Figure 7.36. In this case 100 per cent of the output voltage is fed back and combined

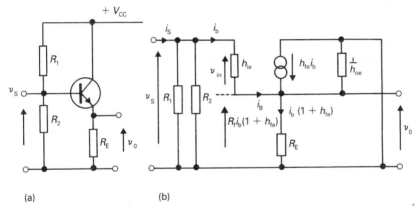

(a)   (b)

Figure 7.36 The emitter-follower: (a) Circuit (b) equivalent circuit. The directions of the arrows show the senses for negative feedback

with the input in antiphase and therefore $\beta = 1$. As shall see, the voltage gain cannot exceed unity and in practical cases is marginally less than 1. The circuit is therefore useless as a voltage amplifier, but is very useful as an impedance transformer; very high input impedances are obtained together with quite low output impedances. The circuit is therefore employed in matching, and coupling high-impedance devices to low-impedance circuits.

Provided the frequency of operation is not extremely high we may use the equivalent circuit of Figure 7.36(b). The direction of the current $h_{fe}i_b$ has been drawn for the negative feedback case (i.e. it is not the general circuit of Figure 7.31(c)). From an examination of the equivalent circuit we see that

$$v_o = R_E i_b \, (1 + h_{fe})$$

$$v_{in} = v_s - R_E i_b \, (1 + h_{fe}) \text{ and } i_b = \frac{v_{in}}{h_{ie}}$$

Therefore

$$v_s = i_b[h_{ie} + R_E(1 + h_{fe})]$$

Hence the closed-loop gain, A', is given by

$$A' = \frac{v_o}{v_s} = \frac{R_E(1 + h_{fe})}{h_{ie} + R_E(1 + h_{fe})}$$

In most cases $h_{fe} \gg 1$ and hence the voltage gain of the emitter follower may be given as

$$A' = \frac{R_E h_{fe}}{h_{ie} + R_E h_{fe}}$$

The expression may be further simplified by remembering that $r_e = h_{ie}/h_{fe}$, so that

$$A' = \frac{R_E}{r_e + R_E}$$

We see that the voltage gain is not quite equal to unity. In practice R_E may be $1\,k\Omega$ or more and r_e is not likely to exceed about $25\,\Omega$. The output signal voltage therefore follows the input signal voltage; hence $v_o \approx v_s$ and $v_o = v_B - 0.6\,V$. The input impedance, v_s/i_b, is equal to the parallel arrangement of R_1, R_2 and the input impedance at the base, say Z_{if}, where $Z_{if} = v_s/i_b$.

$$i_b = \frac{v_{in}}{h_{ie}} = \frac{v_s - R_E i_b(1 + h_{fe})}{h_{ie}}$$

Dividing each side by i_b gives

$$h_{ie} = Z_{if} - R_E(1 + h_{fe})$$

hence

$$Z_{if} = h_{ie} + R_E(1 + h_{fe})$$

Since h_{fe} is usually much greater than unity and R_E is often in practice about the same as h_{ie}, the input impedance, near enough, is $h_{fe}R_E$. If, say, $R_E = 1.5\,k\Omega$ and $h_{fe} = 200$ we see that the input impedance is $300\,k\Omega$. The input impedance as seen by the signal source is then the parallel arrangement of R_1, R_2 and $300\,k\Omega$. Therefore, the complete input impedance is dominated by the parallel arrangement of R_1 and R_2. This means that we do not capitalize fully on the high input impedance at the base, and so the circuit of Figure 7.36(a) is not a good practical one. The way in which we may overcome this difficulty is considered after we have discussed the output impedance.

The output impedance can be found in the usual way by replacing the voltage source with its internal impedance, Z_s say, and applying a voltage, v_2, to the output terminals. An expression is then found for the current, i_2, flowing into the output terminals from which $Z_{of} = v_2/i_2$.

Let the internal impedance of the source be combined with R_1 and R_2 to form Z then

$$v_2 = -i_b(Z + h_{ie}) \qquad \text{(See Figure 7.37)}$$

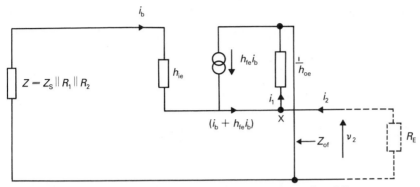

Figure 7.37 Equivalent circuit to find the output impedance of an emitter-follower

R_E may be temporarily removed; its contribution to the output impedance may be taken into account later by adding it in parallel with Z_{of}. Applying Kirchhoff's current law at the point x,

$$i_b(1 + h_{fe}) - i_1 + i_2 = 0 \text{ and } i_1 = v_2 h_{oe}$$

therefore

$$\frac{-v_2(1 + h_{fe})}{Z + h_{ie}} - v_2 h_{oe} + i_2 = 0$$

Dividing throughout by v_2 and remembering that $v_2/i_2 = Z_{of}$

$$-\frac{1 + h_{fe}}{Z + h_{ie}} - h_{oe} + \frac{1}{Z_{of}} = 0$$

therefore

$$Z_{of} = \frac{Z + h_{ie}}{1 + h_{fe} + h_{oe}(Z + h_{ie})}$$

Since h_{oe} is of the order of 10^{-5}, practical values of Z and h_{ie} allow the statement that

$$h_{fe} \gg (1 + h_{oe}[Z + h_{ie}])$$

and a good approximation for Z_{of} is

$$Z_{of} = \frac{Z + h_{ie}}{h_{fe}} = \frac{Z}{h_{fe}} + r_e$$

The output impedance is thus seen to be strongly influenced by the value of Z. If the bias circuit is modified so as to present a high impedance to alternating signals, then the output impedance depends strongly on the internal impedance of the signal source. This is likely to be high since, presumably, the object of using an emitter-follower is to couple a high-impedance driving circuit into the system. The actual output impedance is much reduced when Z is divided by h_{fe}. (r_e in any case is known to be very small.) From the foregoing we see that the output impedance of an emitter-follower is given by $Z_{of} = (Z_s/h_{fe})$ in parallel with R_E, and that it is an advantage to use transistors with high h_{fe} values.

The biasing arrangements for an emitter-follower are determined by the particular circumstances in which the follower is to be used. We could simply omit resistor R_2 in the biasing chain of Figure 7.36(a) and adjust the value of R_1 accordingly. This arrangement is usually frowned upon, but it does have the merit of preserving the large input impedance to the circuit. The voltage levels at the base and emitter are, however, strongly dependent upon h_{fe} and this is a severe disadvantage in some applications. If this form of biasing is used the input and output must be isolated with capacitors and the quiescent voltage level at the emitter must be of little consequence. This means that the input signal amplitudes must be restricted to small values. Under these circumstances the user may as well resort to a standard FET amplifier in order to obtain the desired high input impedance.

Where the signal amplitudes may be large it is necessary to ensure that the emitter voltage is biased to half the supply voltage. From a design point of view we need then to be independent of h_{fe} variations with temperature or transistor sample. Under these circumstances the potential divider of Figure 7.36 must be used. Since the emitter voltage is equal to the base voltage, less one diode (base–emitter junction) drop, the emitter voltage is determined by the potential at the junction of R_1 and R_2. The input impedance is then largely determined by the resistance value of the parallel value of R_1 and R_2. This can be made much larger than the 2 or 3 kΩ of the transistor itself when used as a common-emitter amplifier, but obviously we are unable to capitalize on the high input impedance of the emitter-follower itself. It is more likely, however, that under these circumstances we wish to take advantage of the low output impedance and load-driving capabilities of the circuit. Figure 7.38 is an example in which two resistors of equal

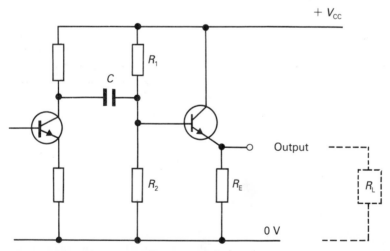

Figure 7.38 Emitter-follower biasing arrangement

value act as a bias potential divider. This arrangement ensures that the base voltage is half of the supply voltage. The emitter voltage is then $(V_{cc}/2) -0.6$, which for most purposes can be taken as about half the supply voltage. If this is not sufficiently accurate an actual R_1, R_2 ratio can easily be calculated. As an example suppose the quiescent current in the emitter-follower is to be 2 mA and a supply voltage of 12 V is being used. $R_E = (V_{CC}/2) \div I_q = 6\,\text{V} \div 2\,\text{mA} = 3\,\text{k}\Omega$. The base voltage must then be $6 + 0.6 = 6.6\,\text{V}$, hence $R_2/R_1 = 1.23$. We could, say, have $R_2 = 100\,\text{k}\Omega$ and $R_1 = 82\,\text{k}\Omega$. The input impedance is then about $39\,\text{k}\Omega$ when $h_{fe} = 100$. (The emitter-follower itself has an input impedance of $h_{fe} \times 3\,\text{k}\Omega$, which is $300\,\text{k}\Omega$.)

To calculate C we note that the lower 3 dB point occurs at a frequency of $1/2\pi 30CR$, i.e. $1/2\pi C\,39\,\text{k}\Omega$. If this point must, for example, be as low as 30 Hz, then $C = 1/2\pi 30 \times 39 \times 10^3 = 0.136$, say 0.15, μF. When the output terminal must be isolated from the load from a d.c. point of view an isolating capacitor must be used here too. The value of the capacitance will depend upon the load, which is likely to be less than R_E; the same expression for a turning frequency can be used which will yield likely values of 1 to 5 μF.

In cases where the load must be directly coupled to the emitter-follower some consideration will need to be given to the quiescent voltage at the emitter. If the load will accept $V_{CC}/2$ there is no problem, but where the quiescent output voltage must be at some other level (commonly 0 V) then the bottom of R_E and R_2 will need to be connected to a negative supply ($= -V_{CC}$ for zero quiescent output voltage). Care will be needed if the load is connected between the output terminal and a zero voltage (i.e. earth) line. R_E and R_L then form a potential divider, and hence the output voltage can swing only as low as $R_L/(R_L + R_E)$ of $-V_{CC}$; this will occur when the input voltage causes the emitter-follower to be cut off, thus reducing the emitter current to zero. The full output voltage swing can then only be $+V$ to $-V_{CC}.R_L/(R_E + R_L)$. For symmetrical swings about 0 V we see that we have a restricted range of $\pm V_{CC}R_L/(R_E + R_L)$.

Regulated power supplies

The emitter-follower is frequently used in regulated power supplies in which use is made of the fact that the emitter voltage is equal to the base voltage minus 0.6V. Simple regulators that depend upon Zener diodes alone are not convenient when

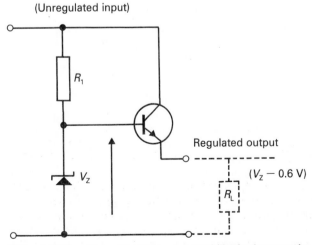

Figure 7.39 Use of an emitter-follower as a stabilized voltage supply

large load current variations are involved. The diode itself must be capable of accommodating these large variations. This situation can be avoided if the Zener diode is isolated from the load current by an emitter-follower as shown in Figure 7.39. Here we see that R_2 is replaced by a Zener diode. The base voltage is thus V_z, the Zener voltage. Since the slope resistance of Zener diodes is commonly only a few tens of ohms, the source resistance R_s is small; the output resistance, being about R_s/h_{fe}, is seen to be very small and hence this regulator approaches an ideal voltage output source. In practice the transistor used for the emitter-follower must be able to accommodate the likely full-load current.

A variation of this technique, which allows not only a variable output voltage but also an improved stabilization of the output voltage, uses a negative feedback arrangement as shown in Figure 7.40. The rectified and smoothed d.c. is fed to the

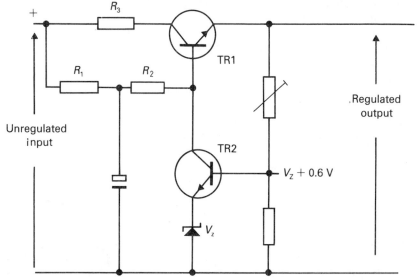

Figure 7.40 The use of an emitter-follower in a stabilized voltage supply in which the degree of stabilization is increased over that obtained with the simple circuit of Figure 7.39

collector of the emitter-follower, TR1 (which must be capable of handling the full load current). The based current of the emitter-follower is supplied by a voltage regulating amplifier consisting of TR2 and a Zener diode. The base of TR2 is connected to the junction of a simple potential divider placed across the regulated output voltage lines. Any change in load conditions, and hence output voltage, affects the base voltage of TR2. This transistor compares the reference voltage of the Zener diode with a suitable fraction of the output voltage. For example, any rises in output voltage increase the base–emitter voltage of TR2. The resulting increase in the collector current causes the collector voltage to fall. Since the collector is connected directly to the base of the emitter-follower, the base voltage of the emitter-follower must fall. This fall very largely offsets the original rise of voltage at the output (emitter) terminal. This negative feedback arrangement thus maintains the output voltage at almost a constant voltage in spite of changes in load conditions. Resistors R_1 and R_2 constitute the collector load of TR2. R_3 is a small-valued resistor that provides some protection for the emitter-follower in the event of short-circuits appearing across the output supply terminals.

The Darlington pair

We have seen how the use of series-voltage feedback increases the input impedance of transistor amplifiers. In particular the emitter-follower can be made to have a very high input impedance at the base. Readers will recall that the input impedance of an emitter-follower is $h_{ie} + R_E(1 + h_{fe})$, while for a common-emitter amplifier with the emitter resistor unbypassed the input impedance is approximately $h_{fe}(r_e + R_E)$. In both cases the expression shows the heavy dependence of the input impedance upon h_{fe}. Practically, it is not yet possible to fabricate transistors with very high (>1000) values of h_{fe}; we usually have to accept values in the low hundreds. A technique for combining two transistors to form a super-h_{fe} transistor equivalent is available; the composite pair of transistors is known as a Darlington pair.

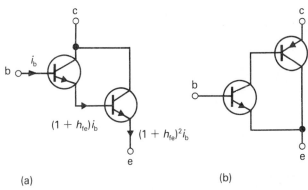

(a)   (b)

Figure 7.41 Compound transistor pairs for the production of high h_{fe} values: (a) Darlington pair of identical transistors; (b) modified arrangement that uses a pnp/npn combination

Figure 7.41(a) shows two npn transistors in such an arrangement. From the diagram it will be seen that the h_{fe} value of the Darlington pair is effectively the product of the h_{fe} value of each transistor making up the pair. Frequently the two transistors are identical, but not always so. For the output stages of a power amplifier it is necessary to use a transistor that can handle large currents. Such a transistor cannot be made also to have a large h_{fe} value. It is possible, however, to form a Darlington pair by combining the power transistor with a high-h_{fe} 'driver' transistor. When the two-diode voltage drop (1.2 V) between base and emitter is inconvenient, as is often the case in integrated circuits, an npn transistor can be combined with a pnp transistor as shown in Figure 7.41(b). We shall see that this arrangement also has its uses when used in power amplifiers that use discrete transistors.

The bootstrap circuit technique

Although series-voltage feedback increases input impedance, the use of conventional potential divider bias networks prevent us from capitalizing on such high input impedances. As we have seen with the emitter-follower, the shunting effect of the bias components reduces considerably the effective input impedance. To obtain satisfactory biasing, the voltage at the junction of the potential divider must remain reasonably constant under varying bias current requirements; this means that the impedance of the biasing network must be low compared with the resistance reflected into the base at zero or very low frequencies ($\approx h_{FE}R_E$). Low values of resistance for the biasing network ensure that the current flowing in the biasing chain is substantially greater than the bias current flowing into the base. The parallel equivalent resistance of the biasing potential divider is usually much lower than $h_{FE}R_E$. An attempt to isolate the base with a high-value resistor (as, for example, R_3 in Figure 7.35(a)) is not really satisfactory since, as the accompanying equivalent circuit shows, the impedance of the biasing circuit becomes too high for satisfactory biasing conditions. The conflicting requirements of high input impedance required to be presented to the driving circuit, and low biasing circuit impedance required for stable biasing can be resolved by a neat and simple modification to the circuit shown in Figure 7.35(a). This modification is shown in Figure 7.42, in which a capacitor is connected between the emitter and the junction of the biasing potential divider. The modification works in cases where the signal frequencies do not extend down into the very low or zero-frequency region.

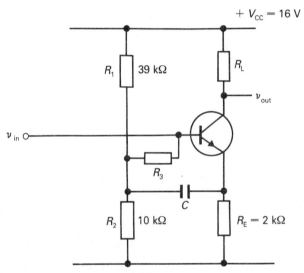

Figure 7.42 The bootstrap amplifier

At zero or very low frequencies (applicable to biasing currents) the presence of C may be ignored. The values of resistance used for R_1, R_2 and R_3 are made low enough to satisfy satisfactory biasing requirements. The input impedance as seen by the driving circuit will be low at these frequencies, but this is of no consequence since low-frequency signals are absent. At signal frequencies the value of C is made high enough so that X_c is negligible in comparison with impedance of the biasing network. Changes of signal voltage at the base are accompanied by similar changes of signal potential at the emitter, and these emitter voltage changes are transferred by the capacitor to the junction of R_1 and R_2. The result is that the net *signal* voltage across R_3 is almost zero, and hence R_3 draws practically no signal current. As seen by the driving circuit, therefore, R_3 appears to have an extremely high a.c. impedance. From a biasing (d.c.) point of view the resistance of R_3 may be quite low.

The circuit configuration is called a bootstrap amplifier because the application of the signal to one end of R_3 causes the voltage at the other end of R_3 to be 'pulled up by the bootstraps'. An expression for the effective a.c. resistance of R_3 may be obtained by noting that the input resistance 'looking into' the base terminal is $h_{ie} + (1 + h_{fe}) (r_e + R'_E)$ where R'_E is the parallel arrangement of R_E, R_1 and R_2. (The presence of C means that from a signal point of view all three components are connected in parallel.) Since h_{ie} is small compared with the remaining part of the expression, then, to a close approximation, this base input resistance is $(1 + h_{fe})(r_e + R'_E)$.

The input base signal current is thus given by

$$i_b = \frac{v_s}{(1 + h_{fe})(r_e + R'_E)}$$

The emitter signal voltage is $(1 + h_{fe})i_b \cdot R'_E$, and thus the voltage difference across R_3 is

$$v_{R_3} = v_s - \frac{v_s R'_E}{r_e + R'_E} = \frac{v_s r_e}{r_e + R'_E}$$

The signal current taken by R_3 is thus $v_s r_e/(r_e + R'_E)R_3$. From the driving circuit's point of view, R_3 has an a.c. resistance given by

$$R_3 \text{ (a.c.)} = \frac{v_s}{v_s r_e/(R'_E + r_e)R_3}$$

i.e.

$$R_3 \text{ (a.c.)} = \frac{r_e + R'_E}{r_e} R_3$$

Thus the a.c. resistance R_3(a.c.) is substantially greater than the d.c. resistance of R_3. For the circuit of Figure 7.42 with a quiescent current of $1\,mA$, $r_e = 25\,\Omega$ (say). If we make $R_3 = 3.3\,k\Omega$ then the d.c. equivalent resistance of the bias network is $(3.3 + 39$ in parallel with $10)\,k\Omega$, which is small compared with the d.c. resistance looking into the base. This gives rise to good bias stability. Since $R'_E = R_E$ in parallel with 10 and 39 $(k\Omega)$ then $R'_E = 1.6\,k\Omega$. The a.c. resistance of R_3 is then the d.c. value of R_3 multiplied by a factor of 65, i.e. $215\,k\Omega$ approximately. Multiplying factors up to 100 or so are practically obtainable with this form of bootstrapping. The total input resistance as seen by the signal source is thus

$$(R_e\text{(a.c.)} + R'_E) \text{ in parallel with } (1 + h_{fe})(r_e + R'_E)$$

The Miller effect

When obtaining expressions for the input impedance of amplifiers that incorporate feedback, we have assumed that the frequency of operation has been low enough to ignore stray and interelectrode capacitances. We have, in effect, been computing the input resistances of the amplifiers. The interelectrode capacitances are small (e.g. the capacitance between the collector and base of a bipolar transistor in operation may be about $3\,pF$). When operating at high frequencies these capacitances should not be ignored. The capacitance between the base (or gate)

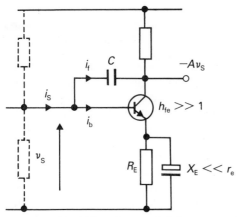

Figure 7.43 The Miller effect in which C represents the interelectrode capacitance between the collector and the base

and emitter (or source) presents no problem and can be added to the stray capacitances of the input circuit. The capacitance between the collector (or drain) and base (or gate) presents a greater problem, however.

From an examination of Figure 7.43 we see that the interelectrode capacitance forms a feedback path; we have in effect voltage-derived current feedback. For the bipolar transistor we assume that the voltage gain is A and that there is a 180° phase reversal.

From the diagram

$$i_s = i_b + i_f$$

$$= \frac{v_s}{h_{fe}\,r_e} + [v_s - (-Av_s)]Y_c$$

where $Y_c = j\omega C$.

The input admittance i_s/v_s is thus increased over what it would have been in the absence of C by $j\omega C\,(1 + A)$. The presence of C between the collector and base thus adds an equivalent input capacitance to ground of $(1 + A)C$. Since A may be up to 100 or more and C may be (say) 3 pF, we must take into account the equivalent 300 pF capacitance added in parallel to the input resistance. (A capacitance of 300 pF has a reactance of only 530 Ω at 1 MHz.)

The reflection of the large equivalent capacitance into the input circuit is known as the Miller effect. So far as amplifiers are concerned it is a nuisance. In the early days of alloy junction bipolar transistors the interelectrode capacitances were comparatively large and the Miller effect imposed a severe limitation on the frequency of operation. This explains the popularity of the common-base amplifier for use as an r.f. amplifier. The Miller effect presents no problem in a common-base amplifier since the collector–base capacitance, C_{cb}, is merely added to the output capacitance, and the base–emitter capacitance merely added to the input capacitance.

The emitter-follower, too, is not affected by the Miller effect since C_{cb} is connected between the base and the zero-signal voltage line. The base–emitter capacitance is connected between the output and input terminals, but since the emitter gain is marginally less than unity this capacitance also is of no consequence.

Power amplifiers

The amplifiers considered so far have been designed to give a maximum voltage or current output with minimum distortion. Although they develop power in their collector load circuits, this power is of little importance. The choice of transistors and associated components is not influenced at all by power considerations except in so far as the components have to be operated within their maximum power ratings. Power amplifiers, on the other hand, are those in which power output is the chief consideration. These are the amplifiers which are designed to operate into loads such as servo motors, potentiometric recorders, moving-coil pen recorders, loudspeakers, meters and other recording devices. The aim is usually to deliver the maximum power into the load, consistent with a reasonably low distortion. The dissipation of power necessarily implies a resistive load since no power can be dissipated in a capacitive or inductive load. There may, however, be a reactive component associated with the load as, for example, in an electric motor. Here, although the power is dissipated in an equivalent resistance, there is always the inductance of the motor coils to be taken into consideration. This affects the design of the power amplifier in which stable operation must be ensured.

In general the actual load has an equivalent resistive value which is not under the control of the designer of the electronic driving circuitry, but is determined by the nature of the output device. Wherever possible, therefore, the power amplifier should be designed to have as low an output impedance as possible. This will ensure that, for a given driving voltage and given load, the power delivered to the load will be close to the maximum. (The maximum power to the load will be delivered when the amplifier output impedance is zero.) Fortunately, negative feedback arrangements allow us to control the output impedance of the power generator and reduce it to extremely low values. With loudspeaker loads especially, low output impedances bring improved transient response; generally, improved overall performance is obtained if the output impedance is made much lower than the load impedance.

In circumstances where the power amplifier's output impedance is fixed at a value that is not very low compared with any likely load impedance then, for a constant driving voltage and given internal impedance, we may find an expression for the power delivered to the load as a function of varying load impedance. By differentiating the expression and equating to zero it can be shown that maximum power is supplied to the load when the load resistance is equal to the output resistance of the amplifier. If the power amplifier's output impedance is complex, and thus of the form $R + jX$, then the load for maximum power transfer is the complex conjugate viz, $R - jX$.

In practice the load impedance will not be able to be varied at will. When both generator and load impedances are fixed we would then have a matching problem. One way of making the load 'look' as though it is variable so far as the amplifier is concerned is to use a transformer. By connecting the primary coil to the amplifier and the secondary coil to the load it can be shown that the impedance reflected into the primary circuit is $(N_p/N_s)^2 R_L$, where N_p and N_s are the number of turns in the primary and secondary coils respectively, and R_L is the actual load assumed to be resistive only. By choosing the correct turns ratio it is always possible to match the load with the internal impedance of the amplifier. Fortunately, we are not often in this difficult situation and thus the use of a troublesome transformer can be avoided.

Efficiency

The efficiency of a power amplifier is defined as the output power divided by the total power supplied via the supply lines. Figure 7.44 shows the graphical position for a Class A amplifier with a resistive load. The absolute maximum collector voltage swing is from zero to the supply voltage V_{CC}. The maximum peak output voltage is therefore $V_{CC}/2$, the r.m.s. value being $V_{CC}/(2\sqrt{2})$. Similarly, the maximum r.m.s. output current is $I_{max}/(2\sqrt{2})$. The output power is therefore $V_{CC}I_{max}/8$. It can be seen from Figure 7.44 that with the operating point in the centre of the load line, $I_{max} = 2I_q$, therefore the output power is $V_{CC}I_q/4$. The input power is that obtained from the supply, which is delivering a mean current of I_q at a supply voltage of V_{CC}. The efficiency is therefore $(V_{CC}I_q/4) \div V_{CC}I_q = 0.25$, i.e. 25 per cent.

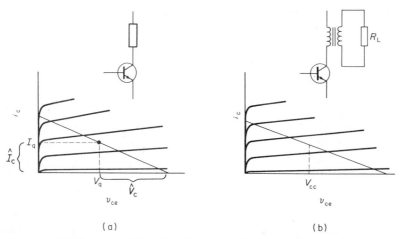

Figure 7.44 (a) Class A amplifier with resistive load; (b) Class A amplifier with transformer-coupled load

In practice, Class A power amplifiers are specifically designed for low distortion with efficiency an important, but secondary, consideration. The amplifiers are, therefore, never driven to their full extent. This means that the output power is less than $V_{CC}I_q/4$. i.e. less than 25 per cent. As the drive voltage is reduced the power delivered to the load is also reduced, but for Class A operation the quiescent current and supply voltage stay fixed; hence the efficiency is reduced as the drive voltage is reduced. In the limit for zero drive the efficiency is reduced to zero.

When the load is coupled to the transistor via a transformer, the peak value of collector voltage change, $\hat{V}_C$, is equal to the supply voltage, if we assume that the d.c. resistance of the primary coil is zero. $\hat{V}_C$ is therefore twice the previous value obtained with a resistive load. The theoretical maximum efficiency is then 50 per cent.

The efficiencies quoted above can be improved upon by operating the transistor under Class B conditions. In this mode, the operating point, Q, is brought down the load line to the cut-off position by the application of suitable bias arrangements. A sinusoidal input signal then produces a series of load current pulses that are half sine waves (Figure 7.45). The power taken from the supply is the supply voltage times the mean current, i.e. $V_{CC} \times 2I_{max}/\pi$. The power into the load is the r.m.s.

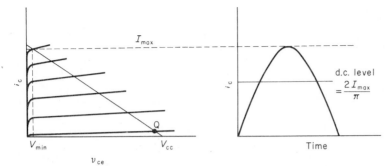

Figure 7.45 Transistor operated under Class B conditions. The maximum theoretical efficiency is almost 78.5 per cent. (The non-linearity at low values of collector current has been ignored)

value of the current, $I_{max}/\sqrt{2}$, times the r.m.s. value of the voltage across the load, i.e. $(V_{CC} - V_{min})/\sqrt{2}$. The efficiency is therefore

$$\frac{I_{max}(V_{CC} - V_{min})}{\sqrt{2} \times \sqrt{2}} \times \frac{\pi}{2I_{max}V_{CC}} = \frac{\pi}{4}\left(1 - \frac{V_{min}}{V_{CC}}\right)$$

V_{min} is very small compared with V_{CC}, and so the efficiency approaches $\pi/4$, i.e. about 78.5 per cent. This, of course, is the theoretical maximum and is achieved only when the amplifier is driven to its fullest extent. Compared with a Class A amplifier there is a considerable improvement; additionally, with low driving voltages we avoid high power dissipation in the transistors. For zero driving voltage no dissipation occurs.

Push-pull amplifiers

It is obvious from Figure 7.45 that it is not possible to use a single transistor as a Class B amplifier because half of the waveform is missing. It is necessary to use two transistors in what is termed a push-pull arrangement. Each transistor conducts for

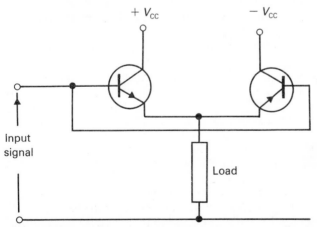

Figure 7.46 Class B prototype power amplifier using a complementary pair (pnp/npn) of output transistors. The circuit is drawn in an unconventional way to emphasize the basic emitter-follower arrangement. $+V_{CC}$ and $-V_{CC}$ need not be numerically equal although often they are. Biasing arrangements are not shown

half a period and thus the complete waveform is restored in the load. The use of emitter-followers and negative feedback enables output stages to be designed that have very low output impedances.

The principle of one form of push-pull output stage is shown, without biasing arrangements, in Figure 7.46. A pair of output transistors, one npn and the other pnp, have their emitters interconnected; the load is connected between the emitters and the zero-voltage line. The bases are also interconnected and fed from the signal source. For positive-going signals the npn transistor is active, while the pnp transistor is cut off; conversely, negative-going signals cause the npn transistor to be cut-off, while the pnp transistor is driven into conduction. Thus, over the complete cycle of input voltage, the npn sources or 'pushes' current into the load for the positive-going half-cycle, and subsequently, during the negative-going half-cycle, the pnp transistor sinks or 'pulls' current from the load. For zero input voltage both transistors are cut off, and hence the quiescent current is zero; this arrangement is therefore a Class B push-pull power amplifier. The power dissipation in the transistors is zero under quiescent conditions, and low for low driving voltages.

Distortion in Class B amplifiers

Although Class B operation has the advantage of high efficiency, and low power dissipation in the transistors when small drive voltages are involved, the attendant distortion is quite severe. The reduced spacing between the output characteristics as the base current is reduced accounts for some of the distortion, but the main factor is what is known as crossover distortion. This distortion arises because of an inherent property of an emitter-follower. The output (emitter) voltage follows the input signal, v_S, but in the active region when the transistor is turned on there is necessarily a $0.6\,V$ diode drop from the base to the emitter. The result is that during that part of the cycle when the signal voltage is less than $0.6\,V$ the transistor current is reduced almost to zero. The signal voltage must subsequently become more negative than $0.6\,V$ before the pnp transistor is fully turned on. As a result the output current waveform becomes severely distorted at the crossover region where active operation is transferred from the npn transistor to the pnp transistor. A similar distortion arises as the signal voltage polarity changes from negative to positive. Figure 7.47 shows the position graphically.

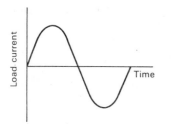

Biased to class A conditions

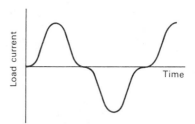

Biased to class B conditions

Figure 7.47 A push-pull amplifier. In Class A the distortion is low and reduces with reducing drive voltages. The distortion with Class B, known as crossover distortion, is due to the curvature of the transfer characteristics near to the cut-off point, and to the fact that for base-emitter voltages of less than $0.6\,V$ the emitter current falls rapidly to zero. For Class B amplifiers the distortion increases with decreasing drive voltages. Many amplifiers are designed for Class AB conditions, where a compromise is reached between the efficiency of the Class B mode and the low distortion of the Class A mode

One way of reducing this form of distortion is to apply sufficient bias to each transistor so as to produce some conduction when the input signal is zero. A popular way of doing this is to modify the circuit of Figure 7.46 in the way shown in Figure 7.48. The transistors are then said to be operating in the Class AB mode, in which a compromise is reached between the efficiency of the Class B mode and the low distortion of the Class A mode. Diodes D1 and D2 are each forward-biased via a current source and sink respectively; these current devices may be no more than a couple of resistors of appropriate values. The signal is fed to the junction of the two diodes. Under quiescent conditions the voltage at the input terminal is zero. Each

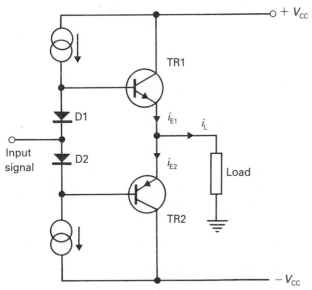

Figure 7.48 Biasing of the output transistors to produce Class AB operation

diode is made to have a voltage drop which is slightly greater than the v_{BE} value of each output transistor. The base–emitter voltage of TR1 is thus slightly positive, and that of TR2 slightly negative. Both output transistors are therefore slightly conducting. In a perfectly symmetrical arrangement each output transistor is conducting equally and, with a matched complementary pair, the joint emitter voltage is zero. We see that one advantage of this type of biasing circuit is the ease with which the output voltage can be adjusted to zero. (Zero output voltage under quiescent conditions is desirable for most loads, and is essential for loudspeaker or servo motor loads.) Any small imbalance in the circuit can be compensated for by adjusting the bias current separately to one or other of the output transistors. This is easy to achieve if the current sources are merely resistors, as is often the case in discrete component amplifiers.

On the application of a signal, the signal voltage is conveyed to the bases of the output transistors. Positive-going signals decrease the forward voltage across D1, thus the resistance of D1 increases and the base voltage of TR1 rises; current is therefore diverted into the base of TR1. The increase in base current brings about an increase in emitter current, and hence an increase in the output voltage. Simultaneously, the forward voltage across D2 increases. The diode current through D2 therefore increases. Since the sum of this current and the base current of the pnp transistor is constant, this latter base current decreases; TR2 is therefore rapidly cut off. Negative-going signals bring about a corresponding result in which the npn transistor is cut off and the pnp transistors is driven into conduction. Because $v_o = v_S + v_{BIAS} - v_{BE}$ and v_{BIAS} is adjusted to be equal to or marginally greater than v_{BE}, changes in signal voltage bring about identical changes in output voltage for values of v_s down to zero volts. Since the output voltage follows the input voltage even when v_S is less than v_{BIAS}, crossover distortion is eliminated.

One further advantage of using the Class AB mode in a push-pull arrangement is the elimination of second harmonic distortion. Second-and higher-order harmonics arise as a result of non-linearity of the transistor output characteristics. The presence of harmonics in the emitter current may be expressed as

$$i_{E1} = I_{DC1} + [I_{e1} \sin\omega t + I_{e2} \sin2\omega t \ldots]$$

where i_{E1} is the total emitter current of TR1, I_{DC1}, the average steady value, and I_{e1}, I_{e2} etc. are the amplitudes of the fundamental, second, third and higher harmonics. The amplitudes of the harmonics are significant, but arise mostly in the region where the collector (and hence emitter) current is low. This, for Class AB operation, is when both transistors are conducting. The emitter current for TR2 is 180° out of phase with that for TR1 for a given input signal because a complementary npn/pnp pair is being used. The emitter current for TR2 is thus

$$i_{E2} = I_{DC2} + I_{e1} \sin(\omega t + \pi) + \ldots = I_{DC2} - I_{e1} \sin\omega t + I_{e2} \sin 2\omega t$$

Since the load current, i_L is given by $i_L = i_{E1} - i_{E2}$ then

$$i_L = 2[I_{e1} \sin\omega t + I_{e3} \sin3\omega t \ldots]$$

We see therefore that the even harmonics are absent. Residual distortion is reduced by the use of negative feedback.

Power amplifiers of the type we have been discussing are not without their problems. Foremost among these are poor thermal stability, no short-circuit protection, and difficulty in obtaining a truly symmetrical arrangement with an npn/pnp pair.

Thermal stability

Power output transistors are required to handle high currents and are designed as such. As a result the transistors need to be physically larger than their small-signal amplifying counterparts. Not only are their h_{fe} values comparatively low, but the increased temperatures produced when high power is being dissipated affects their v_{BE} characteristics in a way discussed in Chapter 5. As a result the collector quiescent current increases; this produces further increases in temperature. In extreme cases a thermal runaway situation may develop. Even though a manufacturer increases the size of the transistor and the metal casing which encloses the device, thermal energy is usually generated at a rate which is too large to be dissipated into the ambient surroundings from the metal case alone. The usual way to solve this problem is to mount the transistor in good thermal contact with a large metal plate. Often this metal plate, known as a heat sink, will need to have fins, and be blackened in order to aid the dissipation of heat.

The design of heat sinks involves the concept of thermal resistance. If the temperature of the device is T_d and the ambient temperature is T_a then the rate of loss of heat, i.e. power transferred to the ambient surroundings, is proportional to the temperature difference; i.e.

$$T_d - T_a \propto P$$

therefore

$$T_d - T_a = \theta P$$

θ, the constant of proportionality, is known as the thermal resistance. If therefore the device is dissipating power, P, and the thermal resistance of the case is θ_c then the temperature of the device will rise to satisfy the above equation. The units of θ_c are degrees $C.W^{-1}$. The value of θ_c for power transistors is rarely small enough to prevent T_d from becoming excessive for the power dissipations involved. We therefore need to reduce the effective value of θ_c by coupling the transistor thermally to a heat sink. The total thermal resistance involved is made up of several thermal resistances in series. The heat is generated at the junctions within the device and so the first resistance involved is that between the device and the case. Manufacturers try to ensure that the thermal bonding of the device to the case is as low as possible. To do this the collector region of the device is bonded directly to the case. The latter is then at the same electrical potential as the collector, and in the vast majority of power transistors the case forms the collector lead. Some electrically insulating layer must then be interposed between the case of the device and the heat sink, because the latter is nearly always at earth or zero potential. The insulating layer itself constitutes a thermal resistance between the case and the heat sink, θ_{CH}, and therefore to reduce this to a minimum the layer must be very thin. Often thin sheets of mica are used, although nowadays suitable flexible synthetics are available. A heat sink compound assists the thermal contact between the device (and heat sink) to the insulating washer. Lastly, we have the thermal resistance between the heat sink and the ambient surroundings, θ_{HA}. Thus the total resistance θ_t is given by $\theta_t = \theta_{DC} + \theta_{CH} + \theta_{HA}$ (θ_{DC} is the thermal resistance between the device and the case).

There is nothing that the user can do about θ_{DC} since this is solely under the manufacturer's control. He therefore usually specifies the case temperature, T_C, that must not be exceeded for a given power dissipation in the device. From the above considerations we see that

$$T_C - T_A = (\theta_{CH} + \theta_{HA}) P$$

where T_C is the case temperature.

If, for example, the ambient temperature were 20°C and the case temperature must not exceed 80°C for a power dissipation in the device of 10 W, then we need to select a suitable insulating washer and heat sink. We see that $80 - 20 = \theta.10$, where $\theta = (\theta_{CH} + \theta_{HA})$, i.e. θ must not exceed 6°C/W. From a manufacturer's catalogue we note that a suitable silicone rubber washer is available with a thermal resistance of 0.33°C/W. (Silicone rubber washers do not need a heat sink compound.) We see therefore that a heat sink is required that has a thermal resistance of less than 5.67°C/W. One with a resistance of 4.0°C/W is readily available and would be suitable.

Although the use of a heat sink assists in preventing excessive rises of temperature in the device, it is essential to adopt additional measures to improve the thermal stability of the circuit. We have already seen in Chapter 5 that the v_{BE} characteristic shifts back towards the origin by about 2.1 mV per °C. This means that, all other factors being constant, a one degree rise in temperature has the same effect on the collector current as a 2.1 mV increase in base–emitter voltage. A rise in temperature of, say, 40°C therefore has the same effect as a rise of 84 mV in the base–emitter voltage. This produces an enormous rise in collector current. It has been shown that when i_{C1} and i_{C2} represent the collector currents at v_{BE1} and v_{BE2} respectively, then

$$v_{BE2} - v_{BE1} = 0.026 \log_e(i_{C2}/i_{C1})$$

We see therefore that a change of 84 mV in v_{BE} results in the collector current being increased by a factor of exp $(84 \times 10^{-3}/0.026)$ =exp (3.23), i.e. = 25. This very large rise can be nullified in power amplifiers by automatically reducing the bias voltage with increasing temperature. This is quite easy to achieve since silicon diodes are used for biasing. The voltage across a silicon diode also falls by 2. mV per °C, hence if the biasing diodes are kept in thermal contact with the output transistors (say by bonding them to the heat sink close to the transistor) then the temperature rises for the transistor and biasing diodes will be almost the same; the diodes will track the output transistors thermally. The voltage across the diodes, and hence the bias voltage, will fall by the necessary amount to compensate for the change of v_{BE} in the transistor. The tracking is not perfect, however, because the temperature of the diodes will be only that of the heat sink near to the transistor; this will be lower than the case temperature, which is itself lower than the base–emitter junction temperature within the device. For this reason small-valued resistors are connected into the emitter circuit to improve the thermal stability. To quantify this improvement let us suppose initially that the diodes are taken out of thermal contact with the output transistors. Assuming that at 20 °C the quiescent current is 30 mA, then, in the absence of any emitter resistor, all of the bias voltage (which we will assume to be 650 mV) appears across the base–emitter junction. If now the temperature rises from 20°C to 60°C then the v_{BE} characteristic shifts towards the origin by 84 mV. If the base–emitter voltage is maintained at its original value, the emitter current must then rise by a factor of 25; the emitter quiescent current now becomes 750 mA.

Let us now assume that we had connected 2 Ω resistors in series with the emitter leads. At 20°C the emitter current is 30 mA and hence the voltage across the emitter resistor will be 60 mV. (The current through the biasing diodes is increased to compensate for this 60 mV drop across the emitter resistor, thus returning the quiescent current to its original value of 30 mA.) The voltage supplied by the diodes now becomes 720 mV of which 650 mV is across the base–emitter junction and 60 mV across the emitter resistor. If the temperature rises to 60°C then, in order to maintain a constant quiescent emitter current of 30 mA, the base–emitter voltage must be reduced by 84 mV to 566 mV. Because the biasing diodes are supplying 720 mV, then to a first approximation the voltage across the emitter resistor must rise to 154 mV. The current through the emitter resistors must therefore rise by a factor of about 2.5. This, however, can only come about if the base current increases.

We may assume therefore that v_{BE}, instead of being 566 mV is, say, some 20 mV higher. Since the diode bias is fixed, less voltage appears across the emitter resistor. The voltage across the emitter then becomes 134 mV, hence the emitter current rises by a factor of 2.3. This is a substantial reduction on the figure of 25 previously obtained without an emitter resistor. When the biasing diodes are returned to thermal contact with the output transistors a further substantial improvement in thermal stability is obtained. This is fortunate since we are able then to reduce the resistance of the emitter resistance without unduly spoiling the thermal stability of the circuit. The kinds of load frequently driven by these power amplifiers have quite a low impedance (say an $8\,\Omega$ loudspeaker requiring $60\text{--}100\,\text{W}$ of audio power). An emitter resistor as high as the $2\,\Omega$ taken in the above example would therefore dissipate a lot of power, and the efficiency would then be severely reduced. Emitter resistors of about 0.33 to $1\,\Omega$ are commonly used, in which case we rely mainly on the thermal tracking of the biasing diodes for adequate thermal stability.

When emitter resistors are used it may be inconvenient to use a total bias voltage that is an integral number of diode voltage drops (i.e. 2 or 3 times $600\,\text{mV}$). In these circumstances the alternative arrangement of Figure 7.49 may be used. From the figure we see that the total bias voltage, V_{BIAS} is the sum of the voltages across R_1 and R_2, i.e.

$$V_{\text{BIAS}} = i_2R_1 + (i_2 - i_B)R_2$$
$$= i_2R_1 + v_{\text{BE}} \text{ (for TR3)}$$

If a transistor with a high h_{fe} is used we may arrange that $i_2 \gg i_B$, hence to a close approximation $i_2 R_2 = v_{\text{BE}}$, and

$$V_{\text{BIAS}} = v_{\text{BE}} \left(1 + \frac{R_1}{R_2}\right)$$

By adjusting the ratio R_1/R_2 we can thus obtain non-integral values of v_{BE}.

It is practically impossible to guarantee that at some time or other an accidental short-circuit will not develop across the output terminals of the amplifier. A conventional metal fuse is of no help in this respect because the output transistors will burn out before the fuse has melted. It is therefore good design to include some

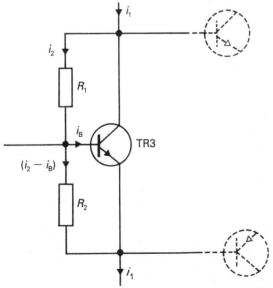

Figure 7.49 Alternative biasing arrangement to produce bias voltages which are not integrals of $600\,\text{mV}$

overload protection circuitry in the amplifier. Without the emitter resistors there is virtually no resistance to limit the short-circuit current to safe values. The presence of emitter resistors is a help in this respect but, as stated above, their resistance value must be kept low if efficiency is to be maintained. Other circuitry should therefore be used to augment the overload protection provided by the emitter resistors. Figure 7.50 shows the basis of a suitable arrangement in which transistors

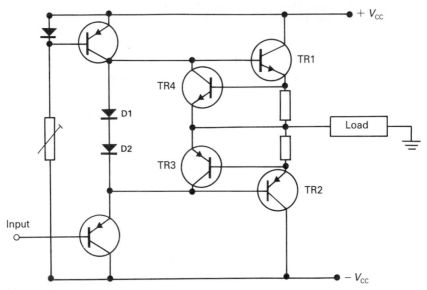

Figure 7.50 Basic overload protection circuit

TR3 and TR4 are turned on by the voltages developed across the emitter resistors when excessive currents flow to a defective load or short-circuit. Once these transistors are turned on they divert base current away from the bases of the output transistors; the emitter currents of the output transistors are thus limited to a safe value. During normal operation TR3 and TR4 are cut off, and thus performance is unaffected.

Quasi-complementary transistors

In all that has been said so far we have assumed that each half of a Class B or Class AB amplifier has made an equal contribution to the performance. In particular, the output transistors must be a matched pair in that the npn transistor characteristics are matched by a corresponding set of characteristics for the pnp transistors. This is difficult to achieve practically since an npn transistor cannot be manufactured in the same run as the corresponding pnp device. Problems of selection also arise. It would obviously be a great advantage if we could employ two npn transistors as the output devices. These could then be easily selected from the same batch, or manufactured as matched pairs. We must, however, make one npn transistor operate as though it were a pnp transistor. This is achieved by using a Darlington arrangement, whereby a pnp transistor serves as the driver device for an npn transistor, as shown in Figure 7.51. The pnp driver transistor is a low-power device and is easy to obtain (unlike high-powered pnp output transistors which are difficult

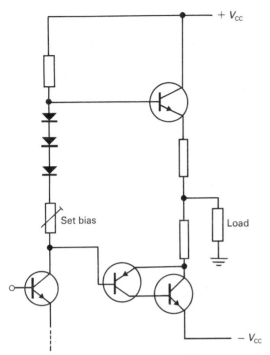

Figure 7.51 Quasi-complementary push-pull amplifier

to manufacture with characteristics that are complementary to an npn device). This arrangement has been described in connection with Figure 7.41(b). In that diagram the npn/pnp combination acted as an npn transistor; it had a single diode voltage drop between base and emitter and had a high h_{fe} figure. In Figure 7.51 we have the same arrangement in principle except that the main current-carrying device is an npn transistor. By driving it with a pnp transistor we still have only a single diode voltage drop between base and emitter, a high h_{fe} figure, but the combination acts as though it were a high-power pnp transistor. The main current-carrying transistors of both Class AB amplifiers in the push-pull arrangement are thus npn transistors which are selected as a matched pair.

Figure 7.52 shows the complete circuit of a high-grade amplifier capable of delivering 70 W into 8 Ω loads. Darlington arrangements are used as output emitter-followers in both sections of the push-pull amplifier. The quasi-pnp output transistor has an extra diode (D9) to provide a two-diode voltage drop to match the npn/pnp used in the other half of the amplifier.

Power FETs are becoming increasingly popular as output devices. Enhancement n-channel VMOS FETs of the type described in Chapter 5 are used as the output transistors; their high voltage and current ratings, low drive power requirements and low 'on' resistances make them attractive alternatives to bipolar transistors in the output position. The absence of thermal runaway confers a special advantage in high-power output stages. It will be recalled that the temperature coefficient under operating conditions is positive (unlike that of the junction transistor). The resistance at any local hot spots within the device therefore increases with temperature. This automatically reduces the local current density; the equalization

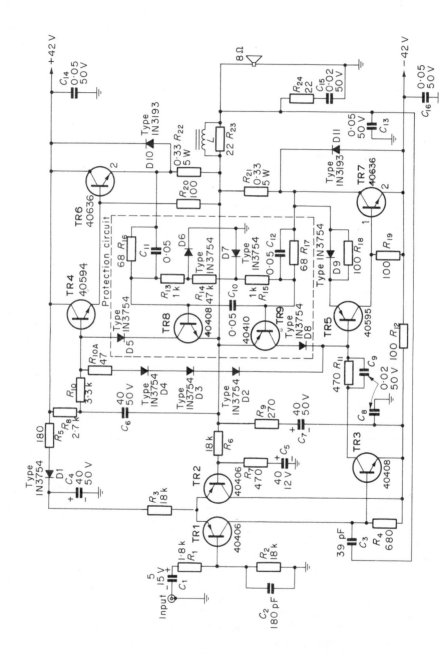

Figure 7.52 A high-grade amplifier capable of delivering 70 W r.m.s. into 8 Ω loads. The circuit (designed by R.C.A. and S.G.S.) uses a quasi-complementary output stage. Note: resistors are ½ W unless otherwise specified; values are in ohms; R_{21} and R_{22} are 5 W wire-round types

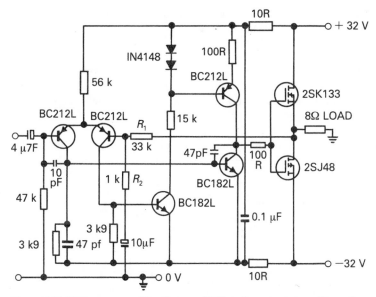

Figure 7.53 25 W power amplifier that uses FETs in the output stage. The voltage gain (determined by R_1 and R_2) is 34. The emitter-coupled input stage is discussed in Chapter 8

of the current throughout the device area results in far fewer device failures than that encountered in bipolar counterparts.

The circuitry for MOSFET output stages follows similar lines to those used for bipolar transistors. For very high powers, n-channel VMOS transistors are employed in a quasi-complementary arrangement, driven by the appropriate bipolar types. For more modest powers, say up to 25 W, the truly complementary arrangement of Figure 7.53 may be used.

Current dumping

One solution to the cross-over distortion problem is to use a technique known as current dumping. The basic principle is illustrated in Figure 7.54. For small driving signals of low voltage the load current is supplied entirely by a highly-linear voltage amplifier via R_1, thus by-passing the main output section. At this time the voltage developed across R_1 is insufficient to forward-bias the emitters of the output section, and thus this section is inoperative. When, however, the input signal is

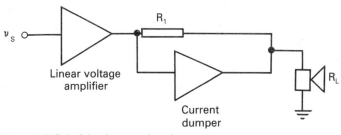

Figure 7.54 Principle of current dumping

large enough to cause the voltage across R_1 to exceed about 0.6 V, the output stage becomes active and takes over the dumping of current into the load circuit. At this point the output current is substantial and the output dumper stage operates in its linear region.

Unfortunately the voltage gain under load rises when the dumper stage takes over because the output impedance of an emitter-follower output stage is much lower than that of the voltage amplifier section. As a result crossover distortion is not entirely eliminated. This problem can be overcome by introducing a relatively small impedance, Z, into the circuit and taking negative feedback to the input of the voltage amplifier from the emitters, as shown in Figure 7.55. With small signals

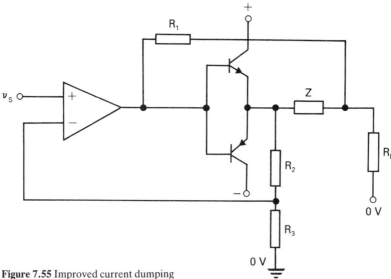

Figure 7.55 Improved current dumping

and the dumper stage inoperative, the relatively small Z is merely in series with R_2 and R_3; the overall gain is then determined by the output impedance of the voltage amplifier, R_1, Z, R_2 and R_3. When the signals are large enough to activate the dumper stage the additional current through Z produces a voltage which is added to the voltage fed back to the input of the voltage amplifier. This increased negative feedback compensates for the higher open-loop gain that would result if Z, R_2 and R_3 were absent. The correct choice of component values may be made with the aid of the following analysis based on the linear equivalent circuit of Figure 7.56. Initially we shall assume that the output impedance of the dumper stage, r, is fixed. In looking for a Norton equivalent circuit we must find an expression for the short-circuit output current, i_{sc}.

Neglecting the current through R_2 and R_3 then

$$i_{sc} = \frac{v_2}{R_1} + \frac{v_2}{r + Z}$$

Now

$$v_2 = \frac{Av_s}{1 + \beta A}$$

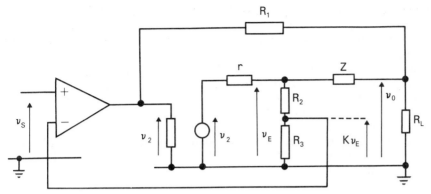

Figure 7.56 Equivalent circuit used for analysis. When finding the Norton short-circuit current i_{sc}, $V = 0$

and under short-circuit output conditions $v_0 = 0$

$$v_E = \frac{Z}{r + Z}v_2 \quad \therefore \quad \beta = \frac{KZ}{r + Z} \quad \text{where } K = \frac{R_3}{R_2 + R_3}$$

Hence

$$v_2 = \frac{Av_s(r + Z)}{r + Z + AKZ}$$

and

$$i_{sc} = \frac{Av_s(r + Z + R_1)}{(r + Z + AKZ)R_1}$$

Although r is non-linear ($\approx r_e \approx 25\,\text{mV/I}_e$—see page 110) depending as it does upon the load current, the above linear analysis is still plausible if we make $(r + Z + R_1) = (r + Z + AKZ)$ because the r term is eliminated on cancellation. This condition is satisfied when

$$Z = \frac{R_1}{AK}$$

and makes i_{sc} a linear function of v_S. The terms R_1 and K are independent of frequency, but A is not (see e.g. Figure 8.19). If the voltage amplifier section is designed to have a first-order low-pass response then $A = A_O/(1 + j\omega/\omega_t)$ where A_O is the low-frequency open-loop gain and ω_t is 2π times the corner or turning frequency. Z may then be expressed as

$$Z = \frac{R_1}{KA_o}\left(1 + j\frac{\omega}{\omega_t}\right) = \frac{R_1}{KA_o} + j\omega\frac{R_1}{\omega_t KA}$$

Thus Z is a resistor $R = R_1/KA_o$ in series with an inductor $L = R_1/(\omega_t KA)$. If these components are realized accurately in practice the amplifier will be free from crossover distortion.

Exercises

1. Explain, with the aid of a diagram, the current amplification mechanism in an npn junction transistor.
2. Explain, with the aid of suitable diagrams and drain characteristics, the construction and mode of operation of
   (a) a p-channel JUGFET
   (b) a p-channel enhancement type of MOSFET.
3. Describe the principle of operation of a depletion-type FET. Explain why this type of FET can be operated in either the depletion mode or the enhancement mode, whereas an enhancement-type FET can be operated only in the enhancement mode.
4. Describe the design procedure you would adopt to produce a single-stage transistor voltage amplifier that will be reasonably immune from changes in ambient temperature and transistor sample. You may assume that no graphical data are available, but that the quiescent current is 2 mA, the supply voltage is 15 V and that the operating frequency range is 40 Hz to 20 kHz. The current gain parameter, h_{fe}, may be taken as being 200.
5. Make a copy of the drain characteristics of Figure 7.2(a) to a larger scale. Use the characteristics to design a single-stage amplifier that uses an n-channel FET. You may assume that the supply voltage is 18 V and that the lowest frequency sine-wave signal is 40 Hz. The input drive voltage is known never to drive the gate more positive than −2.0 V.

   Assuming that the output resistance of your amplifier is the same as the drain load resistor and that the total stray capacitance at the output terminal is 20 pF, sketch the frequency response of your amplifier.

   If square waves with repetition rates as low as 40 Hz are to be amplified with a sag not exceeding 2 per cent, how must the input circuit of your amplifier be modified?
6. Explain the principle of negative feedback as applied to electronic amplifiers. Show how the use of negative feedback improves the performance of amplifiers in respect of bandwidth, harmonic distortion and phase distortion. (You are expected to derive the general formula for the gain of a feedback amplifier, and use the result to illustrate your answer with typical numerical examples.)
7. Show that the closed-loop gain of a feedback amplifier, A, is given by:

$$A' = \frac{A}{1 - \beta A}$$

   where A is the open-loop gain and β is the feedback fraction.
   Calculate the closed-loop gain of an amplifier having $A = 10^4 \angle 135$ and $\beta = 10^{-2} \angle 0$.
   State any conclusions you may draw from your calculations.
8. An amplifier has an open-loop gain of 60 dB. When feedback is applied the resulting closed-loop gain is 40 dB. If the open-loop gain now changes to 54 dB, calculate the new gain with feedback.
   (Ans. 39 dB)
9. The circuit of a source-follower is shown in Figure 7.57. Estimate the input impedance of the arrangement if $R_g = 1 M\Omega$, $R_s = 1.8 k\Omega$, $g_m = 5 mA/V$ at $i_D = 2 mA$ and the gate current may be taken as zero.

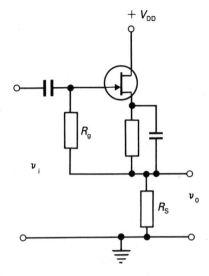

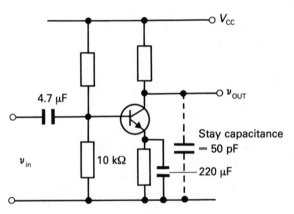

Figure 7.57 The reactance of the capacitors is low enough at the frequency of operation to regard the components as being short-circuits

Figure 7.58

10. Figure 7.58 shows the circuit of a common-emitter transistor amplifier. At the quiescent point $v_{BE} = 600\,\text{mV}$, $I_{CBO} = 10\,\mu\text{A}$ and $v_{CE} = 4.0\,\text{V}$. With device and/or temperature changes h_{fe} (min) $= 50$ and the maximum value is not specified. The minimum voltage between the collector and ground is to be 5 V. Estimate suitable values for the resistors. Estimate the input impedance of the circuit and sketch the frequency and phase response of the amplifier.

8

Analogue integrated circuits

The invention of integrated circuits is perhaps the most important advance in electronics to date. There is little doubt that the historian of the future will see the development of silicon device technology as the greatest single factor in the advance of the twentieth century. Economic, reliable and compact systems – from watches and hearing aids to powerful digital computers – are no longer in the realm of science fiction. Without integrated circuits the progress made in space research, supersonic aircraft, communications systems and automation would not have been possible. These tiny devices are now being incorporated into almost every form of modern machinery. Integrated circuits are rapidly becoming the key to measurement and control, which are the bases of all technology.

Silicon integrated circuits can be constructed to perform switching and memory functions, and to respond to logical digital signals. These functions are described in Part 3 of the book. In this chapter we are concerned with the class known as analogue ICs. Analogue devices are those in which the output is a continuous function of the input signal. These devices include linear amplifiers, in which the output voltage, over the operating range, is linearly related to the input voltage. Such devices can be used to operate upon input voltages in a way that is analogous to the methods used by mathematicians to solve differential equations. Such amplifiers are then called operational amplifiers. The majority of analogue circuits include operational amplifiers that use negative feedback.

The most important type of integrated circuit is the monolithic variety – monolithic because the entire circuit is fabricated within a single block of silicon crystal. Many millions of this type of circuit are sold throughout the world each year. The circuit designers have now progressed so far that almost every type of electronic circuit can be bought as a system within a single package. Complete amplifiers as well as timers, logic gates, memories and computer systems for pocket calculators, clocks and watches can all be bought in single packages. The most common containers are the TO-5 type and the dual-in-line (DIL) type.

The TO-5 is a metal can used frequently for packaging single transistors. The can consists of a small cylinder, and the lead wires are brought out in the same way as for a transistor. Naturally, there are many more lead wires; usually eight leads arranged in a circular pattern are used.

The dual-in-line type (which is much more common) consists of a plastic bar in which the integrated circuit is embedded. The bar is often almost 20 mm long, about 5 mm wide and barely 2.5 mm thick. The leads are brought out along the two sides, hence the term 'dual-in-line'. For most integrated circuits there are four,

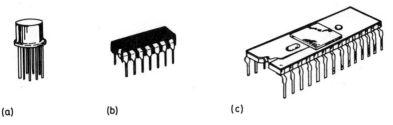

(a)   (b)   (c)

Figure 8.1 Integrated circuit packages: (a) TO-5 metal can; (b) 14-pin plastic dual-in-line (DIL); (c) multiple-pin chip package for calculators, digital clocks and microcomputers, etc.

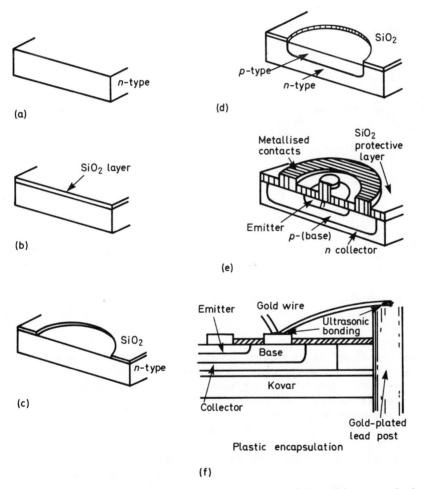

Figure 8.2 Some of the stages in the production of an integrated circuit: (a) n-type region in epitaxial layer; (b) an epitaxial layer of silicon dioxide grows on the surface when the wafer is heated in steam; (c) the oxide layer is coated with photo resist and irradiated with u.v. light through a mask negative. Hydrofluoric acid etches away unwanted portions to cut a window; (d) diffusing in boron from the vapour atmosphere in a furnace converts some of the n-type to p-type; (e) subsequent window cuttings and diffusions produce the final circuit, of which only a single transistor is shown; (f) stitching of chip to connectors

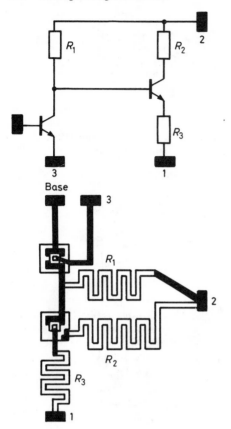

Figure 8.3 Realization of portion of a circuit on a silicon chip. The diagram represents a top view of the chip

seven or eight leads along each side making either an 8-pin or 16-pin DIL package. For clock chips, calculator chips and the more complicated systems 22, 24 or more leads are common. Figure 8.1 shows diagrams of the common types of package.

The manufacture of these circuits is an extension of the planar epitaxial process. When discrete transistors are made by this method large numbers of identical transistors are made on a single slice. Subsequently the slices are cut and the individual transistors mounted in their separate packages. In order to build a circuit it is necessary to reconnect the transistors electrically and incorporate the passive components. It is not difficult to see how the idea was conceived of building the entire circuit on a single chip. By doing so the separation and subsequent reconnections of the transistors can be avoided. The various steps in the production of an integrated circuit are shown in Figures 8.2 and 8.3.

Operation amplifiers–basic circuitry

Operational amplifiers are usually configured as negative feedback, high-gain, directly coupled (d.c.) differential amplifiers. The open-loop gain is very high (often >100 dB) and usually it is desirable to be able to operate with signals down to zero frequencies. Two inputs are available. When a signal is applied to the inverting input, the output is an amplified version of the input, and should be 180° out of phase with the input. On circuit diagrams this input is usually marked with a negative sign; this has nothing to do with polarity and merely indicates the phase reversal of 180°. (The problems associated with phase shifts of less than 180° are discussed later.) The other input terminal is the non-inverting terminal; signals applied to this terminal are amplified and yield an in-phase output signal. The non-inverting terminal is marked on circuit diagrams with a plus sign to show that the output and input signals are in phase. The output voltage of an operational amplifier should preferably be zero when the input voltages are both zero. This is one reason for powering the amplifier with symmetrical positive and negative supply voltages. When energized in this way the output voltage can swing positively and negatively with respect to the earth of zero-potential line.

Differential amplifiers

The input stage of an operational amplifier is based upon a matched pair of transistors connected as an emitter-coupled amplifier. This configuration has important advantages over the single-transistor stage. The thermal properties of an emitter-coupled amplifier are superior to those of the amplifiers so far discussed. Having two inputs, the differential amplifier is able to amplify the difference in voltage between the two input signals. For wanted differential (i.e. out-of phase) signals the amplification can be made very high (often > 100 dB). Signals that are in phase are called common-mode signals, and the amplifier is able to discriminate against them effectively. This is an important advantage since, in a practical situation, unwanted noise signals are often induced on the input lines. These noise signals are, however, common-mode signals and are largely rejected by the emitter-coupled configuration; the wanted differential signal may be much smaller than the noise signal, but experiences a much greater amplification than the noise signal. We can thus extract the wanted signal effectively from a 'noisy' background.

Figure 8.4 shows the circuit diagram of the prototype emitter-coupled amplifier. Two matched transistors have their emitters connected to one end of a resistor R_E; the other end of R_E is connected to the negative supply rail. The action of the circuit is as follows. If signals are applied to terminals 1 and 2 simultaneously then two possibilities arise. If the signals are out-of-phase we have a differential input (e.g. when a generator is connected across the input terminals). As the potential on one input terminal rises the potential on the other terminal falls by the same amount. The current in one transistor increases and that in the other transistor decreases by the same amount. The current through R_E therefore remains substantially unchanged as does the potential at B. The voltages at the output terminals are 180° out-of-phase with their respective inputs. The output voltage across the output terminals is $v_{o1} - v_{o2}$. Since these are out of phase, then if v_{o1} is positive, v_{o2} is negative. The magnitude of the output voltage is therefore $2|v_{o1}|$. If, however, the signals at the input terminals are in phase, i.e. common-mode signals, then the output voltages v_{o1} and v_{o2} are in phase; the output voltage is still $v_{o1} - v_{o2}$, but since v_{o1} and v_{o2} now have the same sign this output voltage is zero. This amounts to a perfect rejection of the common-mode signal. In practice some imbalance in the circuit is inevitable and hence v_{o1} is not quite equal to v_{o2}.

The rejection of the common-mode signal accounts in part for the popularity and usefulness of the circuit. In IC form the transistors are physically close, and since

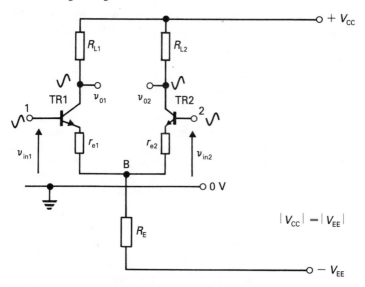

Figure 8.4 The emitter-coupled amplifier with differential input signals. r_{e1} and r_{e2} represent the internal dynamic emitter resistances of the transistors ($r_e = h_{ie}/h_{fe}$)

they were fabricated under identical conditions their parameters match exceedingly well.

Consider now the immunity such an arrangement has from the effects of temperature variations and supply voltage variations. Any effect that increases the current in TR1 will apply equally to TR2. The current rise in TR2 will be almost identical with that in TR1. Although v_{o1} and v_{o2} both decrease as a result of the increase in collector currents, the decrease will be the same in both cases and consequently the output voltage $v_{o1} - v_{o2}$ will not be affected. With discrete components it is impossible to achieve such close matching of the transistor parameters; nor is it possible to arrange such a close physical proximity. Discrete component emitter-coupled amplifiers are therefore inferior to their IC counterparts.

The circuit performance of the emitter-coupled amplifier can be analysed with the help of an equivalent circuit. Since in practice h_{re} and h_{oe} are very small, a first approach to the analysis can be made by replacing the transistors with a simplified version of their equivalent circuits, as shown in Figure 8.5.

The basic equations for the circuit are

$$v_{in1} = i_{b1}[h_{ie1} + R_E(1 + h_{fe1})] + i_{b2}R_E(1 + h_{fe2}) \tag{8.1}$$

$$v_{in2} = i_{b1}R_E(1 + h_{fe1}) + i_{b2}[h_{ie2} + R_E(1 + h_{fe2})] \tag{8.2}$$

$$v_{o1} = -h_{fe1}R_{L1}i_{b1} \tag{8.3}$$

$$v_{o2} = -h_{fe2}R_{L2}i_{b2} \tag{8.4}$$

It is possible to derive many different expressions from the above equations. Various gain formulae can be derived depending upon whether the input signals are applied to only one input or two, in phase or 180° out of phase, whether the output is taken from only one collector, or whether a differential output is wanted, and so

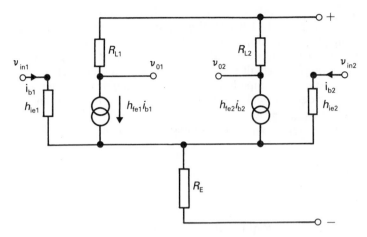

Figure 8.5 Equivalent circuit for the emitter-coupled amplifier

on. Input impedances for different modes of operation can also be derived. We shall confine ourselves here to deriving those results that lead to an appreciation of some of the design features of linear ICs.

To enjoy the benefits of freedom from thermal drift the output should be taken from across both collectors. Let us assume a differential input. The differential gain is given by

$$A_d = \frac{v_{o1} - v_{o2}}{v_{in1} - v_{in2}}$$

From Equations 8.1 and 8.2

$$v_{in1} - v_{in2} = i_{b1}h_{ie1} - i_{b2}h_{ie2} \tag{8.5}$$

therefore

$$A_d = \frac{-h_{fe1}R_{L1}i_{b1} + h_{fe2}R_{L2}i_{b2}}{i_{b1}h_{ie1} - i_{b2}h_{ie2}}$$

For an IC the two transistors are virtually identical and so $h_{ie1} = h_{ie2} = h_{ie}$ and therefore $|i_{b1}| = |i_{b2}|$; hence $i_{b1} = -i_{b2}$ for differential signals. Also $h_{fe1} = h_{fe2} = h_{fe}$ and $R_{L1} = R_{L2} = R_L$ thus

$$A_d = -\frac{h_{fe}R_L}{h_{ie}} = -\frac{R_L}{r_e} \text{ (since } r_e = \frac{h_{ie}}{h_{fe}}\text{)}$$

The differential gain is thus seen to be the same as that for a conventional common-emitter amplifier. The gain is also seen to be independent of h_{fe}.

The input impedance for the differential mode is found from Equation 8.5. For the differential mode $i_{b1} = -i_{b2} = i_b$, say, hence

$$(v_{in1} - v_{in2}) = i_b(h_{ie1} + h_{ie2}) = 2h_{ie}i_b$$

The differential input impedance, $(v_{in1} - v_{in2})/i_b$, equals $2h_{fe}r_e$. Typically, h_{fe} may be 100 and r_e say $50\,\Omega$ with each transistor taking a quiescent current of $500\,\mu\text{A}$. The input impedance is thus $10\,\text{k}\Omega$. This can be increased by using Darlington arrangements to increase h_{fe} considerably (say one hundredfold). The input impedance would then be increased to $1.0\,M\Omega$. We see therefore that although the gain is independent of h_{fe}, the input impedance is not.

Another way of increasing the input impedance is to augment the r_e of each transistor (r_{e1} and r_{e2}) by inserting small-valued resistors into the emitter lead of each transistor. The increased input impedance is then obtained at the expense of the gain. If the resistor added to each emitter is represented by R_1 then the differential gain becomes $-R_L/(r_e + R_1)$.

For common-mode signals, $v_{in1} = v_{in2}$. The common-mode gain is given by

$$A_{cm} = \frac{v_{o1} - v_{o2}}{v_{in}}$$

where $v_{in} = v_{in1} = v_{in2}$.

Also

$$v_{in} = v_{in1} = i_{b1}h_{ie1} + R_E[(1 + h_{fe1})i_{b1} + (1 + h_{fe2})i_{b2}]$$

Therefore

$$A_{cm} = \frac{-h_{fe1}R_{L1}i_{b1} + h_{fe2}R_{L2}i_{b2}}{i_{b1}h_{ie1} + R_E[(1 + h_{fe1})i_{b1} + (1 + h_{fe2})i_{b2}]} \tag{8.7}$$

Taking h_{fe1} and h_{fe2} as being much larger than unity, and $h_{ie1}i_{b1}$ as being much smaller than $R_E i_{b1}$ or $R_E i_{b2}$ then

$$v_{in} = R_E \left[i_{b1}h_{fe1} + i_{b2}h_{fe2}\right]$$

thus

$$A_{cm} = \frac{-h_{fe1}R_{L1}i_{b1} + h_{fe2}R_{L2}i_{b2}}{R_E(i_{b1}h_{fe1} + i_{b2}h_{fe2})} \tag{8.8}$$

Ideally the common-mode gain should be zero since we require zero output for common-mode input signals. The expression for A_{CM} shows that this could be achieved theoretically if we had perfect balance, for then $h_{fe1} = h_{fe2}$ and $R_{L1} = R_{L2}$; the numerator would then be zero. In practice, with modern ICs, the balance, although very good, is never quite perfect; we see therefore that R_E should be as large as possible. Unfortunately, if R_E is an actual resistor the voltage drop across it will be large and there will be little of the supply voltage left for amplifier operation. This difficulty can be overcome by using a transistor as a constant current source. R_E would then be effectively replaced by a dynamic impedance whose value approached infinity. The common-mode gain is then very small. The circuit may then be as shown in Figure 8.6.

Single-ended differential amplifiers

Although a differential amplifier responds to the difference between two input signals, it is useful to have only a single output voltage. Such single-ended outputs can then be used to activate other 'conventional' circuits. The easiest way of obtaining a single-ended output is merely to take the output signal from only one of the collectors. If for example we take the output from the collector of TR2 then R_{L1} is not needed and the prototype emitter-coupled amplifier may be modified to that shown in Figure 8.7.

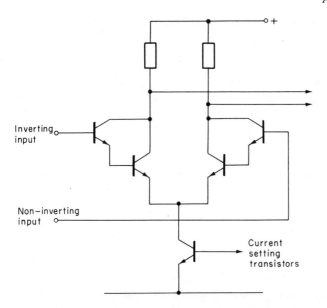

Figure 8.6 One form of input circuit for an operational amplifier

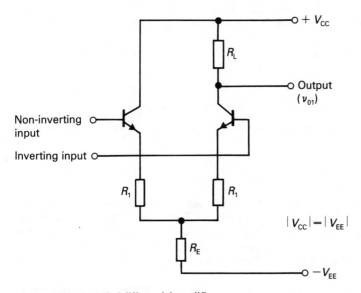

Figure 8.7 Single-ended differential amplifier

The gains of the amplifier can still be obtained from Equations 8.1 to 8.6. The differential gain with $R_1 = 0$ is

$$A_\mathrm{d} = \frac{-h_{\mathrm{fe2}}R_\mathrm{L}i_{\mathrm{b2}}}{i_{\mathrm{b1}}h_{\mathrm{ie1}} - i_{\mathrm{b2}}h_{\mathrm{ie2}}}$$

For a balanced arrangement $i_{b1} = -i_{b2}$, $h_{ie1} = h_{ie2} = h_{ie}$ and $h_{fe2} = h_{fe}$, say, hence

$$A_d = \frac{-h_{fe}R_L}{2h_{ie}} = -\frac{R_L}{2r_e}$$

The gain, as expected, is thus only half of that obtained with a differential output. When finite values of R_1 are included to increase the input impedance, the differential gain becomes

$$A_d = -\frac{R_L}{2(r_e + R_1)}$$

The common-mode gain, again with $R_1 = 0$, is obtained from Equation 8.7. In the balanced condition

$$A_{cm} = \frac{-h_{fe}R_L}{h_{ie} + 2R_E(1 + h_{fe})} \approx \frac{-R_L}{r_e + 2R_E}$$

When $R_1 \neq 0$ this becomes

$$A_{cm} = -R_L/(r_e + R_1 + 2R_E)$$

We thus need very large values of R_E in order to obtain low values of common-mode gain. As before, a large dynamic value of R_E is obtained by replacing R_E with a transistor; then $A_{cm} \approx -R_L/(2R_E)$.

The *common-mode rejection ratio*, CMRR, is a measure of the amplifier's ability to reject in-phase input signals; the CMRR is the ratio between the differential voltage amplification and the common-mode amplification, i.e. A_d/A_{cm}. The ratio is expressed in decibel units; a common figure for modern integrated circuits is 90–100 dB.

For the single-ended output differential amplifier

$$\text{CMRR} = (r_e + R_1 + 2R_E)/2(r_e + R_1)$$

Since R_E is so large in practice, then $(r_e + R_1) \ll R_E$. The CMRR is therefore given approximately by

$$\text{CMRR} = \frac{R_E}{r_e + R_1}$$

Let us see how the above expressions influence the design of a single-ended differential amplifier of the type shown in Figure 8.7. We will assume that an actual resistor is used for R_E, that the supply voltages are ± 15 V, and that matched pairs of transistors with $h_{fe} = 100$ are available. The input impedance is to be not less than 200 kΩ, the CMRR not less than 40 dB, and the differential gain about 30. Can such a specification be achieved?

The input impedance $= 2h_{fe} (r_e + R_1)$, therefore $(r_e + R_1) = 1$ kΩ. It is usual to make the quiescent collector voltage equal to $V_{cc}/2$ to allow for large output voltage swings; the voltage across R_L in our case is thus 7.5 V. For a differential gain of 30, $R_L = 60$ kΩ, hence the quiescent current must be 7.5 V/60 kΩ, i.e. 125 μA. Since $r_e = 26$ mV/I_q (mA), $r_e \approx 200$ Ω. This makes $R_1 = 800$ Ω (say 820 Ω for the nearest preferred value). The common-mode rejection ratio is to be not less than 40 dB, i.e. 100, hence $R_E = 100$ kΩ. We see therefore that with a resistor for R_E we are in difficulties. The quiescent current through one transistor is 125 μA, therefore the current through $R_E = 0.25$ mA. The voltage across R_E is therefore

25 V; but only 15 V is available, therefore it is not possible to realize a comprehensive arbitrary specification; the latter must be relaxed in some way in order to effect a suitable compromise. We could for example use a negative supply of -25 V, but this would not be a very satisfactory solution on practical grounds. We could use the dynamic impedance of a transistor instead of R_E and have 15 V across it, but we would then lose the simplicity of the circuit because of the additional components required.

With integrated circuit technology this would be no problem, and we would then have the advantage of a much larger CMRR. With discrete components, however, we would have to relax the specification in some respect. We could then proceed as follows.

Since we know that about 15 V will be dropped across R_E we could choose a reasonable quiescent current for the transistor, say $200\,\mu$A. The quiescent current through R_E is then $400\,\mu$A, hence $R_E = 15\,$V$/400\,\mu$A which is, say, $39\,$kΩ as a preferred value. For $I_q = 20\,\mu$A, $r_e = 130\,\Omega$. For an input impedance of $200\,$kΩ, ($r_e + R_1$) $= 1\,$kΩ and so $R_1 = 870\,\Omega$ (we would probably choose $910\,\Omega$). The CMRR would then be $39\,$k$\Omega/1.04\,$kΩ which is just about $31\,$dB. $R_L = 7.5\,$V$/200\,\mu$A $= 37.5\,$kΩ and we would have to accept a differential gain of about 18 or 19 with a CMRR of about $31\,$dB. We could, of course, trade off some of the input impedance by reducing R_1; this would increase the gain and the CMRR at the expense of the high input impedance. The reader is invited to investigate the effects of altering the initial quiescent current and input impedance, noting the resulting changes in gain and CMRR. Realistic values of gain and CMRR must be specified initially. For example, the maximum value of CMRR with 15 V across R_E is 300, i.e. just 49 dB. We can estimate the maximum value of CMRR in the following way. The voltage across R_E is given by $V_{EE} = 2I_q R_E$, where I_q is the quiescent current through one transistor. $r_e = 0.026(V)/I_q$ (A). The maximum CMRR is when $R_1 = 0$. i.e. CMRR (max) $= R_E/r_e$, therefore

$$\text{CMRR (max)} = \frac{R_E I_q}{0.026} = \frac{V_{EE}}{0.052} \approx 20\,V_{EE}$$

i.e. the maximum common-mode rejection ratio is 20 times the voltage across R_E.

Similarly the maximum gain can be found if it is assumed that half of V_{CC} is dropped across R_L.

$$R_L I_q = \frac{V_{CC}}{2} = \frac{R_L}{r_e} \times 0.026$$

therefore

$$\frac{R_L}{r_e} \approx 20\,V_{CC}$$

The maximum differential gain is obtained when $R_1 = 0$ therefore

$$A_d\,(\text{max}) = R_L/r_e \approx 20V_{CC}$$

Active loads and biasing circuits

It has already been pointed out in the last chapter that resistors and capacitors are expensive to produce in integrated circuit form, mainly because of the area they occupy on the chip. Whenever possible, therefore, circuit functions are realized using transistors alone. The large values of R_E required for an emitter-coupled amplifier preclude the use of an actual resistor.

We have already seen how discrete components are used to bias a transistor amplifier, the problem being to maintain collector currents constant for all likely variations of temperature. To avoid the use of large-valued resistors and capacitors, extensive use is now made of current mirrors. Such mirrors can be made to serve as the equivalent of large-valued resistors; they are therefore useful as active loads.

The principles of a basic current mirror have been discussed in the previous chapter in connection with Figure 7.16. By using this circuit, the emitter current of an emitter-coupled input amplifier can be kept constant; we thus have a very high effective dynamic value for R_E without using a large-valued resistor. Figure 8.8 shows how the Widlar stabilization circuit can be used to produce the constant 'tail' current of a differential amplifier.

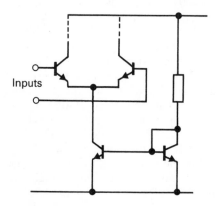

Inputs

Figure 8.8 The use of a current mirror to provide the emitter load of a differential amplifier (see also Figure 7.16)

The technique can also be used to produce active loads of high dynamic impedance for the collector circuits of amplifiers. Often a modified Widlar current mirror, known as the Wilson mirror, is used for this purpose. Not only does the Wilson mirror use no resistors, but a further advantage lies in the fact that the ratio of the current and its 'mirror' counterpart is even less dependent on h_{FE}. We see this from the analyses of the two circuits. Figures 8.9 (a) and (b) make the comparison. In the Widlar circuit I_1 is $h_{FE}i_B$, and its mirror counterpart, I_2, is $2i_B + h_{FE}\, i_B$ therefore $I_1/I_2 = h_{FE}/(2 + h_{FE})$. If, say, $h_{FE} = 50$ then $I_1 = 96$ per cent of I_2. In the Wilson circuit the emitter current of TR3 is $i_B (2 + h_{FE})$. This is equal to the base current of TR3 plus h_{FE} times the base current. The base current of TR3 must therefore be $i_B (2 + h_{FE})/(1 + h_{FE})$.

From this, $I_1 = i_B h_{FE} + i_B (2 + h_{FE})/(1 + h_{FE})$ and $I_2 = h_{FE}i_B (2 + h_{FE})/(1 + h_{FE})$; hence

$$\frac{I_1}{I_2} = \frac{h_{FE}^2 + 2h_{FE} + 2}{h_{FE}^2 + 2h_{FE}}$$

(We are assuming throughout that transistors TR1 and TR2 are identical.) If again we assume $h_{FE} = 50$ then $I_1 = 100.1$ per cent of I_2. We see therefore that for an h_{FE} of 50 the two currents are within about 0.1 per cent of each other for the Widlar

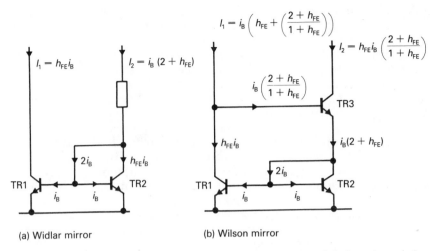

Figure 8.9 The Widlar and Wilson current mirrors compared to show their dependence on h_{FE}

circuit, but within 0.1 per cent of each other for the Wilson circuit. Furthermore, by taking different values of h_{FE} for each circuit it will be found that the Wilson circuit is less affected by changes of h_{FE}.

The Wilson circuit, using pnp transistors, can be used as an active load for an npn pair of emitter-coupled transistors. Voltage gains of several thousand are achievable in a single stage. As we shall see later, high gain is a very desirable feature in an operational amplifier. One form of input stage for an operational amplifier is shown in basic form in Figure 8.10. The output is taken from one of the collectors. If the very high gain of the stage is to be realized the load impedance connected to the collector must be very high, hence the use of a Darlington transistor. This latter transistor can be suitably biased by the use of multiple transistor current mirrors. Here both the differential amplifier and the Darlington transistor are supplied by constant current sources.

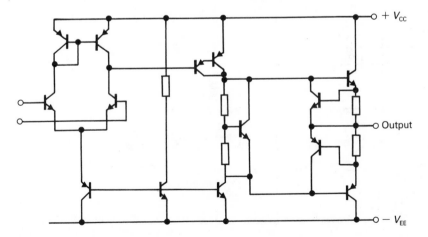

Figure 8.10 Basic circuit of one form of operational amplifier

The single-ended output stage is required to have a low output impedance and reasonably high drive capability. A conventional complementary pair of output transistors are connected in the power amplifier mode discussed in the last chapter. The circuit of Figure 8.10 therefore shows the basic simplified circuit of a complete operational amplifier. In practice, additional circuitry is added to protect the amplifier from short-circuits between the output terminal and either the power supply lines or zero voltage line. Early types of op-amp suffered from latch-up, a condition whereby the potential at the output terminal would become fixed at or near to either V_{CC} or V_{EE}. Circuitry in modern op-amps prevents this.

Perhaps the most famous and popular operational amplifier to date is the 741. This is an amplifier with a typical gain of 106 dB (minimum 86 dB) and comprises a differential input stage, an intermediate stage of amplification, and a complementary npn/pnp emitter-follower output stage. It has an on-chip feedback capacitor that gives it what is termed internal compensation. These high-gain amplifiers when used as such are all operated with negative feedback. The problem of stability then arises, which necessitates the use of compensation. The term is defined and discussed later in the chapter.

The input stage of a 741 is different from that shown in Figure 8.10. We still have a pair of transistors connected as a differential amplifier, but instead of taking the output from a collector, emitter-follower techniques are used. The loads consist of two transistors, working in the common-base mode, used in connection with a further pair, as shown in Figure 8.11. The input resistance of such an arrangement is about $1 \text{ M}\Omega$. Current mirrors are used for biasing as explained above. An intermediate driving stage is designed primarily to present a high impedance to the input stage while being able to drive a 'conventional' push-pull complementary pair output stage. Short-circuit protection is also incorporated.

As will be seen from Figure 8.11 the emitters of TR1 and TR2 (which form part of the emitter-follower loads) are connected to pins on the DIL package. The user therefore has access to these two points in the circuit, which are called the offset

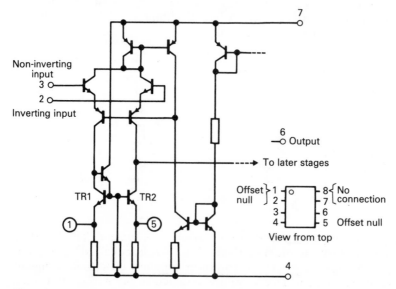

Figure 8.11 Input stage for a 741 operational amplifier. Numbers refer to lead pins of an 8-pin DIP

null terminals. The reason for making these points accessible is to enable the user to apply external voltages to them so as to force the output voltage to be zero when the voltage across the inverting and non-inverting terminals is zero. In this way we can compensate for any small imbalance that may be present in the differential amplifiers.

When input impedances greater than $2\,\mathrm{M\Omega}$ are required the user can obtain a version of the 741 that uses FETs for the input differential pair. The input impedance is then raised to $10^{12}\,\Omega$.

Applications of analogue ICs

Analogue integrated circuits of the type we have been discussing can be used as voltage comparators, timers, multipliers, phase-locked loops and voltage regulators. The most common application is, however, as an operational amplifier in which the IC is combined with passive components to form a negative feedback amplifier.

A linear IC amplifier with feedback and input impedances is shown in Figure 8.12. This arrangement is often called an operational amplifier (op-amp) since it can operate on input voltages in a manner analogous to operations performed by mathematicians. With suitable feedback and input impedances the op-amp can add, subtract, differentiate and integrate voltages applied to the input terminals. A large number of electronic instruments can therefore be made using operational amplifiers as a basis for the design. An important incidental advantage is that the user is freed from the chore of designing in detail the electric circuitry of the basic amplifier, and can concentrate instead on the design of the overall system.

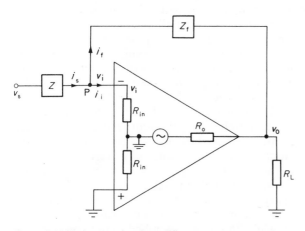

Figure 8.12 Basic operational amplifier

In circuit diagrams the IC amplifier is represented by a triangle. The inverting and non-inverting terminals are often marked $-$ and $+$ respectively. The output is at the apex of the triangle. Connections to power supplies, offset null terminals (and compensation terminals if necessary) are assumed to be present, but often omitted on the diagrams for reasons of clarity.

To obtain an expression for the overall gain of the amplifier we note that when v_s is positive, v_o is negative

$$v_i = i_i R_{in} = R_{in}(i_s - i_f)$$

$$v_o = -Av_i = -AR_{in}(i_s - i_f)$$

$$i_s = \frac{v_s - v_i}{Z_i} \text{ and } i_f = \frac{v_i - v_o}{Z_f}$$

$$\therefore v_o = -AR_{in}\left(\frac{v_s - v_i}{Z_i} - \frac{v_i - v_o}{Z_f}\right)$$

$$\therefore \frac{v_o}{v_s} = -\frac{1}{Z_i} \bigg/ \left[\frac{1}{Z_f} + \frac{1}{A}\left(\frac{1}{R_{in}} + \frac{1}{Z_i} + \frac{1}{Z_f}\right)\right]$$

When A is very large

$$\frac{v_o}{v_s} = A' \approx -\frac{Z_f}{Z_i}$$

We see that the use of the approximate formula does not involve much error when A is large. For example, when $A = 10\,000$ (i.e. 80 dB), $R_{in} = 1\,M\Omega$, $Z_i = 1\,k\Omega$ and $Z_f = 100\,k\Omega$ then the percentage error when using the approximate expression is only about 1 per cent. For a 741 op-amp., A is, say, 106 dB and hence the error will be less than 0.1 per cent. Since it is highly unlikely that the resistors will be known to better than 0.1 per cent, we are justified in using the approximate formula in most practical cases. The influence of R_{in} is seen to be small.

When Z_i and Z_f are both resistances $A' = -R_f/R_i$. This means that the gain of the amplifier can be very accurately determined. The accuracy is determined only by the quality and tolerance of the resistors used. Provided the open-loop gain is very large, amplifier characteristics do not affect the closed-loop gain.

Because the open-loop gain of the amplifier is so large, practical output voltages are produced when the voltage difference across the inverting and non-inverting terminals is only a few tens of microvolts. If, therefore, the non-inverting terminal is held at zero potential, as shown in Figure 8.12, then the potential at the point P is always close to zero; the point P is then called a virtual earth. (For reasons given below, the non-inverting terminal is not often connected directly to the zero voltage line. The potential at the non-inverting terminal may therefore not be quite at zero potential. Feedback makes the potential at the inverting terminal track the potential at the non-inverting terminal so as to maintain the voltage difference at almost zero. The absolute value of the potential may therefore be several millivolts relative to the zero potential line, but still small enough to call P a virtual earth point.) We see that for practical purposes the input impedance at the signal input terminal is R_i. This is sometimes inconveniently low. We can then use the configuration of Figure 8.13(b), in which the signal is applied to the non-inverting terminal and the gain is controlled by selecting a suitable proportion of the output voltage and feeding it back to the amplifier in antiphase; this is achieved by feeding back to the inverting terminal. The output voltage is in phase with the signal voltage. The feedback fraction, β, is $R_i/(R_i + R_f)$. Assuming simple negative feedback and a huge open-loop gain, the closed-loop gain is $1/\beta$, i.e. $(R_i + R_f)/R_i$. We see therefore that the effective gain is different from that obtained when the signal is fed, via R_i, to the inverting terminal.

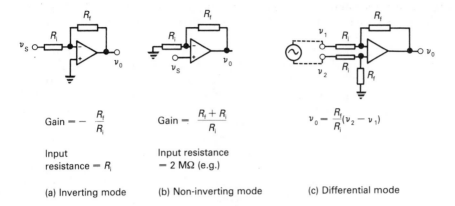

$$\text{Gain} = -\frac{R_f}{R_i}$$

$$\text{Gain} = \frac{R_f + R_i}{R_i}$$

$$v_0 = \frac{R_f}{R_i}(v_2 - v_1)$$

Input
resistance $= R_i$

Input resistance
$= 2\,\text{M}\Omega$ (e.g.)

(a) Inverting mode   (b) Non-inverting mode   (c) Differential mode

Figure 8.13 Three basic configurations for an operational amplifier. In (b) when $R_f = 0$ the voltage gain is 1 and the arrangement is known as a voltage follower

By making $R_f = 0$ the voltage gain is unity. We then have 100 per cent negative feedback. The output voltage 'follows' the input voltage very closely and hence this circuit is known as a voltage-follower; it is the op-amp equivalent of the emitter-follower. The circuit is used extensively as a buffer, or impedance transformer. Special amplifiers have been developed for use as voltage-followers; they have input impedances in excess of $10^3\,\text{M}\Omega$ and output impedances of only $1\,\Omega$, or so.

When using the amplifier in the differential mode, signals are fed simultaneously to both inputs. For example, the signal source, instead of being connected between one of the terminals and the zero voltage line, may be connected directly across the two input terminals. In order to preserve balance in the circuit, and to have identical gains at both input terminals the arrangement of Figure 8.13(c) is used. By using an input attenuator at the non-inverting terminal the effective gain at that signal input terminal is reduced by a factor of $R_f/(R_i + R_f)$ to R_f/R_i. By using values of R_i and R_f in the attenuator we maintain the essential balance of the circuit. Each amplifier input terminal 'sees' the same effective external resistance. The bias currents into each terminal are therefore equal; this helps to reduce the offset voltage at the output.

Before discussing some applications it may be useful to consider some of the terms used in connection with IC analogue amplifiers.

Open-loop gain is the differential voltage gain for a given load and at a given frequency and temperature. Manufacturers quote the gain at zero frequency since this is usually the largest figure in the range. Modern ICs have open-loop gains of about 10^5; the 741 gain figure is typically 2×10^5, i.e. 106 dB, but any sample may have a gain as low as 86 dB.

Common-mode rejection ratio has already been discussed; 90–100 dB are figures to be expected with modern units.

Input offset current is the difference between the currents flowing into the input terminals when the output voltage is zero. This is about 10^{-7} A for bipolar transistors and of the order of 10^{-12} A for FET inputs. It is temperature-dependent.

Input offset voltage. In an ideal, perfectly balanced, amplifier the output voltage is zero if the voltage between the input terminals is zero. In a practical amplifier the imbalances must be trimmed out to bring the output voltage to zero. The input offset voltage varies with time and temperature, giving rise to drift of the output voltage.

Slewing rate defines the maximum rate of change of the output voltage when a step voltage signal is applied to the input terminals. The rate is limited by the compensating circuits used. (Compensation is described in a later section.) For the internally compensated general purpose op-amp the slew rate is 0.5 V/μs. A general purpose FET amplifier (TL081 type) has a slew rate of 13 V/μs. Special high-speed types have rates of several hundreds of volts per microsecond.

Input bias current, which is temperature-dependent, may result in a non-zero output voltage when the input voltage is zero. Each input terminals should 'see' the same d.c. impedance to earth. This means that if, for example, we were using the inverting mode of Figure 8.13(a) then the non-inverting terminal, instead of being connected directly to earth, should be connected via a resistance whose value is the parallel combination of R_i and R_f. Other factors permitting, the feedback resistance should be kept low.

Bandwidth. The gain/frequency characteristics are to be discussed shortly. From these characteristics it will be seen that the open-loop gain is not constant with frequency; it falls as the frequency rises because of capacitive effects. The lower limit of frequency is zero for op-amps; this means that the upper 3 dB point can be quoted as the bandwidth. Alternatively, some manufacturers quote the frequency at which the gain falls to unity.

The *supply voltage rejection ratio* is the change in output voltage divided by the change in supply voltage that produces it; the input terminals are short-circuited when measurements are made. For a 1 V change in supply voltage the corresponding change in output voltage may be expected to be as low as 10 μV with many amplifiers; the rejection ratio is often expressed in dB (in this case 100 dB). The 741 has a typical s.v.r.r. of 90 dB.

Supply voltage range. Analogue ICs are usually operated from two supply rails (apart from the zero voltage line). The voltage on these rails must not exceed the manufacturer's limit otherwise damage will result. The 741 can tolerate ±18 V as an absolute maximum, but normal operating voltages do not usually exceed ±15 V. The device will not operate satisfactorily below ±3 V.

Input voltage range. The input voltage must never be allowed to exceed the supply line voltages. Although the 741 is short-circuit protected at its output, it is vulnerable if input voltages are applied when the supply lines are not energized.

Output voltage range. The maximum output voltage available from op-amps is usually a volt or two less than the supply voltages. Attempts to obtain more output will saturate the amplifier and clip the top or bottom (or both) of the output waveform.

Frequency compensation

We have seen that the application of negative feedback to high-gain amplifiers produces a circuit with characteristics that are largely independent of amplifier characteristics. The overall performance is almost entirely dependent on the nature of the feedback elements. Substantial improvements are obtained in gain stability, bandwidth, distortion and phase response; the improvements are in direct proportion to the amount of feedback used. Provided the open-loop gain is large enough, it is possible to tailor the closed-loop gain to almost any degree of accuracy. The application of large amounts of feedback does, however, lead to problems not so far discussed. The most important of these problems involves amplifier stability.

In the examples discussed previously the feedback has been referred to as simple negative feedback. The term 'simple' implies that the feedback signal is always 180° out of phase with the input signal. Under these circumstances the open-loop gain, A, and the feedback fraction, β, may be regarded as simple algebraic quantities. When using feedback with amplifiers that have very high gains and wide bandwidths, special attention must be paid to the open-loop gain and phase characteristics. It is no longer possible to regard A and β as simple algebraic quantities; both are complex in that phase as well as magnitude must be taken into consideration.

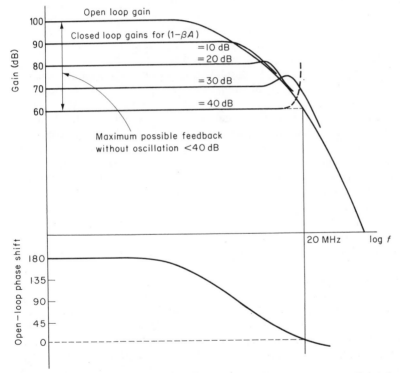

Figure 8.14 Open-loop gain and phase shift for a typical linear IC. It is assumed that the signal is applied to the inverting input terminal, hence the phase shift at low frequencies is 180°

Figure 8.14 shows the open-loop gain as a function of frequency and the corresponding phase shift for a typical linear integrated circuit. Assuming the input signal to be applied to the inverting terminal, the phase shift at low frequencies is 180°. When feedback is applied to the amplifier the gain is given by the general formula

$$A' = \frac{A}{1 - \beta A}$$

To emphasize the complex nature of A and β this may be written:

$$|A'|\angle\phi = \frac{|A|\angle\theta}{1 - |\beta|\angle\alpha|A|\angle\theta}$$

where θ and α are the phase angles of the amplifier and feedback circuit respectively. (It will be recalled that a complex quantity can be written as $X + jY$. Plotting this on a phasor diagram we see that $X + jY$ can be represented as $r\angle\phi$, where r is the distance from the origin, i.e. $\sqrt{(X^2 + Y^2)}$, and ϕ is $\tan^{-1}(Y/X)$. Let us assume the usual case when the phase shift in β is zero. Then

$$|A'|\angle\phi = \frac{|A|\angle\theta}{1 - |\beta A|\angle\theta}$$

For negative feedback $\theta = 180°$ and $A' = -A/(1 + \beta A)$. So long as the feedback stays negative, stable operation is achieved. However, as the frequency of operation increases, inspection of Figure 8.14 shows that θ is then no longer 180°. At 20 MHz $\theta = 0$, hence $A' = A/1 - \beta A)$. It can be seen that if the loop gain βA is equal to unity at 20 MHz, A' becomes infinitely large and uncontrollable oscillations result. The frequency of the oscillation will be 20 MHz.

ICs incorporate transistors made by modern planar epitaxial techniques; they therefore have considerable gain at high frequencies. The open-loop gain characteristic shown in Figure 8.14 for a typical IC shows that at 20 MHz, when the phase shift has altered from 180° to 0°, the gain is as much as 60 dB. Clearly we are restricted in our choice of values for β since it is easy to have $\beta A = 1$ at a frequency that gives zero phase shift. The absolute maximum value of β can be found by noting the value of the open-loop gain at zero phase shift. In our example this is 60 dB (i.e. 10^3), hence the maximum value of β is 10^{-3}. An alternative way of looking at this is to restate the closed-loop gain formula as follows:

$$A' = \frac{A}{1 - \beta A}$$

Therefore

$$20 \log A' = 20 \log A - 20 \log (1 - A)$$

i.e.

$$A' \text{ (dB)} = A \text{ (dB)} - (1 - \beta A) \text{ (dB)} \tag{8.9}$$

$(1 - \beta A)$ is the amount of feedback. This is shown graphically in Figure 8.14. We see that the absolute maximum value of feedback is found by drawing a vertical line through that frequency which gives zero phase shift. A horizontal line through the intersection of the vertical line and open-loop gain shows the maximum permissible closed-loop gain. The difference between the open-loop gain in dB and the closed-loop gain in dB is the amount of feedback. This is confirmed by a rearrangement of Equation 8.9.

A problem arises if the required closed-loop gain is less than the minimum permissible. Unless the open-loop frequency response of the IC is modified, unstable operation will be encountered.

We can account for the fall-off in the open-loop gain by supposing the IC to consist of stages as shown in Figure 8.15. Usually it is sufficient to have three stages. For this purpose each stage is considered to be a perfect amplifier

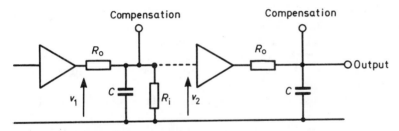

Figure 8.15 Various stages of an IC amplifier

(represented by the triangle) having an output resistance R_o and a stray capacitance to ground of C. The input resistance of the next stage is represented by R_i. Now $v_2/v_1 = Z/(Z + R_o)$, where $Z = R_i/(1 + j\omega C R_i)$. In ICs C is very small and therefore $j\omega CR$ is small enough to be neglected over a wide band of frequencies. At low frequencies $j\omega CR \ll 1$ and $v_2/v_1 = R_i/(R_i + R_o)$ which is independent of frequency. Eventually, however, as the frequency is raised, $j\omega CR$ becomes significant and a fall of gain, accompanied by phase shift, occurs. When large amounts of feedback are required to meet a given gain specification, the fall in gain may not occur at a low enough frequency, and the possibility of instability arises. Manufacturers therefore provide compensation terminals to which compensation components may be attached. The compensation network is quite often no more than a single capacitor. This capacitor has the effect of increasing C, thus ensuring that the gain falls off at lower frequencies than would otherwise be the case. Figure 8.16 shows the effect on the open-loop gain of connecting various compensating circuits. As the effective time constant between the stages of the amplifier is increased, the 'roll-off' in the response occurs at ever lower frequencies, thus permitting the application of greater amounts of feedback before instability conditions arise.

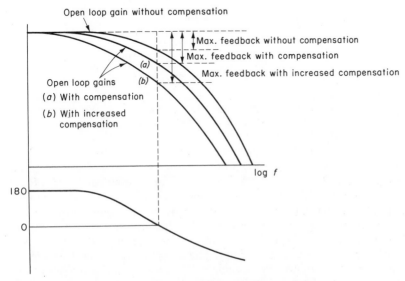

Figure 8.16 Effect of compensation on open-loop gain characteristics

In applying feedback it is always advisable to avoid using the maximum theoretical amount of feedback. Some margin of safety must be allowed. This margin is specified as a phase margin and values of about 30° are usually considered adequate. To understand the meaning of statements such as 'the phase margin should be at least 30°', we need to consider the way in which the loop gain, βA, varies with frequency.

The denominator of the closed-loop gain expression is $(1 - \beta A)$. If $|\beta A| = 1$ and the phase angle, $\theta = 0$ then $A' = \infty$, mathematically. The quantity βA is complex and is of the form $X + jY$; if $\beta A = (1 + jO)$ then instability results. The condition $\beta A = 1 + jO$ is known as the Barkhausen criterion for oscillation and is often illustrated diagrammatically by a Nyquist plot. Such a plot is shown in Figure 8.17. The magnitude and phase of βA are plotted on what is called the j-plane. For our case, the initial conditions at low frequencies are that β is positive and the open-loop gain is $A < 180°$, i.e. $-A$. As the frequency increases, β remains constant, but both $|A|$ and the phase shift θ, alter. $|A|$ is reduced, and θ moves from 180° to 0°. In Figure 8.17(a) we see that $|A|$ falls quickly enough as ω increases for the trajectory to avoid enclosing the critical point (1,0). Even when θ is zero the magnitude of the loop gain, $|\beta A|$, is not large enough to sustain oscillations. Somewhere along the Nyquist plot $|\beta A| = 1$. The value of the angle at that point is known as the phase margin. As previously stated, adequately stable operation is obtained if this angle is about 30°. Figures in excess of 30° improve the stability, but it can be shown that the transient response of the amplifier is degraded.

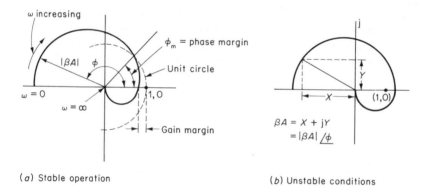

(a) Stable operation   (b) Unstable conditions

Figure 8.17 Plots of the complex quantity βA for frequencies from $0 \rightarrow \infty$. Such trajectories are known as Nyquist plots: (a) Shows a stable position in which there are both phase and gain margins; (b) shows an unstable situation, since there is one frequency where the loop gain is sufficient to sustain oscillations

Figure 8.18 shows two typical amplifier circuits using the S.G.S. μA709 op-amp. With the compensation components shown the gain may be as low as unity without instability. For unity gain, $R_f = R_i = 20\,k\Omega$. Since the open-loop gain of this IC at zero frequency is about 4×10^4, this means that as much as 90 dB of feedback may be applied without running into stability problems. It will be realized that heavy compensation results in a restricted bandwidth. For the amplifier shown in Figure 8.18(a) the bandwidth at the -3 dB points is 500 kHz; the slewing rate is 0.3 V μs^{-1} and the noise voltage is 0.03 mV peak to peak. The slewing rate can be improved at the expense of the noise figure by reducing C_1 and C_2 and using input lag compensation as shown in Figure 8.18(b). With these modifications, the slewing

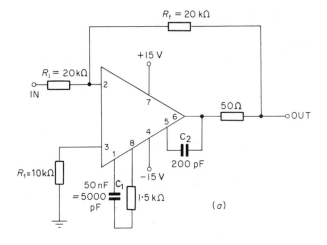

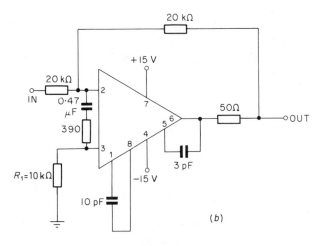

Figure 8.18 Unity-gain amplifier using S.G.S. μA709 ICs. Value of $R_1 = R_f$ in parallel with R_i. The resistances to earth at both the inverting and non-inverting input terminals are then identical; the offset voltage at the output terminal is thereby minimized. Note the compensation to produce stability with optimum bandwidth

rate (which is a measure of transient response) is increased to $22\,\mathrm{V\,\mu s^{-1}}$, but the noise figure is now 20 mV peak-to-peak.

For 741 and other types of internally compensated op-amps the user has no control over the bandwidth for a given closed-loop gain. In the case of a 741 a 30 pF capacitor is used for compensation, and is built into the chip. 30 pF is a large capacitance in this context and the resulting time constant dominates all others. The first break point is therefore at a very low frequency (about 10 Hz); thereafter the open-loop response falls at 6 dB per octave as shown in Figure 8.19. The whole of the open-loop response curve, which would appear in the absence of compensation, is thus pulled back towards the ordinate, and hence at unity (closed-loop) gain the loop gain is too small to sustain oscillations, i.e. $\beta A < 1$.

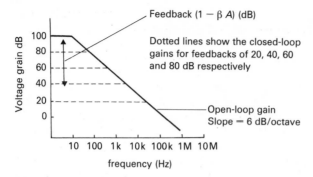

Figure 8.19 Open- and closed-loop responses for a 741 operational amplifier. Since the curves are similar to those for a first-order low-pass filter (see Figure 4.14 and associated text) the above responses may be expressed as $A = A_o/(1 + jw/w_t)$. A_o is the open-loop gain at zero frequency

The summing amplifier

We have already discussed in connection with Figure 8.13 the standard basic arrangements of integrated circuit amplifiers and associated feedback and input impedances. If additional resistors are added to the input as shown in Figure 8.20, then, by using Kirchhoff's Laws, it is easy to show that

$$v_o = - \left(\frac{v_1}{R_1} + \frac{v_2}{R_2} + \frac{v_3}{R_3} \right) \times R_f$$

When $R_1 = R_2 = R_3 = R_1$

$$v_0 = -(v_1 + v_2 + v_3)$$

We are thus able to add voltages in this way. We also have the opportunity of applying factors to any of the input voltages merely by adjusting the relative values of the input resistances and the feedback resistance. If the minus sign is inconvenient all that is necessary is to take the output voltage and feed it to a sign reverser; this sign reverser is merely a simple arrangement of a feedback resistor and an input resistor of equal resistance.

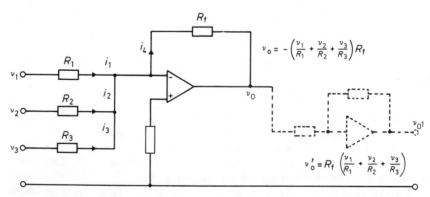

Figure 8.20 The summing amplifier

With multiple inputs some of the freedom of choice of R_f is lost since each input signal must be amplified by a specified amount and must ideally 'see' a specified input impedance. At times R_f may be inconveniently large (say $>10\,\mathrm{M\Omega}$). Such resistors are not readily available, and are often associated with undesirable stray capacitance. Under these circumstances the feedback impedance may take the form shown in Figure 8.21. By comparing the corresponding elements of the A-matrices for each network it is easy to show the equivalence of the two top networks in the figure; the transformation is familiar to electrical engineers as a star-delta conversion. When the actual circuit of Figure 8.21(a) is used as a feedback path for an operational amplifier, it is equivalent to using the network of (b), in which R_c is merely connected to the low-impedance output of the amplifier and R_a is equivalent to a second input to the amplifier in which the input voltage is zero.

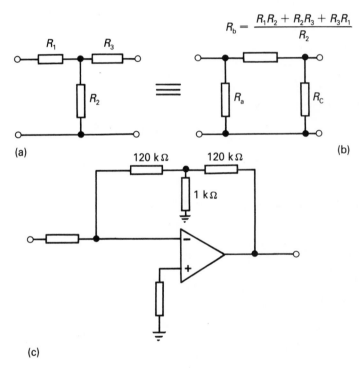

$$R_b = \frac{R_1 R_2 + R_2 R_3 + R_3 R_1}{R_2}$$

(a)

(b)

120 kΩ   120 kΩ

1 kΩ

(c)

Figure 8.21 Method of producing high effective feedback resistances from commonly available low-valued resistors

R_a and R_c therefore play no part in the feedback process; R_b is the effective equivalent feedback resistance. When $R_1 = R_3$ then

$$R_b = 2R_1 + R_1^2/R_2$$

In the example given, commonly available values of 120 kΩ, 120 kΩ and 1 kΩ may be for R_1, R_3 and R_2 respectively. This gives an effective feedback resistance of 14.64 MΩ. High gains are therefore possible even with high input resistances.

The integrating amplifier

If the feedback impedance, Z_f, consists of a capacitor instead of a resistor we can analyse the circuit by considering Figure 8.22(a). It is assumed that A is very large and that the input current to the amplifier itself is negligibly small (which is valid for the high input impedances in modern IC op-amps). Remembering that the charge on a capacitor, q, is related to the voltage across the plates by the relation $q = CV$, then $dq/dt = Cdv/dt$. Since $dq/dt = i_f$ then

$$i_f = \frac{dq}{dt} = C\frac{dv}{dt}$$

$$\frac{v_s - v_i}{R} = C\frac{d(v_i - v_o)}{dt}$$

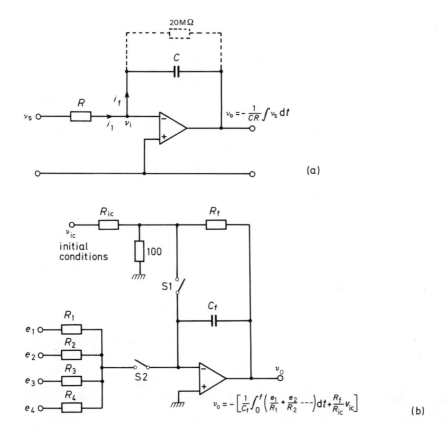

(a)

(b)

Figure 8.22 (a) Use of an operational amplifier as an integrator; (b) to achieve accurate integration, low-loss dielectric capacitors should be used. (Mylar or polystyrene dielectrics are satisfactory.) To reset the integrator S1 is closed and S2 open. To commence integration S1 is open while simultaneously S2 is closed. When both switches are open, a 'hold' condition is maintained. V_{ic}, R_f and R_{ic} set the initial conditions, thus giving the constant of integration

In practice A is extremely large and therefore v_i is vanishingly small, and can be ignored. We then have

$$\frac{v_s}{R} = - C \frac{dv_o}{dt}$$

therefore

$$v_o = -\frac{1}{CR} \int v_s dt \text{ i.e. } v_o(s) = \frac{1}{sCR} v_s(s)$$

That is to say, the output voltage is proportional to the time integral of the input voltage. The time constant, CR, is a scaling factor and can be adjusted just as can the ratio $R_f/R_{1,2,3}$ in the case of a summing amplifier.

In practice any offset voltages that may be present are also integrated along with v_s. This limits the time that the integration can be effected. Even with a small offset voltage, v say, the output will contain a term $\int v dt = vt$ if v is constant. (v is the equivalent input voltage that would produce the same offset in a perfect amplifier.) Eventually the output voltage will rise, or fall, to the value of the positive or negative supply rail voltage respectively. To avoid this the d.c. gain is often defined by placing a high value resistor across the capacitor, as shown by the dotted component in Figure 8.22(a).

Figure 8.22(b) shows a basic integrator of the type that would be suitable for an analogue computer. The switching action is performed by reed relays.

When long integration times are involved it is necessary to reduce offset currents and voltages to a minimum so that the resulting drift in the output voltage with time is negligibly small. One technique for reducing drift is to use a chopper-stabilized operational amplifier.

Chopper-stabilized amplifier

This type of amplifier has an extremely low d.c. offset voltage and bias current. These useful characteristics are achieved by first extracting the steady and low-frequency components of the input signal using a low-pass filter. The output from the filter is then periodically shorted to the zero voltage line by means of a switch operated by an oscillator. Nowadays it is convenient to use a FET as a switch. This process is known as chopping. The chopped low-frequency components are then amplified by an a.c. amplifier which, of course, has zero drift since it is a.c. coupled. The ouput from the a.c. amplifier is then demodulated.

Demodulation is akin to rectification in a power supply except that instead of using a diode to give unidirectional pulses, synchronous switches (FETs) are used. These are operated by a square-wave oscillator so that when one FET is on, the other is off. Hence in Figure 8.23, one FET is shown chopping the input signal; the other switch is in the demodulator and chops off all negative-going pulses coming from the output of the a.c. amplifier.

The demodulated output is smoothed via a low-pass filter and the filter output is then further amplifier in a d.c. amplifier. High-frequency components of the input signal, which are filtered out at the input of the chopper amplifier, are coupled directly to the d.c. amplifier. The resulting d.c. offsets, and drift associated with the final output voltage, are thus reduced by a factor equal to the gain of the chopper

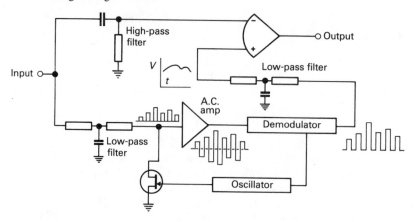

Figure 8.23 Principles of chopper-stabilized operational amplifier

channel. Figure 8.23 shows a simplified circuit of the 3291 chopper-stabilized amplifier of Burr–Brown. This amplifier is virtually immune from component ageing, temperature changes and power supply variation, and is ideal for use as an integrator where long integration times are involved.

The differentiating circuit

A circuit that achieves differentiation is similar to Figure 8.22(a) but with the capacitor and input resistor interchanged. Then

$$v_o(t) = - CR. \frac{dv_s}{dt}$$

$$\text{i.e.} \quad v_o(s) = -sCRv_s(s)$$

$$\text{and} \quad v_o(j\omega) = -j\omega CRv_s(j\omega)$$

Differentiators must be designed with care otherwise noise and instability problems arise. We see from the equivalent expressions for v_o that the differentiator's frequency response rises continually with frequency. This means that the upper frequency noise signals experience considerable amplification. In analogue computing, differentiators are best avoided altogether. It is usually possible to manipulate the differential equations involved so that only integrators and summers are required.

Miscellaneous circuits

By arranging more complicated networks for Z_i and Z_f we can modify the overall response of the operational amplifier in almost any desired way. We may, for example, be required to amplify signals in a narrow band of frequencies, or alternatively to amplify signals of all frequencies but one; for the reproduction of gramophone records it is necessary to have a falling response to meet R.I.A.A. specifications, and additionally a variable response is required for tone control purposes. The designing of suitable circuits is greatly simplified today by the use of ICs together with the appropriate input and feedback networks.

Some examples of useful circuits are given in Figures 8.24 to 8.31.

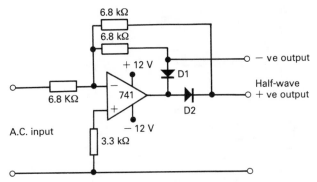

Figure 8.24 Half-wave rectifier. D1 and D2 are general-purpose silicon diodes

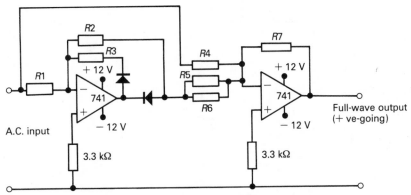

Figure 8.25 A full-wave rectifier for small-amplitude sinusoidal input voltages. R1–R7 should be closely matched 10 kΩ high-stability resistors

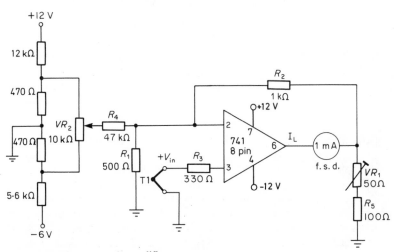

Figure 8.26 Thermocouple amplifier

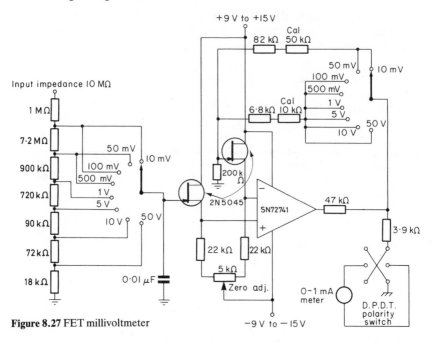

Figure 8.27 FET millivoltmeter

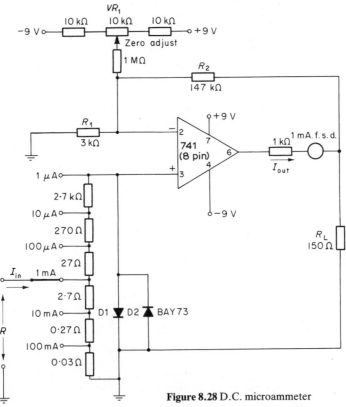

Figure 8.28 D.C. microammeter

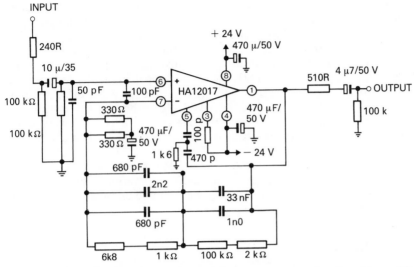

Figure 8.29 The HA12017 is a low-noise IC intended for use as a very low distortion amplifier (THD<0.002 per cent for 10 V output at 1 kHz). The above circuit is intended for high-quality magnetic amplifiers in which the necessary R.I.A.A. characteristic is followed accurately

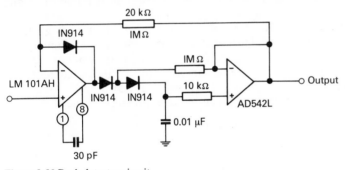

Figure 8.30 Peak detector circuit

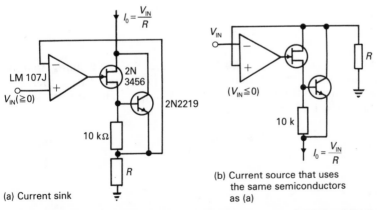

Figure 8.31 Precision current source and sink. For small currents 2N2219 and 10 kΩ resistor may be omitted; output is then taken from the source of the JFET. The type of JFET is not critical

Rectifier circuits that use diodes alone suffer from non-linearity of the diode characteristics for signal amplitudes less than about 600 mV. The circuit of Figure 8.24 avoids this. For extremely low input voltages, diodes D1 and D2 are both non-conducting. There is thus no feedback, and therefore the full open-loop gain is applied to the signal. Once the magnitude of a positive-going input signal becomes large enough to initiate conduction in D1, the closed-loop gain rapidly reduces until we have unity-gain amplifier conditions. During this time D2 is cut off. An almost perfect negative-going half sine wave is thus available at the negative output. For negative-going input signals, D2 conducts and D1 is cut off; a half sine wave that is positive-going, and of the same amplitude of the input signal, is then available at the positive output.

A full-wave rectifier for small-amplitude sine waves is shown in Figure 8.25. It is left as an exercise for the reader to explain the mode of operation of the circuit. The second IC is used as a two-input adder; the resistors should be closely matched and preferably metox types.

Logarithmic amplifiers

There are numerous occasions in scientific work where it is convenient to generate a voltage which is a logarithmic function of an input signal. Such amplifiers are useful in the fields of acoustics, nuclear physics, medical instrumentation and process control.

We have seen that the transfer function of an operational amplifier is given by the ratio of feedback to input impedances. A non-linear feedback element will obviously give a non-linear transfer function. The non-linear element for this application is conveniently a bipolar transistor since the collector current is a logarithmic function of the base–emitter voltage over several decades. We have already seen that the collector current is given by an expression of the form

$$i_C = I_{L1} \exp (e v_{BE}/kT)$$

hence

$$v_{BE} = K_1 \log i_C/I_{LI}$$

where

$$K_1 = \frac{kT}{e}$$

A prototype logarithmic amplifier is shown in Figure 8.32. Assuming that no current enters the inverting terminal then $i_S = i_c$. (An FET type of IC is therefore necessary for this application if we are to have an output ranging over several decades.) Since the point P is a virtual earth

$$i_s = \frac{v_s}{R_1} = i_C$$

Therefore

$$v_o = - v_{BE} = - K_1 \log \frac{v_s}{R_1 I_{L1}}$$

v_0 is seen to be a logarithmic function of v_S.

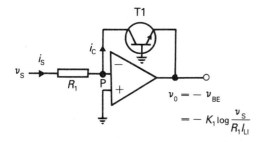

Figure 8.32 Prototype logarithmic amplifier

In practice the circuit of Figure 8.32 needs to be modified before a useful arrangement is obtained. The most important modification is the inclusion of temperature compensation. The collector current of the 'feedback' transistor, although an exponential function of v_{BE}, is also temperature-dependent. We can stabilize the collector current against temperature changes by using the circuit of Figure 8.33.

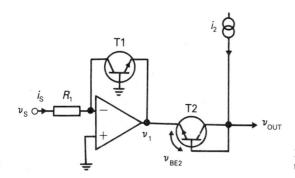

Figure 8.33 Logarithmic converter with temperature compensation

From the relations above (replacing v_0 by v_1), for transistor T1

$$v_1 = -K_1 \log \frac{v_s}{R_1 I_{L1}}$$

For transistor T2 the collector current i_2 is given by

$$i_2 = I_{L2} \exp (ev_{BE2}/kT)$$

Therefore

$$v_{BE2} = \frac{kT}{e} \log \frac{i_2}{I_{L2}}$$

Hence

$$v_{OUT} = v_1 + v_{BE2}$$

$$= -\frac{kT}{e} \log \frac{v_s}{R_1 I_{L1}} + \frac{kT}{e} \log \frac{i_2}{I_{L2}}$$

$$= -\frac{kT}{e} \log \frac{v_s}{R_1 I_{L1}} \cdot \frac{I_{L2}}{i_2}$$

Now if we use a matched pair of transistors in the same thermal environment, $I_{L1} = I_{L2}$; also $kT/e = 26\,\text{mV}$ at 20°C, therefore

$$v_{OUT}\,(\text{mv}) = -26\,\log\frac{i_s}{i_2}$$

where $i_S = v_S/R_1$.

It is easy to keep i_2 constant by using a constant current source (e.g. a larger resistance or the circuit of Figure 8.31(b)). We thus have an output voltage that is proportional to the logarithm of the input current. This input current should preferably be supplied from a constant current source. If the circuit of Figure 8.31(b) is used, the input current is programmed by R; we then have an output voltage proportional to the logarithm of the input voltage. If the input voltage (and hence input current) changes by a decade, $\log\,(i_S/i_2)$ changes by $\log_e 10$ for constant i_2. The output voltage is therefore $26\,\log_e 10$, i.e. $60\,\text{mV/decade}$. This output can be amplified to give any desired output level per decade. For a final output of $-1\,\text{V}$ per decade we need a gain of $1\,\text{V}/60\,\text{mV} = 16.7$. A final circuit may therefore be similar to Figure 8.34. From a practical constructional point of view it may be

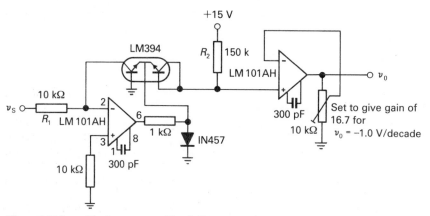

Figure 8.34 Logarithmic converter. The diode protects the LM394 if v_s should inadvertently become negative. R_1 and R_2 may be replaced by current sources

convenient to buy a unit in which the set of necessary ICs and transistors together with a constant current source are mounted within single encapsulation (e.g. Burr–Brown 4127 or similar). The only necessary external components are then the programming resistors, trimmers and power supply.

Instrumentation amplifiers

Many measurements of physical quantities can be made with the aid of the fundamental instrumentation circuit shown in Figure 8.35. Here we have a 741 operational amplifier configured as a differential amplifier giving an output voltage that is an amplified version of the voltage across a Wheatstone bridge arrangement. For measuring resistance the classic relationship $R_1/R_2 = R_3/R_4$ applies for a null condition; if three of the resistor values are known the fourth can be calculated. The 741 is then being used as a null detector, i.e. $v_{IN} = 0$. When the bridge is not in

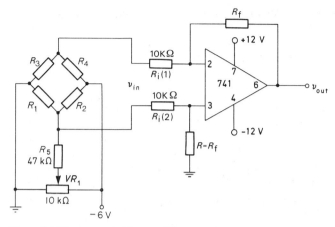

Figure 8.35 Resistance bridge amplifier

balance the above relationship does not hold; then v_{IN} is the difference of potential that exists between the junctions of R_1 and R_2, and R_3 and R_4. If the bridge has an energizing voltage, E across it (e.g. $-6\,\mathrm{V}$ in Figure 8.35) then

$$v_{IN} = E\left(\frac{R_1}{R_1 + R_2} - \frac{R_3}{R_3 + R_4}\right)$$

(We are assuming that the detector circuit impedance is high compared with the bridge resistors.)

To measure a physical quantity we must first obtain or devise a suitable transducer whose resistance is a known function of the physical quantity. Thus by using thermistors, light-dependent resistors or strain-gauges, for example, the output voltage from the bridge is a function of temperature, light intensity or mechanical strain respectively. Basically the measurement system depends only upon our ingenuity in devising suitable resistive transducers.

The advantage of using a resistor bridge lies in the fact that the output voltage bears a linear relationship to the change of resistance provided the latter is small compared with the resistance value, i.e. $\delta R/R$ is small. It is usually convenient to make the resistance values of the 'arms' equal to the transducer resistance at a convenient value of the physical quantity. For example, if we were measuring temperature over the range $0°$ to $100°C$ then each resistor in the bridge would be made equal to R, where R is the resistance value of the transducer at $50°C$. Let us make R_4 the transducer whose resistance is $R \pm \delta R$ at temperatures different from $50°C$. The output from the bridge is the input to the differential amplifier therefore

$$v_{IN} = E\left(\frac{1}{2} - \frac{1}{2 \pm \dfrac{\delta R}{R}}\right) = \frac{E}{2}\left(1 - \frac{1}{1 \pm \dfrac{\delta R}{2R}}\right)$$

When $\delta R/2R$ is small we may invoke the binomial theorem and take only the first two terms; then

$$v_{IN} = \frac{E}{2}\left(1 - \left[1 \pm \frac{\delta R}{2R}\right]^{-1}\right) = \pm\frac{E\delta R}{4R}$$

We see therefore that v_{IN} is a linear function of δR. If, in addition, δR is a linear function of the change in physical quantity then a linear measurement system exists. Even when $\delta R/R$ is not very small the bridge exerts a linearizing influence.

For non-critical temperature measurements a thermistor may be used as R_4. The change of resistance is quite large over an extended temperature range and thus the instrument would need to be calibrated. For more accurate measurements a transducer is needed in which $\delta R/R$ is very small over the range of interest. It is obvious that as $\delta R/R \rightarrow 0$, the output from the bridge becomes very small. In addition, many practical measurement systems involve a transducer position that is remote from the electronic measuring apparatus. Inevitably the transducer leads pick up a good deal of induced noise unless special precautions are taken. In many instrumentation systems we are therefore presented with the problem of measuring very small signals in the presence of comparatively high noise or other voltages. Fortunately, standing voltages and/or induced noise voltages are very often common-mode signals. An amplifier with very good common-mode rejection properties is therefore needed. We may take as an example the circuit of Figure 8.35 in which the gain ($R_f/10\,\text{k}\Omega$) is say 100. If it is assumed that the physical quantity being measured changes the transducer resistance by 1 per cent (i.e. $\delta R/R = 0.01$) then the wanted output signal from the bridge is 15 mV. The output voltage from the amplifier is then ideally 1.5 V. Unfortunately we have a common-mode standing input voltage of -3 V, which will give an additional output in a less than ideal amplifier. Suppose we require the error in the wanted 1.5 V output voltage to be not greater than 1 per cent (a fairly modest request in a high-grade instrumentation amplifier) then the output error voltage would be 15 mV. The input common-mode signal is 3 V and hence the common-mode gain would need to be less than 5×10^{-3}. Since the differential gain is 10^2 the CMRR must be $10^2/5 \times 10^{-3} = 2 \times 10^4$ i.e. 86 dB at least. Clearly, a 741 type of amplifier would be unsatisfactory since these types have a worst-case CMRR of 70 dB; the rejection properties of a 741 are not good enough for instrumentation purposes. We would need to go to a 725 type of amplifier for this application. In general when high-grade amplification is required for instrumentation purposes it is better to use a special instrumentation amplifier (e.g. AD 521 or AD 522).

Unlike an ordinary operational amplifier, an instrumentation amplifier is dedicated to amplifying accurately the difference voltage between the two input terminals. To this end an instrumentation amplifier is constructed so as to have low drift, an extremely high impedance between the two input terminals and high common-mode rejection ratios ($>100\,\text{dB}$). The differential gain is usually in the range 1–1000, and the amplifier is single-ended, with a very low output impedance. The amplifier gain is usually set by the user with a single external resistor.

Since the primary aim of the amplifier is to yield an output voltage that is strictly proportional only to the voltage difference between the input terminals, rejecting almost completely any common-mode signals it may seem convenient and attractive to choose a high-grade operational amplifier that is sold as an 'instrumentation' amplifier. For example, a 725CN has a CMRR of 115 dB and open-loop gain of 127 dB (worst-case 110 dB and 108 dB respectively). The trouble with this excellent specification is that often it cannot be realized in practice; it is the characteristics of the overall instrumentation system that counts. With standard operational amplifier techniques we must make compromises in the design. One of the difficulties is associated with input impedance. With the usual differential configuration the gain is determined by R_f/R_i (Figure 8.35). For practical values of

R_f, large gains require small values of R_i; on the other hand, high input impedances require R_i to be large.

The effect of mismatch in the input resistors can be seen by comparing the different gains at each of the input terminals. The output voltage is

$$v_o = -\frac{R_f}{R_{i(1)}}v_1 + \frac{R_f}{R_{i(2)}}v_2$$

If we employed an almost perfect IC having an extremely high CMRR, the overall CMRR would then be governed by the mismatch in the input resistors. When $R_{i(1)} = R_{i(2)}$ then $v_o = 0$ for common-mode signals; the gain for each input is G, say. If, however, $R_{i(1)}$ is slightly higher than the specified value and, for a worst case, $R_{i(1)}$ were lower by a similar amount then

$$v_0 = -(G - \delta G)v_1 + (G + \delta G)v_2$$

For common-mode signals $v_1 = v_2 = v_{cm}$ hence $v_{0(cm)} = 2\delta G v_{cm}$, i.e. the common-mode gain = $2\delta G$.

The CMRR is therefore $G/2\delta G$. We see therefore that if $R_{i(1)}$ were to be 0.05 per cent higher than the specified value, and $R_{i(2)}$ to be 0.05 per cent lower (i.e. the mismatch is 0.1 per cent), $\delta G/G = 0.05$ per cent and hence the CMRR would be 1/0.1 per cent, i.e. 1000, or 60 dB. An excellent CMRR in the op-amp would therefore be degraded to 60 dB.

It might seem that a solution would be to trim the two input resistances for perfect equality. The procedure would not necessarily improve matters since the effective values of the input resistors are $R_{i(1)}$ and $R_{i(2)}$, to which must be added the equivalent source impedances. In the case of a bridge source, trimming would be very difficult to achieve practically, especially as the resistance of a transducer alters with changes in the physical quantity being measured. The only way in which to preserve high values of CMRRs in instrumentation amplifiers is to have very high input impedances. The effect of changes in source impedances would then be relatively unimportant.

One way of achieving a high input impedance is to use voltage-followers in each channel of the input stage. A third IC can then be used to combine the outputs from the voltage followers and give a final single-ended output. This third IC would then have to supply all of the necessary gain and achieve all of the common-mode rejection. To avoid this, a preferred input stage, as shown in Figure 8.36, is used.

Since the open-loop gain of an op-amp is very large, the difference between the voltages at its input terminals is extremely small when practical output voltages are involved. The voltages at the input terminals may then be taken to be equal when considered in connection with the voltages in the rest of the circuit. Thus for the circuit in Figure 8.36 $v_1 = v_{S1}$, and $v_2 = v_{S2}$. Thus

$$v_{01} - v_{02} = i(R_1 + 2R_2)$$

where i is the current through the resistor chain.

Now

$$i = (v_1 - v_2)/R_1 = (v_{S1} - v_{S2})/R_1$$

therefore the differential gain is

$$\frac{v_{o1} - v_{o2}}{v_{s1} - v_{s2}} = 1 + 2\frac{R_2}{R_1}$$

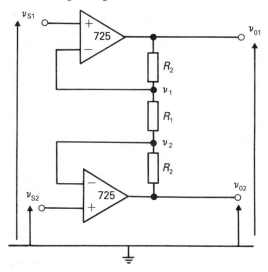

Figure 8.36 Input stage of instrumentation amplifier that uses two high-grade operational amplifiers. (All voltages are referenced to the zero-voltage line)

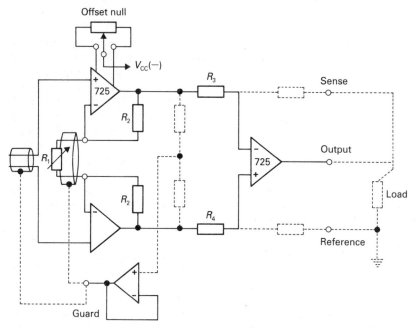

Figure 8.37 Basis of a commercial instrumentation amplifier, which can be purchased as a complete single package

We can therefore achieve large differential gains in the input stage by a suitable choice of a single resistor R_1. Simultaneously, large input impedances are evident at each input terminal. The impedance between the input terminals, in practice, is several hundred megohms.

If throughout the circuit all of the voltages had been referred to some voltage level other than zero the analysis would be unaffected and the same result would have been obtained for the differential gain. With a different reference voltage level we are effectively adding a common-mode signal to the input. The same common-mode signal would also appear superimposed on v_{01} and v_{02}. The common-mode gain of the arrangement shown is therefore unity. The ratio of differential to common-mode signal at the output of the stage is considerably greater than that at the input.

The residual common-mode signal is eliminated by a third operational amplifier, which also yields a single-ended output. An instrumentation amplifier therefore has the configuration illustrated by Figure 8.37. Commercial units often incorporate the additional facilities shown as dotted lines in the diagram. The leads to the signal transducer (and to R_1 if this component cannot be conveniently mounted in the amplifier housing) needing to be shielded to minimize noise pick-up. To avoid undesirable cable capacitive and leakage effects, the shield should be maintained at the common-mode potential of the input signal. Since this potential also exists at the v_{01} and v_{02} terminals, it is convenient to use a voltage-follower connected as shown. The guard terminal is then supplied from a low-impedance source. Sense and reference terminals are also often provided. These involve feedback in the third operational amplifier and nullify the effects of wiring losses to the load.

Active filters

The majority of analogue signals encountered in electronics may be regarded as a synthesis of component sinusoidal signals. These components each have different amplitudes and frequencies and bear differing phase relationships to one another. In some cases, e.g. high-fidelity sound reproduction and video amplifiers, the whole of the frequency spectrum of the composite signal must be processed without introducing distortion. There are occasions, however, when only a part of the information is required. In radio transmissions, for example, we wish our electronic system to select only one carrier wave that may be present along with those which are temporarily unwanted. The separation is performed by means of an electronic filter.

Filters are transmission systems that are designed to alter the frequency spectra of signals presented to the input terminals. In practice the filter is usually required to modify the frequency spectrum in one of four ways.

(a) High-pass filters are required to transmit without attenuation all components of a signal having a frequency above that of a specified value, known as the cut-off frequency, f_c; no components with frequencies below the cut-off value should be transmitted.
(b) Low-pass filters operate upon signals so as to eliminate all components with frequencies above a specified cut-off value; below the cut-off frequency all components should pass through the filter without attenuation.
(c) Band-pass filters pass only those components within a specified range of frequencies.
(d) Band-reject filters reject all components within a specified range of frequencies, passing all others without attenuation.

Figure 8.38 shows the ideal amplitude/frequency response characteristics of the four types of filter.

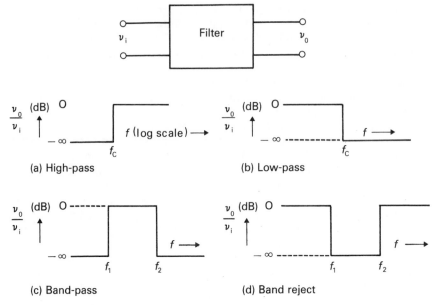

Figure 8.38 Ideal characteristics of the four basic types of filter

When only high-frequency components are present in the signal, satisfactory filters can often be constructed using inductors and capacitors. When the frequencies are low, however, inductors are impractical because large inductance values are required. This implies large physical size, inherent resistor losses and a tendancy to hum and noise pick-up. Corresponding responses may be obtained without inductors by using capacitors and resistors in conjunction with operational amplifiers; the combination is known as an active filter. The sensitivity of filter performance to tolerance in component values, especially the change in open-loop gain and phase shift of the op-amp with frequency, generally makes active filters inferior to their LC counterparts. The choice of IC is important; in many applications the use of an internally compensated amplifier (such as the 741) is unsatisfactory. The emergence in recent years of op-amps with superior performance has alleviated the position.

Filters are classified according to their order. The input and output voltages are related, in linear systems, by a linear differential equation. The filter order is defined as the highest order of differential coefficient present in the equation. For linear systems it is convenient to transform the differential equations into their Laplace counterpart. The ratio v_o/v_s expressed as a function of the Laplace operator, s, is called the *transfer function* of the system.

The performance of an active filter is described mathematically by its transfer function. As we have seen in the first part of the book, transfer functions may be written as the quotient of two polynominals in the Laplace variable. Take, for example, a series LCR circuit energized by a sinusoidal voltage

$$L \frac{di}{dt} + iR + \frac{1}{C} \int i dt = E \sin \omega t = v_i(t)$$

Therefore

$$\bar{\imath}\left(sL + R + \frac{1}{sC}\right) = \bar{v}_i$$

If the output is taken across the capacitor then

$$\bar{v}_o = \frac{\bar{\imath}}{sC} = \frac{\bar{v}_i}{sC\left(sL + R + \dfrac{1}{sC}\right)}$$

therefore

$$\frac{\bar{v}_o}{\bar{v}_i} = \frac{1}{s^2 LC + sCR + 1}$$

The transfer function is $1/(s^2 LC + sCR + 1)$; it describes a second-order system since the highest power of s in the denominator is 2. In general, analyses of filter systems yield transfer functions of the form

$$F(s) = \frac{\alpha_n s^n + \alpha_{n-1} s^{n-1} \ldots \alpha_o}{a_n s^n + a_{n-1} s^{n-1} \ldots a_o}$$

where α and a are coefficients. The highest power of s in the denominator defines the order of the filter. (It is the order of the denominator that corresponds to the order of the differential equation giving rise to the transfer function – see example below.) High-order filters $(n>3)$ are made by cascading prototype sections of order 1 or 2.

The transfer function polynomials may be factorized and hence $F(s)$ may be written as

$$F(s) = \frac{(s + z_1)(s + z_2)(s + z_3) \ldots}{(s + p_1)(s + p_2)(s + p_3) \ldots}$$

When $s = -z_1, -z_2$ and/or $-z_3$ etc, then $F(s) = 0$; z_1, z_2, z_3, etc are therefore called *zeros*. When $s = -p_1, -p_2$ and $-p_3$, etc., then $F(s) = \infty$; p_1, p_2, p_3, etc. are therefore called *poles*. Since the denominator of a transfer function defines the order of a filter, and an nth-order polynominal has n poles, then an nth-order filter can be called an n-pole filter. The terms poles and zeros arise from considering $s = \sigma + j\omega$, and plotting the real and imaginary parts of $F(s)$ in the usual $(a + jb)$ way on what is called the s-plane. Points z_1, z_2 etc. correspond to $F(s) = 0$, while p_1, p_2 are those points at which $F(s)$ stands out from the plane like a long pole because $F(s) \to \infty$.

The frequency response of an ideal filter is unrealizable, and we must therefore be content with a suitable approximation. The quality of this approximation depends upon the order of the filter used. Since the complexity of a filter increases with the order, it is advisable to choose the lowest order that will meet the specification. A high-pass filter, for example, could be realized by applying the input signal to a simple series CR circuit, taking the output across the resistor. The transfer function for the filter is $sCR/(sCR + 1)$. This is a first-order filter with a cut-off frequency of $1/2\pi CR$. Above f_c the pass characteristics are satisfactory, but the cut-off rate at f_c is very different from the required ideal. Below f_c the cut-off rate is only 6 dB/octave, which is usually too slow for filter purposes.

In practice, the lowest order that gives reasonable approximations to ideal characteristics is the second. Many control systems and active filters may be approximately described by the second-order differential equation

$$\frac{d^2y}{dt^2} + 2\zeta\omega_o\frac{dy}{dt} + \omega_o^2 y = \omega_o^2 x(t) = a\ddot{y} + b\dot{y} + cy$$

y is the output quantity and $x(t)$ the input quantity; $x(t)$ is some unspecified function of time. In control and mechanical systems ζ (zeta) is called the damping factor and $1/(\zeta\omega_o)$ is the time constant of the system. ω_d (the damped angular frequency of oscillation) $= \omega_o \sqrt{(1 - \zeta^2)}$ where $\omega_o = 2\pi \times$ the undamped natural frequency.

In active filters ζ is related to the Q-factor of the filter and $\omega_o = 2\pi f_c$ for low- and high-pass filters.

In Laplace transform terms the differential equation becomes

$$(s^2 + 2\zeta \omega_o s + \omega_o^2)\bar{y} = \omega_o^2\bar{x}$$

where $\bar{y}$ and $\bar{x}$ are the Laplace transforms of $y(t)$ and $x(t)$ respectively. The transfer function of the system in this example is thus $\omega_o^2/(s^2 + 2\zeta \omega_o s + \omega_o^2)$. If we compare the denominator of this expression with that of the general polynominal form, and write $a = a_2$, $b = a_1$ and $c = a_o$ then

$$\omega_o = \sqrt{\left(\frac{c}{a}\right)}$$

and

$$Q = \frac{\sqrt{(ac)}}{b} = \frac{1}{2\zeta}$$

The transfer function for this example could thus be given as

$$F(s) = \frac{\omega_o^2}{s^2 + \omega_o s/Q + \omega_o^2}$$

In general we can recognize second-order filter characteristics from four transfer functions, viz.

(1) Low-pass

$$F_{LP}(s) = \frac{A_1\omega_o^2}{s^2 + \dfrac{\omega_o s}{Q} + \omega_o^2}$$

(2) High-pass

$$F_{HP}(s) = \frac{A_2 s^2}{s^2 + \omega_o s/Q + \omega_o^2}$$

(3) Band-pass

$$F_{BP}(s) = \frac{A_3(\omega_o/Q)s}{s^2 + \omega_o s/Q + \omega_o^2}$$

(4) Band-stop

$$F_{BS}(s) = \frac{A_4(s^2 + \omega_o^2)}{s^2 + \omega_o s/Q + \omega_o^2}$$

where A_1 to A_4 are gain factors. The frequency response of each filter is easily deduced by replacing s with $j\omega$ and noting what happens to the transfer function when $\omega \rightarrow O$, $\omega \rightarrow \infty$ and any values of ω of special interest. For example, in high-pass filters

$$F_{HP}(j\omega) = \frac{-A_2\omega^2}{-\omega^2 + j\dfrac{\omega_o}{Q}\omega + \omega_o^2}$$

from which we see when $\omega \rightarrow 0$, $F_{HP}(j\omega) \rightarrow -A_2(\omega/\omega_o)^2$, i.e. there is a fall of 12 dB per octave with 180° phase shift. At high frequencies $\omega \rightarrow \infty$ and $F_{HP}(j\omega) \rightarrow A_2$, i.e. is constant. (We have used this technique earlier in the book when dealing with Bode diagrams.)

We see that when designing or analysing an active filter we first must obtain the transfer function of the system and then examine the function in order to reach some basic conclusions.

The range of active filters is huge, and it would be impossible in the space available here to give other than an appreciation to the principles involved.

Let us initially take a simple and popular configuration which leads to Sallen and Key filters (Sallen, R.P. and Key,S.L. 'A practical method of designing RC active filters', *I.R.E. Trans* CT-2,1955,74–85).

This paper contains the first proposals for the design of simple second-order filters based on positive feedback circuits. These filters have found widespread acceptance, especially since inexpensive high-grade d.c. amplifiers have become available in IC form. The basic general circuit is shown in Figure 8.39. Y_1 to Y_4 are the admittances of the respective two-terminal networks which contain linear passive components. The general method of obtaining the transfer function of the system is to assign arbitrary symbols for the voltages at each node. We are here using an op-amp as a voltage-follower; the gain is therefore unity. The amplifier is assumed to have ideal characteristics, mainly that the open-loop gain is extremely

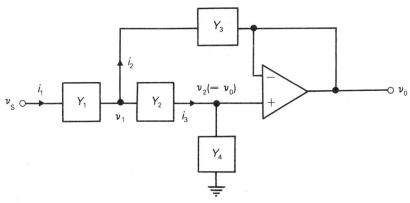

Figure 8.39 Basic general circuit for filters based on the Sallen and Key arrangement

large and remains so over the range of frequencies of interest; the input impedance at each input terminal is assumed to be infinitely large. We then proceed to establish linear equations for the system by using Kirchhoff's current Law. From Figure 8.39

$$i_1 = i_2 + i_3$$

$$(v_s - v_1)Y_1 = (v_1 - v_0)Y_3 + (v_1 - v_2)Y_2$$

Since we are using a unity gain amplifier with an infinite input impedance, $v_2 = v_0$ and $i_3 = v_0 Y_4$; therefore

$$(v_s - v_1)Y_1 = (v_1 - v_0)Y_3 + v_0 Y_4$$

Also Y_2 and Y_4 form a potential divider, hence

$$v_0 = v_1 Y_2/(Y_2 + Y_4) \text{ i.e. } v_1 = v_0(Y_2 + Y_4)/Y_2$$

Substituting in the above expression and rearranging terms

$$v_s Y_1 = v_1(Y_1 + Y_3) + v_0(Y_4 - Y_3)$$

$$v_s Y_1 Y_2 = v_0(Y_1 Y_2 + Y_3 Y_4 + Y_4 Y_1 + Y_2 Y_4)$$

Let us take the case where $Y_1 = Y_2 = 1/R$ and $Y_3 = Y_4 = sC$, then

$$v_s = v_0(s^2 C^2 R^2 + 2 s \, CR + 1)$$

$$\frac{v_0}{v_s} = \frac{1}{s^2 C^2 R^2 + 2sCR + 1}$$

This is a second-order filter in which $\omega_0 = 1/(CR)$; the filter has a Q-value of 0.5.

By considering $s \equiv j\omega$ we see that we have a low-pass filter with $f_c = 1/(2\pi CR)$ and an ultimate slope at high frequencies of 12 dB/octave. From the Q-value, however, we see that the transition from *pass* to *reject* regions is not very sharp.

If we need a sharper cut-off it will be necessary to increase the Q-value. This is achieved by altering the gain of the amplifier. Appropriate value for R_f and R_i as shown in Figure 8.13(b) must then be selected. If the gain of the amplifier is then K, a similar analysis to the one above yields a transfer function given by

$$\frac{v_0}{v_s} = \frac{K}{s^2 C^2 R^2 + sCR(3 - K) + 1}$$

We see that the cut-off frequency is unaffected by the gain, while $Q = 1/(3 - K)$ and is therefore a function of the gain. The Q-factor is seen to be sensitive to variations of gain, especially for values of K close to 3. When K varies from 1 to 3, Q varies from 0.5 to ∞.

The sensitivity of filter performance to changes in component values constitutes one of the major problems associated with the design of active filters. Sensitivity factors can be defined to enable quantitative assessments, and comparisons, to be made of filter performance. In general a sensitivity factor may be defined as

$$S_b^a = \frac{\delta a/a}{\delta b/b} = \frac{b}{a} \frac{\partial a}{\partial b}$$

(The partial derivative is used because all other variables are assumed to be temporarily constant.)

S_b^a is the sensitivity of the filter property 'a' to variations in component parameter 'b'; thus if $a = \omega_o$ and $b = K$ then $S_K^{\omega_o}$ represents the sensitivity of the parameter ω_o to changes in the gain K. For example, in the case of the above filter $S_K^{\omega_o} = 0$ because the break (or cut-off) frequency is independent of K. Q, however, does depend upon K.

$$Q = \frac{1}{3 - K} \therefore \frac{dQ}{dK} = \frac{1}{(3 - K)^2}$$

$$S_K^Q = \frac{K}{Q} \cdot \frac{\partial Q}{\partial K} = \frac{K}{3 - K}$$

The filter is thus extremely sensitive to variations of K when K is close to 3. The effect that changes of K have on the transfer function is illustrated in Figure 8.40. Since $Q = 1/(3 - K) = 1/2\zeta$ then $K = 3 - 2\zeta$. In Figure 8.40 the magnitude of the

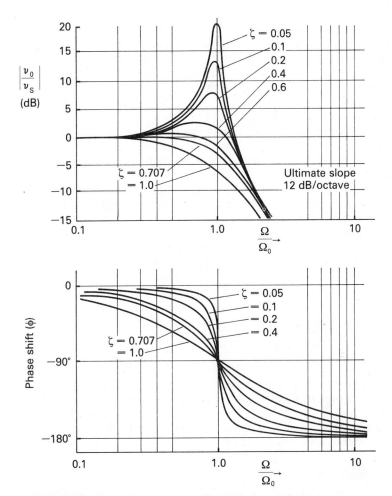

Figure 8.40 Magnitude of the transfer function and phase shift as functions of frequency for a second-order filter

transfer function has been plotted for different values of ζ. We see that when $\omega = \omega_o$ there is marked peaking of the frequency response for values of ζ lower than 0.1 ($K>2.8$; $Q = 5$). At the transition from the pass band to the stop band the phase shift changes rapidly for Q values in excess of 5. The filter characteristics at high Q-values are therefore far from ideal. On the other hand for low Q-values (say $Q = 1 = K$) the phase transition is smooth enough, but the cut-off of the frequency response is not sharp enough.

An enormous amount of investigation has taken place by many workers in order to improve the characteristics of practical filters. Well-known names are Butterworth, Chebyshev and Bessel.

When a name is applied to a filter, that name does not refer to a specific circuit arrangement, but indicates the type of response, and the characteristics of the filter. Each worker in the field has made proposals intended to optimize some aspect of filter performance. Unfortunately, improvements in one aspect of performance are usually gained at the expense of other properties. The circuit designs therefore become compromises in which improvements are limited by the amount of degradation that can be tolerated in the other characteristics. Three properties are uppermost in the filter designer's mind. First, the amplitude/frequency characteristic must ideally be flat in the pass band; second, the transition from the pass band to the stop band must be extremely sharp; and third, distortion of complex waveforms must be minimized. In this respect it will be recalled that a complete waveform may be considered as being a synthesis of harmonic components. If the time delay through the filter is different for the different components, then the composite wave will emerge from the filter with its waveform distorted. To ensure that the time delays are the same for all frequencies in the pass band, then the filter must have a phase response that changes linearly with frequency.

Since ideal filter characteristics cannot in practice be realized, suitable approximations to ideal transfer functions must be found. From the approximate expressions, the amplitude/frequency characteristic of the practical filter is obtained by substituting $j\omega$ for s and then evaluating the magnitude of transfer function at different frequencies. The phase characteristics are deduced from the ordinary and quadrature (real and imaginary) components of the transfer function.

Low-pass filter characteristics and design have received a good deal of attention because transformations, tables and computer programs are available to enable the designer to convert low-pass filter data into high-pass and band-pass form. The general low-pass filter characteristics can be described by

$$F(s) = \frac{1}{a_n s^n + a_{n-1} s^{n-1} \ldots a_o}$$

By choosing sufficient terms we can approach as near as we wish to the ideal low-pass characteristic in which we have a flat amplitude/frequency response in the pass band and an extremely sharp fall in response to the stop or reject band. In practice we must find a polynomial that approximates the response closely enough for our needs and which is physically realizable without undue complication.

The Butterworth approximation

This approximation is based upon the Butterworth polynomials, the first four of which are:

n	$F(s)$
1	$\dfrac{s}{\omega_o} + 1$
2	$\left(\dfrac{s}{\omega_o}\right)^2 + 1.414 \left(\dfrac{s}{\omega_o}\right) + 1$
3	$\left(\dfrac{s}{\omega_o}\right)^3 + 2 \left(\dfrac{s}{\omega_o}\right)^2 + 2 \left(\dfrac{s}{\omega_o}\right) + 1$
4	$\left(\dfrac{s}{\omega_o}\right)^4 + 2.613 \left(\dfrac{s}{\omega_o}\right)^3 + 3.414 \left(\dfrac{s}{\omega_o}\right)^2 + 2.613 \left(\dfrac{s}{\omega_o}\right) + 1$

The magnitude of the polynomials is given by

$$B_n = \sqrt{\left[1 + (\frac{\omega}{\omega_o})^{2n}\right]}$$

For low-pass filters

$$|F(j\omega)| + 1/\sqrt{\left[1 + (\frac{\omega}{\omega_o})^{2n}\right]}$$

A sketch of this function for low values of n is given in Figure 8.41. Irrespective of the value of n, when $\omega = \omega_o$ then $|F(j\omega)| = 1/\sqrt{2}$, i.e. ω_o defines the turnover or cut-off frequency; at this angular frequency $|F(j\omega)|$ is 3 dB down on its value at low frequencies. For values of ω much less than ω_o the term (ω/ω_o) becomes small especially for large values of n. For high-order filters the response in the pass region is maximally flat. This constitutes the major advantage of filters with a Butterworth response. Flatness of the frequency response in the pass region is obtained at the expense of phase properties. For frequencies where $\omega > \omega_o$ the value of $|F(j\omega)|$ falls off rapidly, especially so for high values of n. Thus we can obtain a rapid transition

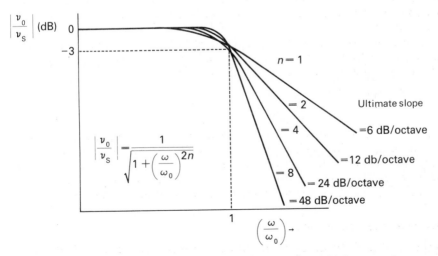

Figure 8.41 Normalized frequency response curve for low-pass filters with Butterworth responses

from the pass to the stop region, but only by using many sections to produce the necessary high order. Note also that the higher the value of n the sharper is the 'knee' of the amplitude/frequency characteristic.

A Sallen and Key second-order filter of the type shown basically in Figure 8.39 can be made to have a Butterworth response by considering the transfer function given previously. For a low-pass filter $Y_1 = Y_2 = 1/R$ and $Y_3 = Y_4 = j\omega C$. Instead of having a unity-gain voltage-follower the gain should be set at 1.586. We note that the transfer function when the gain is K is given by

$$\frac{v_o}{v_s} = \frac{K}{s^2 C^2 R^2 + sCR(3 - K) + 1}$$

By making $CR = 1/\omega_o$ (hence CR determines the desired cut-off frequency) the denominator of the transfer function corresponds to the second-order Butterworth polynomial when $3 - K = 1.414$ (i.e. $\sqrt{2}$); hence $K = 1.586$. Since $K = 1.586$, then $\zeta = 1/\sqrt{2} = 0.707$; the frequency response and corresponding phase response can be seen from the appropriate curve in Figure 8.40.

When very sharp transitions from pass to stop regions are required, it is necessary to use a large number of filter sections. The output impedance of operational amplifiers is quite low and therefore sufficient identical sections can be cascaded to achieve the required cut-off rate.

If we are not too concerned about a very flat response in the pass-region, and if an extremely sharp cut-off is essential, a more economical way of achieving the necessary filter response is to abandon Butterworth polynomials and substitute Chebyshev polynomials.

The Chebyshev approximation

This approximation to an ideal low-pass filter characteristic concentrates on sharpness of cut-off. Very sharp cut-off rates can be achieved by devising an appropriate set of polynomials, while simultaneously choosing a form of filter function that confines the variations of response in the pass region to within specified limits. The Chebyshev approximation is

$$|F(j\omega)| = \frac{v_o}{v_i} = 1/\sqrt{[1 + \epsilon^2 T_n^2]}$$

ϵ is a constant the significance of which will be explained shortly. T_n is a Chebyshev polynomial (T is used because an alternative spelling is Tchebyschev; C is also a frequently used symbol). For filter purposes it is convenient to express the polynomial as a function of ω/ω_o (or f/f_o).

$$|F(j\omega)| = \left|\frac{v_o}{v_i}\right| = 1/\sqrt{\left[1 + \epsilon^2 T_n^2\right]}$$

Five of the polynomials are given below.

Order	T_n
1	$\left(\dfrac{\omega}{\omega_o}\right)$
2	$2\left(\dfrac{\omega}{\omega_o}\right)^2 - 1$

3	$4 \left(\dfrac{\omega}{\omega_o}\right)^3 - 3 \left(\dfrac{\omega}{\omega_o}\right)$
4	$8 \left(\dfrac{\omega}{\omega_o}\right)^4 - 8 \left(\dfrac{\omega}{\omega_o}\right)^2 + 1$
5	$16 \left(\dfrac{\omega}{\omega_o}\right)^5 - 20 \left(\dfrac{\omega}{\omega_o}\right)^3 + 5 \left(\dfrac{\omega}{\omega_o}\right)$

From an inspection of the polynomials we see that the value of $T_n{}^2$ increases very rapidly for $\omega > \omega$. The reciprocal function for v_o/v_i therefore falls rapidly at frequencies in excess of the break frequency, especially so for the higher order polynomials.

It is, of course, easy to obtain very rapid falls in response merely by choosing large enough coefficients for the polynomial. Unfortunately, difficulties arise when we must simultaneously have a suitable response in the pass band. Arbitrary choices of large values for the coefficients will, in general give pass band characteristics that are wildly different from the ideal response. In Chebyshev low-pass filters, however, it will be seen that in the pass band ($0<\omega<\omega_o$) the Chebyshev polynomials are always such that $-1 \leqslant T_n \leqslant 1$.

Substituting various values of ω will show that T_n oscillates between -1 and $+1$. In the absence of ϵ, the filter function is

$$\left| \frac{v_o}{v_i} \right| = \frac{1}{\sqrt{[1 + T_n{}^2]}}$$

Even at the maximum excursions of T_n the ripple in the pass band is never more than $3\,dB$ down, i.e.

$$\frac{1}{\sqrt{2}} \leqslant \left| \frac{v_o}{v_i} \right| \leqslant 1$$

The ripple can be further reduced by introducing the term ϵ into the filter function, and choosing a suitable value for ϵ. Then

$$\left| \frac{v_o}{v_i} \right| = \frac{1}{1 + \epsilon^2 T_n{}^2}$$

In the pass, when $0<\omega<\omega_o$

$$\frac{1}{\sqrt{(1 + \epsilon^2)}} \leqslant \left| \frac{v_o}{v_i} \right| \leqslant 1$$

Plots of the Chebyshev filter function for various orders are shown in Figure 8.42. The ripple $(1-\gamma)$ is equal throughout the pass band and for this reason Chebyshev filters are known as equiripple filters. For $f = 0$ those filters whose order is odd have $|v_o/v_i| = 1$; even-order filters have $|v_o/v_i| = \gamma$ at $f = 0$. From a consideration of the filter function we see that

$$\gamma \leqslant \frac{v_o}{v_i} \leqslant 1 \qquad \text{i.e. } \gamma = 1/\sqrt{(1 + \epsilon^2)}$$

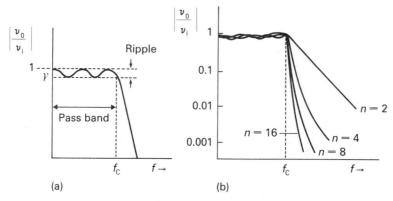

Figure 8.42 Chebyshev filter responses. Odd-order filters start at $|v_o/v_i| = 1$ as in (a). For even-order filters $|v_o/v_i|$ starts at γ; in (b) the ripple is about 2 dB

If the ripple is not to exceed 1 dB, for example, then $\gamma = 0.89$ and hence $\epsilon^2 = 0.2589$. If we once again use the circuit arrangements of Figure 8.39, which is a second-order system, then for a Chebyshev response

$$\left|\frac{v_o}{v_i}\right| = \frac{1}{\sqrt{\left[1 + 0.2589\left(2\left(\frac{\omega}{\omega_o}\right)^2 - 1\right)^2\right]}}$$

This will have its maximum value when $[2(\omega/\omega_o)^2 - 1] = 0$ i.e. $\omega = 0.707\omega_o$. When $\omega = 0$ or ω_o then $|v_o/v_i| = 1/\sqrt{1.2589}$, i.e. 0.89. The Chebyshev response is sketched in Figure 8.43 in which the left-hand ordinate shows the figures for the Chebyshev function.

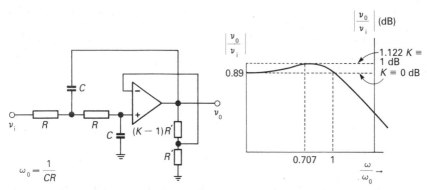

Figure 8.43 Second-order filter with Chebyshev response. (For calculation of K, and explanation of response curve, see text)

The transfer function for the circuit of Figure 8.43 is

$$F(s) = \frac{K}{s^2 C^2 R^2 + sCR(3 - K) + 1}$$

Therefore

$$|F(j\omega)| = \left|\frac{v_o}{v_i}\right| = \frac{K}{\sqrt{\left\{\left[1 - \left(\frac{\omega}{\omega_o}\right)^2\right]^2 + \left[\left(\frac{\omega}{\omega_o}\right)(3 - K)\right]^2\right\}}}$$

To make this correspond to a Chebyshev response we need to estimate a suitable value for K. The ordinate values also need to be changed. This is no problem since we may use a decibel scale and position the 0 dB point at any convenient situation. Noting that the maximum value of the Chebyshev function occurs when $\omega/\omega_o = 0.707$, then for our circuit

$$\left|\frac{v_o}{v_i}\right|_{max} = \frac{K}{\sqrt{[0.25 + 0.5(3 - K)^2]}} = 1.122K$$

This yields a value of $K = 1.96$; the values of R' and $R'(K - 1)$ for a Sallen and Key type second-order filter, having a Chebyshev response with a ripple not exceeding 1 dB, can then be estimated.

Bessel filters

When transient response is of prime importance the Butterworth and Chebyshev type filters are unsatisfactory. The overshoot with a step input signal for second-order filters is 4 per cent and 11 per cent for Butterworth and Chebyshev (0.4 dB ripple) filters respectively. The corresponding overshoots for sixth-order filters are 14 per cent and 21 per cent; for a sixth-order Chebyshev filter with a 1.6 dB ripple, the overshoot is 32 per cent. For best transient response we need a filter that has a phase shift in the pass band that is linear with frequency; the Bessel filter, though not ideal, comes closest to the necessary characteristics. Because this type of response approximates a constant time delay in the pass band, the filters handle transient waveforms with the minimum of distortion. Figure 8.44 shows the amplitude and phase response. We see that for higher order filters the response is well down at the cut-off frequency. The amplitude response in the pass band is therefore poor. For a second-order filter with a Bessel response, using the Sallen and Key filter of Figure 8.43, the gain. K, must be 1.27.

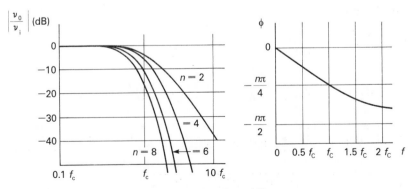

Figure 8.44 Amplitude and phase response for a Bessel filter

High-order filters

As can be seen from the sketches of the amplitude responses of the various types of filter, higher order filters are required when the cut-off needs to be sharp. As mentioned previously, prototype second-order sections are cascaded in order to produce high-order filters. When designing complete filters it is desirable to retain the same definitions for ω_o and Q for each section of the complete filter. We therefore have $\omega_o = 1/CR = 2\pi f_o$; f_o is the cut-off frequency ($=f_c$) for low- and high-pass second-order filters. For band-pass filters $f_o = \sqrt{(f_1 f_2)}$ where f_1 and f_2 are the upper and lower 3 dB points. Q for band-pass filters $= f_o/(3 \text{ dB bandwidth})$. (These points relating to Q have already been discussed in Part 1 of the book.) For low- and high-pass filters, Q is related to the damping factor; as previously stated, $Q = 1/2(\zeta)$ for second-order filters. For Butterworth responses, since all curves pass through the 3 dB point irrespective of n (the filter order), the resistors in every section may be identical, as may be the capacitors; hence $CR = 1/2\pi f_c$ where f_c is the desired 3 dB cut-off frequency. For orders of filter higher than 2 we do, however, need to adjust the gain for each section in order to obtain the correct overall response. To avoid the tedious analyses involved when high orders are involved, most designers consult tables of filter factors. An example is given in Table 8.1, which contains data supplied by Burr–Brown in their general catalogue. This firm produces high-grade instrumentation products, and we will have occasion below to refer to them again. In the case of Chebyshev and Bessel filters, not only is it necessary to trim the gain for each section but the time constant, CR, must also be adjusted for each section. Retaining the notation f_c for the cut-off frequency, then CR for a particular section is given by $1(2\pi f_c f_n)$ where f_n is a normalizing factor. The cut-off frequency is the -3 dB point for Bessel filters, while for Chebyshev filters f_c is the frequency at which the amplitude response passes through the ripple band and enters the stop band. If, therefore, we wish, for example, to design a six-pole low-pass Chebyshev filter we would cascade three second-order sections. Let us suppose that the ripple must not exceed 2 dB and that we are going to use the Sallen and Key low-pass circuit. It has been shown that $Q = 1/(3 - K)$, hence the gains for each of the three sections will be given by $K = 3 - 1/Q$, i.e. 1.891, 2.648 and 2.904 respectively. If the required cut-off frequency is f_c

Table 8.1 Low-pass filter parameters (figures are taken from Burr–Brown catalogue)

| Number of poles | Butterworth | | Bessel | | Chebyshev | | | |
| | | | | | 0.5 dB ripple | | 2 dB ripple | |
	$fn^{(1)}$	Q	$fn^{(1)}$	Q	$fn^{(2)}$	Q	$fn^{(2)}$	Q
2	1.0	0.70711	1.2742	0.57735	1.23134	0.86372	0.907227	1.1286
4	1.0	0.54118	1.43241	0.52193	0.597002	0.70511	0.470711	0.9294
	1.0	1.3065	1.60594	0.80554	1.031270	2.9406	0.963678	4.59388
6	1.0	0.51763	1.60653	0.51032	0.396229	0.68364	0.31611	0.9016
	1.0	0.70711	1.69186	0.61120	0.768121	1.8104	0.730027	2.84426
	1.0	1.93349	1.90782	1.0233	1.011446	6.5128	0.982828	10.4616
8	1.0	0.50980	1.78143	0.50599	0.296736	0.67657	0.237699	0.89236
	1.0	0.60134	1.83514	0.55961	0.598874	1.6107	0.571925	2.5327
	1.0	0.89998	1.95645	0.71085	0.861007	3.4657	0.842486	5.58354
	1.0	2.5629	2.19237	1.2257	1.005984	11.5305	0.990142	18.6873

(1) -3 dB frequency
(2) Frequency at which amplitude response passes through the ripple band.

then the time constants, CR, for each section are $1/(2\pi.0.31611f_c)$, $1/(2\pi.0.730027.f_c)$ and $1/(2\pi0.0.982828f_c)$ respectively.

It will be obvious from examination that the figures are produced by a computer, and therefore imply high arithmetic accuracy. In practice, component values are not known to this degree of accuracy; the final filters must therefore be trimmed to produce the desired characteristics. When fabricated in thin or thick film, laser trimming is often used.

High-pass filters can be realized using the Sallen and Key circuit. With reference to Figure 8.39, Y_1 and Y_2 are each capacitors ($=sC$) and $Y_3 = Y_4 = 1/R$. The capacitors and resistors are thus interchanged when changing from a low-pass to a high-pass filter. For Butterworth responses Q-values given in the table are also applicable to high-pass filters (and hence the K-values are unaffected) and $CR = 1/(2\pi f_c)$ as before. For Chebyshev and Bessel filters, however, the normalizing factors are the reciprocals of the low-pass normalizing factors, i.e.

$$f_n \text{ (high-pass)} = \frac{1}{f_n \text{ (low-pass)}}$$

Band-pass and band-stop filters can be made by cascading high-pass and low-pass filter sections. For band-pass filters, f_c (low-pass) $> f_c$ (high-pass); for band-stop filters, f_c (low-pass) $< f_c$ (high-pass). Very often, however, band-pass and band-stop filters are required to have a high Q-value. These are difficult to realize in practice because of the sensitivity of K and Q to component value tolerances. In these cases we really require the high Q, and comparative insensitivity to component values, that are typical of LC filters; unfortunately we are forced to use active filters when low frequencies are involved. The problem can be overcome by using a different circuit arrangement from that used by Sallen and Key. The circuit shown in Figure 8.45 is due to Delyiannis and is suitable for high-Q circuits (T.Delyiannis, 'High Q-factor circuits with reduced sensitivity', *Electronic Letters* 4 Dec. 1968, p.577). This circuit is another example of a voltage-controlled amplifier being energized from a voltage source. The Sallen and Key filters are also voltage-controlled, voltage-source (VCVS) filters.

By using the same techniques as those used previously we see that

$$(v_S - v_A)Y_1 = (v_A - v_o)Y_3 + (v_A - v_1)Y_2$$
$$v_o = (v_2 - v_1)A$$

where A is the magnitude of the open-loop gain of the IC.

$$v_2 = v_oR_3/(R_3 + R_4) = Mv_o$$
$$\text{where } M = \frac{R_3}{R_3 + R_4}$$

$v_o = (Mv_o - v_1)A$, i.e. $v_1 = Mv_o$ when $|A| \to \infty$.
$$v_AY_2 + v_oY_4 = v_1(Y_2 + Y_4)$$

From these equations v_A and v_1 may be eliminated, yielding

$$v_sY_1Y_2 = v_o[(Y_1 + Y_2 + Y_3)(MY_2 + Y_4[M - 1]) - Y_2Y_2 - MY_2^2]$$

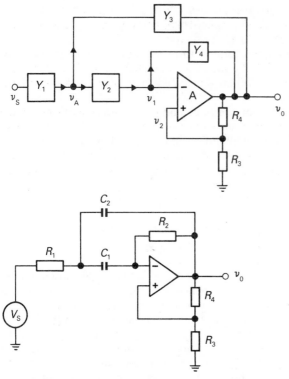

Figure 8.45 Active filter circuit (due to Delyiannis) for high Q-factors with reduced sensitivity to component tolerances

When $Y_2 = Y_3 = sC$, which is usually the case,

$$\frac{v_o}{v_s} = \frac{sCR_2}{(M-1)\left[s^2C^2R_1R_2 + sC\left(2R_1 + \dfrac{MR_2}{M-1}\right) + 1\right]}$$

from which we can see that

$$\omega_o = \frac{1}{C\sqrt{(R_1R_2)}} \text{ and } Q = \frac{\sqrt{(R_1R_2)}}{\left[2R_1 + \dfrac{MR_2}{M-1}\right]}$$

Having set ω_o with C, R_1 and R_2 the Q-value can then be set by M. A suitable potentiometer may be used for R_3, R_4. Although from the expressions it may seem desirable to make $R_1 = R_2$ we would lose some freedom of action. The sensitivity of ω_o and Q to component tolerances can be minimized by a suitable choice of R_1, R_2, R_3 and R_4.

Universal active filters

From a user's point of view, it is attractive to be able to have a single versatile circuit block that can be configured to give low- ,high- and band-pass filters with minimum alteration of associated components. Burr–Brown, National, and Analogue Devices all produce such filter blocks. Extensive design data are supplied to enable the designer to assemble a filter easily.

Universal active filters use the techniques of analogue computing to achieve the desired characteristics. In effect they operate in an analogous way on the differential equations that describe the filter performance. Usually the available blocks are intended to be configured as second-order filters; they therefore contain two integrators together with two other op-amps, which can serve as summers or sign reversers. Filters using this arrangement are also known as state-variable filters.

The basic configuration is shown in Figure 8.46. The method of analysis involves finding expressions for v_1 to v_5 and then obtaining the appropriate ratios (v_3/v_s, v_4/v_s and v_5/v_s) for the transfer functions.

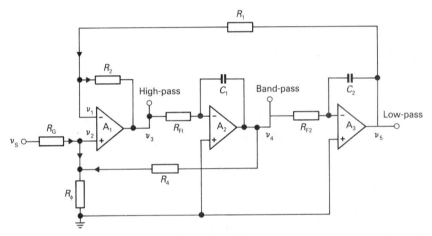

Figure 8.46 One configuration for a universal active filter based on the Burr–Brown UAF41 integrated circuit. These arrangements of operation amplifiers are known as state-variable filters

The necessary equations are

$$(v_s - v_2)Y_G + (v_4 - v_2)Y_4 = v_2 Y_Q$$

i.e.

$$v_s Y_G + v_4 Y_4 = v_2(Y_G + Y_Q + Y_4)$$
$$v_3 = A_1(v_2 - v_1)$$

From these we obtain

$$\frac{A_1(Y_1 + Y_2)Y_G}{Y_Q + Y_G + Y_4}v_s = (Y_1 + Y_2 + A_1 T_2)v_3 - \frac{A_1(Y_1 + Y_2)Y_4}{Y_Q + Y_G + Y_4}v_4 + A_1 Y_1 v_5$$

Since $A_1 Y_2 \gg (Y_1 + Y_2)$ we may write $(Y_1 + Y_2 + A_1 Y_2)$ as $A_1 Y_2$.

The A's in each term then cancel and we obtain

$$mv_s = Y_2 v_3 - nv_4 + Y_1 v_5$$

where $m = (Y_1 + Y_2)Y_G/(Y_Q + Y_G + Y_4)$ and $n = (Y_1 + Y_2)Y_4/(Y_Q + Y_G + Y_4)$.

Assuming that A_2 and A_3 are both large at the frequencies of interest, $v_4 = -v_3/(sC_1R_{F1})$ and $v_5 = -v_4/(sC_2R_{F2}) = v_3/(s^2C_1C_2R_{F1}R_{F2})$. Therefore

$$mv_s = v_3[Y_2 + n/(sC_1R_{F1}) + Y_1/(s^2C_1C_2R_{F1}R_{F2})]$$

and hence the transfer function for the first op-amp is

$$\frac{v_3}{v_s} = \frac{mC_1C_2R_{F1}R_{F2}s^2}{s^2C_1C_2R_{F1}R_{F2}Y_2 + snC_2R_{F2} + Y_1}$$

$$= \frac{mR_2s^2}{s^2 + \dfrac{nR_2}{C_1R_1}s + \dfrac{R_2}{C_1C_2R_{F1}R_{F2}R_1}}$$

$$= \frac{A_{HP}s^2}{s^2 + \dfrac{\omega_o}{Q}s + \omega_o^2}$$

From which we see that we have a high-pass filter with the following properties:

$$\omega_o^2 = \frac{R_2}{C_1C_2R_{F1}R_{F2}R_1}$$

$$Q = \frac{C_1R_{F1}\omega_o}{nR_2} = \frac{R_4\left(\dfrac{1}{R_Q} + \dfrac{1}{R_G} + \dfrac{1}{R_4}\right)}{1 + \dfrac{R_2}{R_1}}\sqrt{\left[\frac{C_1R_{F1}R_2}{C_2R_{F2}R_1}\right]}$$

$$A_{HP} = mR_2 = \frac{\left(1 + \dfrac{R_2}{R_1}\right)}{R_G\left(\dfrac{1}{R_Q} + \dfrac{1}{R_G} + \dfrac{1}{R_4}\right)}$$

Having obtained a high-pass filter voltage at v_3 we can easily see how to obtain the band-pass and low-pass outputs.

The numerator of the transfer function of a high-pass second-order filter contains an s^2 term. We can reduce this by one order by passing the high-pass output through an integrator since an integrator divides the its input by sC_1R_{F1}.

A further division by sC_2R_{F2} in a second integrator reduces the numerator of the function of v_3 to a term not containing s. We thus have an output from the first integrator, v_4, that has a band-pass characteristic, and simultaneously a low-pass output is available from the second integrator.

The gyrator

We have mentioned previously that high-Q LC filters can be designed so as to be relatively insensitive to departures of component values from their nominal values. For use at low frequencies, inductors are inconvenient; fortunately, now that operational amplifiers are easy and cheap to produce, we can simulate the properties of an inductor by using a capacitor and a couple of integrated circuits. Such a circuit arrangement is known as a gyrator, and is shown in Figure 8.47.

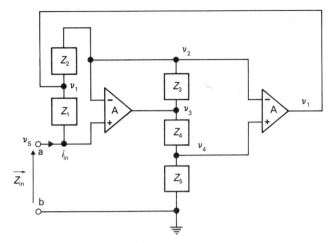

Figure 8.47 The gyrator circuit

In analysing the circuit we will assume that the open-loop gain of each amplifier is very large, and that each input terminal has an infinite input impedance. The aim of the analysis is to produce an expression for the input impedance $Z_{in} = v_s/i_{in}$. Taking the same point of view as that expressed in connection with the analysis of the circuit given in Figure 8.36, $v_2 = v_s = v_4$. (In almost all analyses involving op-amps we may assume that the difference between the voltages at the inverting and non-inverting terminals of an IC is zero.)

$$v_3 = \left(1 + \frac{Z_4}{Z_5}\right) v_s \therefore i_{Z_3} = \frac{v_s Z_4}{Z_3 Z_5} = i_{Z_2}$$

where i_{Z2} and i_{Z3} are the currents through Z_2 and Z_3 respectively. Hence

$$v_1 = v_s - Z_2 i_{Z_2} = v_s \left[1 - \frac{Z_2 Z_4}{Z_3 Z_5}\right]$$

Since $v_s - v_1 = i_{in} Z_1$

$$\frac{v_s}{i_{in}} = \frac{Z_1 Z_3 Z_5}{Z_2 Z_4}$$

If we make $Z_1 = R_1$, $Z_2 = R_2$, $Z_3 = R_3$ and $Z_5 = R_5$ we have

$$Z_{in} = \frac{R_1 R_3 R_5}{R_2 Z_4} = \frac{N}{Z_4}$$

$$\text{where } N = \frac{R_1 R_3 R_5}{R_2}$$

$$\text{Thus, if } Z_4 = \frac{1}{sC}$$

$$Z_{in} = sNC \equiv j\omega NC$$

An external circuit, therefore, 'looking into' terminals ab, appears to see an inductance of NC. We are thus able to design standard LC filters in which, at the lower frequencies, the gyrator replaces the inductor. It is very convenient to produce a gyrator in integrated circuit form since three of the resistors need not be accurately known; after selecting a suitable value for C, NC can be obtained accurately by using a single external trimming resistor.

Exercises

1. An IC op-amp together with feedback and input impedances may be represented by the circuit diagram shown in Figure 8.48. Shown that the gain of such an arrangement is given by

$$A' = -\frac{1}{Z_i} \left/ \left[\frac{1}{Z_f} + \frac{1}{A} \left(\frac{1}{R_{in}} + \frac{1}{Z_i} + \frac{1}{Z_f} \right) \right] \right.$$

where A is the open-loop gain and the other symbols have their usual meaning.

By making some simplifying assumptions, show how such an arrangement may be used:
   (a) to add two voltages, and
   (b) to integrate a voltage.

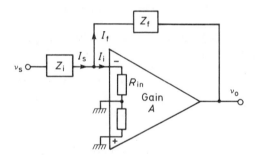

Figure 8.48

2. An analogue integrated circuit may be regarded as consisting of three stages of the type shown in Figure 8.15. The resistance of R_i is usually so large that it can be neglected. R_o and C for each section form a low-pass filter; each of the three filters have different time constants. Show that the overall open-loop voltage gain, A_{OL}, is given by

$$A_{OL} = \frac{A}{\left(1 + \dfrac{f}{f_1} \right) \left(1 + \dfrac{f}{f_2} \right) \left(1 + \dfrac{f}{f_3} \right)}$$

where A is the gain at zero frequency, f_1, f_2 and f_3 are the turning, or break, frequencies of the three low-pass filters respectively, and f is the frequency of operation. Plot the amplitude and phase of this expression against $\log f$, and compare the results with the graphs shown in Figure 8.14.

3. Figure 8.49 shows an arrangement that can be used as part of an integrated circuit amplifier. Find an expression for the ratio v_o/v_i. You may neglect v_{BE} and the base currents; assume that all of the transistors are identical.

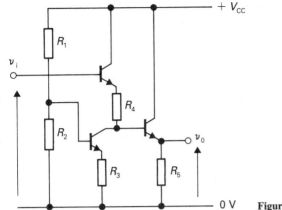

+ V_{cc}

v_i

R_1

R_4

R_2

v_0

R_3

R_5

0 V   **Figure 8.49**

4. An operational amplifier with an open-loop voltage gain of A, and infinite input impedance, has an input resistor and feedback resistor connected as shown in Figure 8.50. State expressions for (a) the closed-loop voltage gain, (b) the impedance between the point X and earth, and (c) the impedance between the point P and earth.

5. Figure 8.51 shows an arrangement of resistors and an operational amplifier. When one of the resistors is a transducer, the resistance of which varies with a physical quantity (e.g. a thermistor, light-dependent resistor or strain gauge), the arrangement forms the basis of a useful instrument. Assuming that the open-loop gain is very large, obtain an expression for the output voltage, v_0, as a function of R_1, R_2, R_3, R_4 and V. Suggest which resistor should be the transducer in order to have a linear relationship between the transducer resistance and the output voltage.

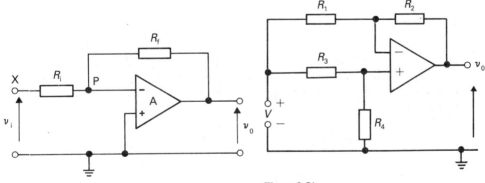

Figure 8.50   **Figure 8.51**

6. Design an inverting operational amplifier that has a gain of 20 and an input resistance of 22 kΩ. What value of resistor should be connected between the non-inverting terminal and ground? If the fractional change in the closed-loop gain is to be no greater than 0.001 of the corresponding fractional change in the open-loop gain, determine the minimum open-loop gain of the operational amplifier.

7. Figure 8.52 shows an operational amplifier being used as an integrator. Plot the frequency response of the integrator, bearing in mind that the open-loop gain is not infinite, but is equal to 60 dB. At what frequency will the actual output of the integrator differ from the ideal output by 1 per cent? (Hint: the input voltage at the 'virtual earth' point may not be taken as zero, but is $-v_o/A$.)

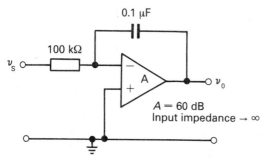

Figure 8.52

8. Figure 8.53 shows the circuit diagram of a dual-feedback active tuned circuit. If $C_1 = C_2 = 0.1\,\mu F$ and $R_1 = R_2 = 10\,k\Omega$ find the value of R_3 that gives a maximum circuit response at 22.5 Hz. What is the Q-value of the circuit at this value of R_3?

Ans.
$$TF = \frac{-s/CR}{s^2 + s\dfrac{2}{CR_3} + \dfrac{2}{C^2RR_3}}$$

($R_1 = R_2 = R$; $C_1 = C_2 = C$)
$R_3 = 1\,M\Omega$
$Q = 10/\sqrt{2}$

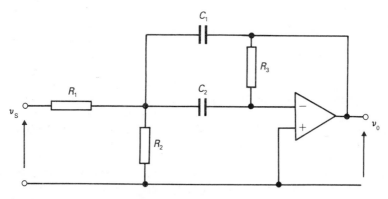

Figure 8.53

9. Modify the amplifier of Figure 8.39 so that it has a gain of K. Show that the resulting Sallen and Key filter has a transfer function given by
$$\frac{v_o}{v_s} = \frac{K}{s^2 C^2 R^2 + sCR(3 - K) + 1}$$
Discuss the consequences of varying K.

9

Harmonic oscillators

An oscillator is an instrument for producing voltages that vary in a regular fashion; the waveforms of the voltages are repeated exactly in equal successive intervals of time. In many cases the waveform of the output voltage is sinusoidal and the oscillator is then called a sine wave generator or harmonic oscillator. (Those instruments that produce repetitive waveforms that are square, triangular or sawtooth in shape are called relaxation oscillators. The term 'relaxation' is used because during the generation of the waveform there is a period of activity in which there is a sharp transition from one state to another. This period is then followed by a relatively quiescent one, after which the cycle is repeated.)

Oscillators can be constructed so as to operate at frequencies as low as one or two cycles an hour or as high as hundreds of megahertz. The selection of a suitable frequency or range of frequencies depends upon the function that the oscillator is required to perform. For the testing of equipment, or as a source of power for energizing a.c. bridges, the frequencies are usually in the l.f. or low r.f. range. Conductivity cells and electrolytic tanks are supplied with energy at frequencies of a few hundred hertz. Radio-frequency oscillators are widely used in the generation of carrier waves for telecommunication systems and in the construction of non-lethal e.h.t. supplies. Industrial heaters of dielectric materials such as wood, glue and plastics depend upon r.f. oscillators. Physiotherapy departments in hospitals use this type of heater in the treatment of bone and tissue disorders. Where the heating of electrically conducting material, such as metal ingots, is involved, induction coils are fed from power oscillators operating at lower frequencies. The material to be heated is placed within the coil and the eddy currents that are induced within the material cause rises in temperature. Both induction heating and dielectric heating have the advantage that the heating is produced within the bulk of the material. These methods of heating do not therefore rely on conduction from a hot surface layer.

For the applications described above, the waveform produced by the oscillator has usually to be sinusoidal or nearly so. In other applications such as cathode-ray oscilloscopes, television receivers, radar equipment, digital computers and automatic industrial controllers, relaxation oscillators are important and necessary sections of the equipment.

The most general method of producing sinusoidal oscillation is to use a feedback amplifier in which the feedback is positive at some desired frequency. The feedback circuit must therefore be frequency selective. From previous work we recall that the gain of a feedback amplifier is given by

$$A' = \frac{A}{1 - \beta A}$$

where A is the gain of the amplifier without feedback and β is the fraction of the output voltage fed back to the input. In the cases we have considered so far the feedback has been negative. Consideration of the gain formula for A' shows that with positive feedback A' is greater than A. (In preliminary discussion in this paragraph it is implicit that the quantities A and β are both real.) For values of βA less than unity the output voltage for a given input voltage increases and βA increases. When $\beta A = 1$, A' is theoretically infinite and any disturbance in the circuit is amplified, fed back in just the right phase to be further amplified, and so on until the amplifier is driven to its full extent. Infinite amplification is impossible in a practical case because of the limits set by the saturation and cut-off points of the transistors. However, even in absence of an input signal any disturbance within the amplifier is able to maintain itself. An amplifier that provides its own input in this way functions as an oscillator. By making the feedback circuit frequency selective we automatically arrange that $\beta A = 1$ at only one selected frequency. Oscillations of only this frequency are, therefore, produced and so the voltage waveform is sinusoidal. In practice, the non-linearity of the oscillator characteristic causes the output waveform to depart from a true sinewave., Non-linearity is unfortunatily essential for amplitude stability since it is impossible in practice to maintain precisely the condition $\beta A = 1$. Non-linearity is deliberately introduced, often in the form of a thermistor or small tungsten lamp, in the circuit that controls the gain of the amplifier.

The conditions under which oscillations are initiated and maintained require that any losses and attentuation in the feedback line must be made good by the amplifier. In mathematical terms the loop gain βA must be unity. The quantity βA is usually complex and is of the form $X + jY$. For oscillations to be maintained $\beta A = 1 + jO$. The real or ordinary part of βA must be unity and quadrature component must be zero. This is equivalent to saying that the feedback is positive. The condition $\beta A = 1 + jO$ is known as the Barkhausen criterion for oscillation and is often illustrated diagrammatically by a Nyquist plot. Figure 9.1 shows two such plots of the ordinary (real) part of βA against the quadrature (imaginary) component for all frequencies from zero to infinity. Figure 9.1(b) shows stable conditions and represents the position that must be achieved in a negative feedback amplifier. Since the plot does not enclose the point $(1, 0)$ there is no frequency at which feedback is positive and the loop gain high enough to sustain an oscillatory condition. Figure 9.1(c) shows the plot of an unstable feedback amplifier in which oscillations can be initiated and sustained. When using a feedback amplifier as an oscillator we deliberately ensure that $\beta A = 1 + jO$ at a selected frequency. This is achieved by using a suitable feedback circuit, examples of which are described later in the chapter. Arranging that the input and output voltages of the amplifier are in phase ensures positive feedback when there is no shift in the feedback path; alternatively a 180° phase shift in the amplifier requires a further 180° phase shift in the feedback circuit. The gain of the amplifier automatically adjusts itself until $|\beta A| = 1$ (i.e. the magnitude of the loop gain – written $|\beta A|$ – is unity). The self-adjustment is achieved by virtue of the non-linear gain of the amplifier.

The mode of operation of this type of oscillator is thus as follows. On switching on the instrument, it may be assumed that a switching surge is present at the output terminals. This surge is fed back to the input circuit via the feedback line and

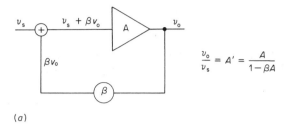

(a)

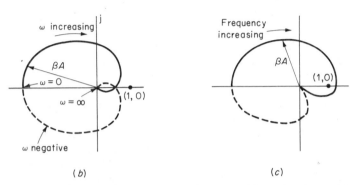

(b)   (c)

Figure 9.1 Nyquist plots of the ordinary component of βA against the quadrature component for all frequencies from zero to infinity: (a) General representation of a feedback amplifier; (b) stable conditions since the Nyquist plot of βA does not include the point (1,0). To satisfy the Nyquist criterion a closed loop must be obtained by introducing the concept of negative frequencies. Such frequencies have only a mathematical, not practical, reality; (c) unstable conditions since the closed loop encloses the point (1,0). Oscillations are initiated at that frequency which makes βA lie along the positive real axis. Oscillations are maintained when $|\beta A| > 1$ initially. The gain subsequently adjusts itself until $\beta A = 1$

subsequently amplified. The larger output is fed back again and further amplified. The result is that the natural responses of the closed-loop system gives rise to an increasing sinusoid at a given frequency until the amplitude is restricted by the non-linearity introduced by the amplifier characteristics and the amplitude control circuitry.

Oscillators for high frequencies

When an oscillator is required to operate at frequencies above about 100 kHz, the feedback and load circuits associated with the transistors consist of tuned circuits. A simple high-frequency oscillator is shown in Figure 9.2. Here we have a standard amplifier arrangement with a biasing potential divider and bypassed emitter resistor. The load, however, is a tuned circuit. To obtain a good sinusoidal waveform the Q-value of the tuned circuit must be as high as possible; this ensures that the oscillating currents in the LC resonant circuit are very much greater than the supply, i.e. maintaining, currents. In this context the resonant circuit is often called a 'tank' circuit. The tank circuit contains the large circulatory currents and is 'topped-up' by the maintaining current to make good the resistive and load losses.

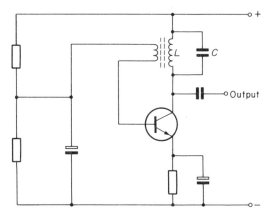

Figure 9.2 Tuned-collector oscillator

Feedback is obtained in this case by coupling the input base circuit to the tank circuit using transformer action. The feedback fraction is dictated by the magnitude of the mutual coupling between the primary coil of the tank circuit and the secondary coil in the base circuit. To avoid too much disturbance of the tank circuit the mutual coupling should be as small as possible consistent with maintaining oscillations.

Positive feedback is obtained by ensuring that the secondary coil leads are connected into the base circuit in the correct way. The output may be taken from the collector via a capacitor. Alternatively, a second secondary coil mutually coupled to the tank circuit may be used.

The design of transistor oscillators is complicated by the fact that loading of the tank circuit adversely affects both the frequency stability and the waveform. A common solution is to provide some sort of impedance isolation between the oscillator and the load. An emitter-follower may well be used since this circuit has a very high input impedance and its low output impedance enables a wide range of loads to be driven. The oscillator designer's worries are reduced by the use of integrated-circuit amplifiers that have been specifically designed for use at radio frequencies. An oscillator based on the S.G.S.L103T2 r.f./i.f. amplifier is shown in Figure 9.3. Here the tank circuit is connected to the high input impedance input terminals, and the low output impedance stage is able to drive loads of a few tens of ohms.

Two popular and well-tried oscillator circuits, that are basically the same, are due to Hartley and Colpitts. Figure 9.4(a) shows the basic arrangement. When Z_2 is a capacitor and Z_1 and Z_3 are inductors, then the arrangement is known as a Hartley oscillator; when Z_2 is an inductor and Z_1 and Z_3 are capacitors the circuit becomes a Colpitts oscillator.

The operation of both the Hartley and Colpitts oscillators may be understood by considering the basic circuit of Figure 9.4(a). Z_1, Z_2 and Z_3 form a resonant circuit in which quite a large oscillating current is circulating. The losses that inevitably occur are made up by the supply of alternating power from the active device. The circulating current in the resonant circuit is many times the supply current when Q-value is high. A tapping is provided along one of the reactive arms at the junction of Z_1 and Z_3. The voltages at each end of the tank circuit are necessarily 180° out of phase with each other relative to the tapping point. The collector

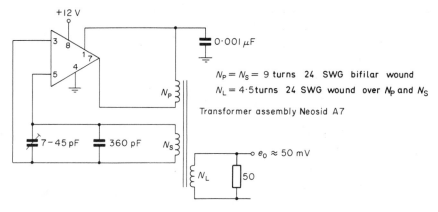

Figure 9.3 10 MHz oscillator using an S.G.S. L103T2 r.f./i.f. amplifier

voltage is 180° out of phase with the base voltage, so by connecting the tapping to the emitter and the two ends of the tank circuit to the collector and base, respectively, positive feedback is introduced into the amplifier, and oscillations are sustained when the loop gain is unity. The tapping point is chosen so that a suitable proportion of the voltage between the two ends of the tank circuit is applied between the base and emitter. As the tapping point is moved towards the collector end of the tank circuit the feedback is increased. Several factors such as the Q of the tank circuit, stray capacitances and the damping imposed by the base circuit and the load make a precise calculation of the tapping point difficult. It is usual to determine the best point experimentally; the analysis shown below gives good approximations to the impedance ratios, and thus indicates a suitable starting point.

In the Hartley circuit there is ambiguity about the tapping point. Although the tapping point on the coil is precise enough, there is also a hidden tapping point due to the stray capacitances that exist across L_1 and L_3 (of Figure 9.4(b)). For low radio frequencies the effect of strays can usually be neglected, but if the operating frequency must be high (say several megahertz), then the effect of the strays becomes important. Since the ratio of the stray capacitances is unlikely to be the same as L_1/L_3, uncertain operation can result. At high radio frequencies, therefore, many workers prefer the Colpitts circuit in which the tapping is effectively along the capacitor. Allowance can then be made for the strays, which are merely added to the two capacitors forming the tapping.

Both circuits can be analysed with the help of the simplified equivalent circuit shown in Figure 9.5. Y_1, Y_2, Y_3 and Y_{ie} are the admittances of Z_1, Z_2, Z_3 and h_{ie}, respectively.

$$i_1 + i_b - i = 0 \text{ and } h_{fe}i_b + i + i_2 = 0$$

Substituting the appropriate voltage-admittance products for the current terms

$$v_i(Y_1 + Y_{ie}) = (v_o - v_i)Y_2$$

hence

$$v_i(Y_1 + Y_2 + Y_{ie}) = v_oY_2$$

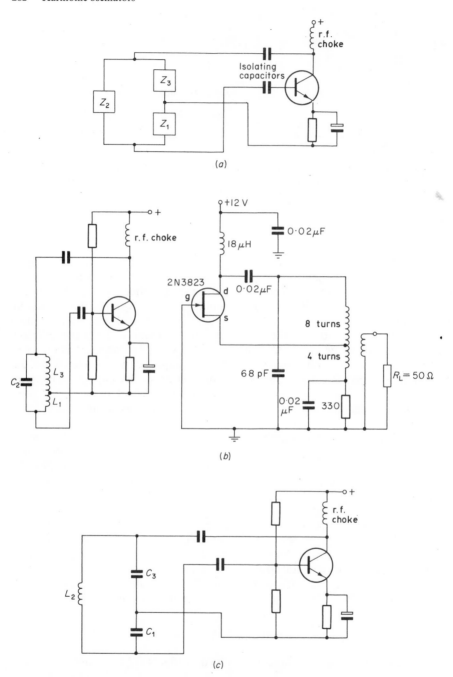

Figure 9.4 High frequency oscillators based on the circuits of Hartley and Colpitts: (a) Basic arrangement; the isolating capacitors have negligible reactance at the oscillator frequency. The bias conditions are fixed by components not shown; (b) two versions of the Hartley oscillator; (c) the Colpitts oscillator

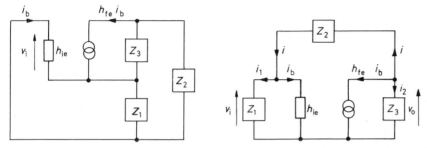

Figure 9.5 Simplified equivalent circuits for the basic arrangement of Figure 9.4(a). Figure 9.4(b) is merely the (a) arrangement redrawn to make the analysis more obvious. h_{oe} and h_{re} are considered small enought to be ignored in the equivalent circuit for the transistor

Also

$$v_i Y_{ie}(h_{fe} + 1) + v_i Y_1 + v_o Y_3 = 0$$

Substituting for v_o and cancelling v_i (since $v_i \neq 0$)

$$Y_2 Y_{ie}(h_{fe} + 1) + Y_1 Y_2 + Y_3(Y_1 + Y_2 + Y_{ie}) = 0 \qquad (9.1)$$

In the case of the Hartley oscillator $Y_2 = j\omega C$, and Z_1 consists of L_1 in parallel with the bias resistors. The relative impedances of L_1 and the bias resistors enable us to ignore these resistors in this analysis; hence $Z_1 = j\omega L_1$. Also Z_3 consists of L_3 in parallel with the collector load; the relative values in this case make us unable to neglect the effect of load initially. Assuming a resistive load, R, then $Y_3 = 1/R + 1/j\omega L_3$. Equation 9.1 becomes

$$j\omega C Y_{ie}(h_{fe} + 1) + \frac{C}{L_1} + \left(\frac{1}{R} + \frac{1}{j\omega L_3}\right)\left(\frac{1}{j\omega L_1} + j\omega C + Y_{ie}\right) = 0 \qquad (9.2)$$

Equating the ordinary (real) terms to zero

$$\frac{C}{L_1} + \frac{Y_{ie}}{R} - \frac{1}{\omega^2 L_1 L_3} + \frac{C}{L_3} = 0$$

hence

$$\omega^2 = \frac{1}{C(L_1 + L_3) + \dfrac{L_1 L_3}{R h_{ie}}}$$

Since $R h_{ie} \gg L_1 L_3$

$$\omega^2 = 1/[C(L_1 + L_3)]$$

hence the frequency of oscillation is given by

$$f = \frac{1}{2\pi \sqrt{[C(L_1 + L_3)]}}$$

In practice L_1 and L_3 constitute a single coil and therefore mutual coupling exists; each inductance value must then be increased by a factor M. Under these circumstances the frequency of oscillation is given by

$$f_o = 1/[2\pi\sqrt{\{C(L_1 + L_3 + 2M)\}}]$$

Where the collector load is an r.f. choke, this load is $j\omega L$, L being the inductance of the choke. An analysis can now be made by making the appropriate substitutions in Equation 9.1.

A similar analysis shows that for the Colpitts oscillator the frequency of oscillation is given by

$$f_o = \frac{1}{2\pi\sqrt{\left\{L\left(\dfrac{C_1 C_3}{C_1 + C_3}\right)\right\}}}$$

Crystal oscillators

There are several applications, notably in telecommunication systems and in laboratory frequency-standard equipment, where highly stable frequencies are required. Frequency drift in oscillators is due to changes in the resonant frequency of the tuned circuit resulting from the variations of component values with temperature. Oscillation at a frequency slightly different from the resonant frequency is caused by the components associated with the tuned circuit, such as leads to transistors, interelectrode capacitances and output loading coils. Changes in the parameters of these components contribute to frequency drift.

In nearly all cases the difficulties can be largely overcome by using tuned circuits with a high Q and constructed from stable components. With ordinary inductors and capacitors, however, Q values greater than a few hundred cannot be obtained; very large improvements in frequency stability can be obtained when a quartz crystal is used as the resonating element, in place of the conventional tuned circuit.

Quartz crystals exhibit piezoelectric properties, that is to say mechanical stresses imposed on the crystal give rise to potential differences across the faces of the crystal, and vice versa. Special cuts are needed relative to the crystallographic axes to produce the best performance. Quartz is chosen for oscillator frequency standards because this material is almost perfectly elastic; if mechanical oscillations are initiated it takes a long time for the oscillations to die away. Quartz crystals, therefore, have a very high mechanical Q. So far as the electrical properties are concerned, a quartz crystal is equivalent to the LC resonant circuit shown in Figure 9.6. The values of L, R, C_1 and C_2 depend upon the physical size of the crystal and the type of cut used. The crystal itself has conducting electrodes sputtered on to two crystal faces. Connecting leads are then joined to the sputtered electrodes. When the leads are connected to a source of oscillating voltage, mechanical vibrations are established in the crystal plate. Provided the frequency of the oscillating voltage is close to a resonant frequency of the crystal plate, the crystal forces the oscillating voltage to assume a resonant frequency determined by the plate. By using the crystal in place of an LC resonant circuit in an oscillator, the frequency is determined almost entirely by the crystal. Q-values in excess of 20 000 are easily obtained while values up to 0.5×10^6 can be achieved with care. The frequency stability of a crystal oscillator is therefore very high. The stability depends upon the temperature, but by using crystal cuts that exhibit extremely small temperature coefficients, frequency variations of no more than one part in 10^4 can easily be achieved. Enclosing the crystal in a thermostatically controlled oven improves the frequency stability considerably so that with care the frequency variation can be reduced to one part in 10^8.

The actual oscillator circuits follow much the same line as those for a conventional oscillator. Some examples are given in Figures 9.7 to 9.9.

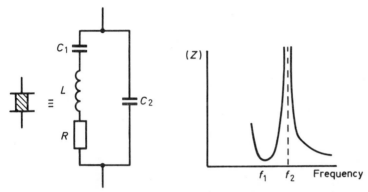

Figure 9.6 The circuit symbol for a crystal together with its equivalent circuit. The graph shows the variation of impedance with frequency in the region of series (f_1) and parallel (f_2) resonant frequencies. The values of L, R, C_1 and C_2 depend upon the individual crystal. One typical sample has values of $L = 5.2\,\text{H}$, $R = 280\,\Omega$, $C_1 = 0.01\,\text{pF}$ and $C_2 = 6\,\text{pF}$. The series resonant frequency is 698 kHz and the Q-value 81 400

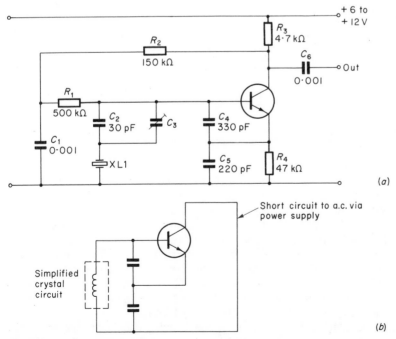

Figure 9.7 Crystal oscillator for use in the range 50 kHz to 1 MHz (a) Complete circuit diagram (b) Simplified equivalent a.c. circuit

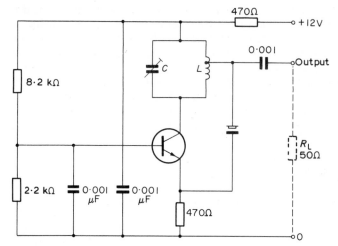

Figure 9.8 Crystal oscillator for use in the range 30 MHz to 200 MHz

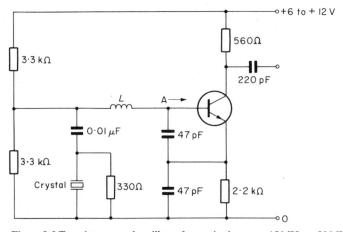

Figure 9.9 Transistor crystal oscillator for use in the range 15 MHz to 30 MHz

Figure 9.7(a) shows an arrangement that is in effect a Colpitts oscillator. R_1 and R_2 provide the bias current, but play no part in the alternating currents involved since there is effective decoupling via C_1 and isolation because of the large value of R_1. The emitter resistor, R_4, provides a d.c. path for the collector current, and forms part of the stabilizing circuitry, but has such a large value that most of the oscillatory component of the emitter current is carried by C_5. The small-valued collector resistor, R_3, does not affect the Colpitts mode of operation, but provides sufficient output voltage to drive later stages via C_6. The equivalent a.c. circuit, shown in Figure 9.7(b), makes the standard Colpitts arrangement obvious. The oscillator is suitable for use in the range 50 kHz to 1 MHz.

Figure 9.8 can be seen to be a variant of the Hartley oscillator, the feedback taking place at only one frequency when the crystal is being used in its series-resonant mode; in this mode the effective resistance is very low. The resonant circuit CL in the collector lead must be adjusted to match the operating frequency of the crystal. The coil should be tapped approximately 1/5th of the total number of turns from the $+ 12\,\mathrm{V}$ end. The circuit is intended for use between 30 MHz and 200 MHz. Figure 9.9 shows a transistorized crystal Colpitts oscillator for use between 15 MHz and 30 MHz. L should be adjusted to resonate at the crystal frequency with the input capacitance at A, i.e. if the crystal is shorted out the circuit should operate as an LC oscillator at approximately the crystal frequency. The circuit is not critical of transistor type; any modern silicon planar transistor for small-signal amplifiers should work satisfactorily.

Negative resistance oscillators

When a charged capacitor is discharged through an inductor the voltage across the capacitor does not fall exponentially unless the resistance of the coil is large. In practical inductors used for oscillators it is always ensured that the Q-factor is large; the resistance of the coil is therefore quite small compared with the reactance of the coil. With such small resistances the current in a parallel LC circuit does not decay exponentially, but oscillates with a decreasing amplitude. The voltage across the capacitor, therefore, has a waveform that corresponds to damped simple harmonic motion; i.e. it has the appearance of a sine wave whose amplitude decays exponentially. The cause of the decay is due to energy losses mainly in the resistance of the coil; at high frequencies part of the loss is due to electromagnetic radiation. We have, in previous sections, seen how these losses can be made good by supplying power in the correct phase from an amplifier. Another way of making good the losses is to place a negative resistance in parallel with the LC arrangement. The negative resistance cancels or neutralizes the positive equivalent resistance that represents the losses in the LC arrangement.

With a normal resistor (i.e. one having positive resistance) any rise in applied voltage is accompanied by a rise in current through the resistor. Devices that exhibit negative resistance are characterized by a fall in current as the applied voltage is increased; conversely, falls in applied voltage are accompanied by rises in current. Two semiconductor devices exhibit this effect over a limited range of applied voltage, namely a tunnel diode and a unijunction transistor. The unijunction transistor is used as a relaxation oscillator and is mentioned later in the chapter.

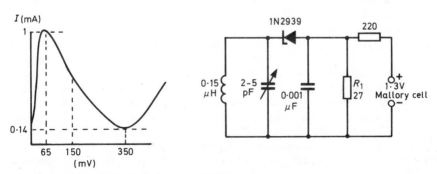

Figure 9.10 Approximate characteristic for the IN2939 tunnel diode and a practical circuit for generating oscillations in the region of 100 MHz

Sinusoidal oscillators can be obtained from a circuit that incorporates a tunnel diode. Typically tunnel-diode oscillators operate in the v.h.f. region, i.e. hundreds of megahertz. Figure 9.10 shows a tunnel-diode oscillator that operates at a frequency of about 100 MHz. For correct operation, the diode should be forward-biased about 150 mV. The bias supply should be stable, a convenient source being a Mallory type cell. The resistor R_1 must be adequately bypassed so that the tunnel diode is connected effectively across the resonant tank circuit. At 100 MHz or thereabouts a capacitance of 0.001μF is adequate. Connected in this way the negative resistance of the tunnel diode neutralizes the losses associated with the tank circuit.

RC oscillators

When oscillators are required to operate at frequencies below about 50 kHz, and especially at audio frequencies, it is inconvenient to use LC circuits. The size of the coil to obtain the necessary inductance is inconveniently large with the result that it is difficult to construct coils with a sufficiently high Q-value. Their bulk makes them unsuitable for use in transistorized equipment, and they are prone to pick up 50 Hz signals. At high frequencies a variable-frequency output is easily obtained by the use of a variable capacitor in the resonant circuit. At low frequencies it is not easy to construct capacitors of sufficiently large value that are also variable. For these reasons oscillators for use at low frequencies are based on combinations of resistance and capacitance.

The principle of operation is the same as that used in LC oscillators; a feedback amplifier is used in which the feedback line consists of a suitable frequency-selective network of resistors and capacitors. Two RC networks are in common use. They are the three-or four-section phase-shift network and the Wien bridge network.

Figure 9.11 shows a phase-shift oscillator using three RC sections in the feedback line. In order to have the collector voltage and base voltage in phase, it is necessary to have a 180° phase shift in the RC network. Since there is a 180° phase shift in the transistor, the collector voltage being 180° out of phase with the base voltage, a further shift of 180° in the feedback network brings the overall phase shift to 360°. It can be shown that a three-section RC network has a 180° phase shift at only one frequency, namely $1/\{2\pi\surd(6)CR\}$, and this, therefore, is the frequency of operation of the oscillator. The attenuation of the network is 29 so the amplifier must have a gain of at least this figure. It is quite easy to obtain a gain of 29 in a single stage. The phase advance for a single isolated RC loop is given by $\phi = \tan^{-1}(1/\omega CR)$ and a consideration of the appropriate phasor diagram shows that $0° < \phi < 90°$. The output voltage across R would, however be very small for values near to 90°, and figures of approximately 60° are used. When two identical CR networks are cascaded, the phase shift is not twice that of a single stage because the first section is loaded by the second. Since it is not possible to obtain a phase shift of 180° with only two sections, the minimum, and therefore usual, number of sections to use is three. More sections can be used; for example, if four sections are used the attenuation at a frequency that gives a 180° phase shift is 18.39 so the gain of the amplifier need not be greater than this. The operating frequency becomes $\surd0.7/(2\pi CR)$. The marginal drop in the attenuation seldom justifies the use of the fourth section.

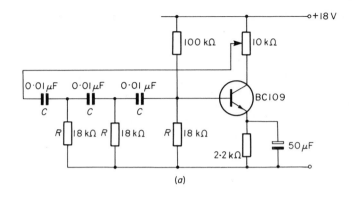

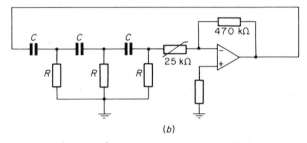

Figure 9.11 Three-section phase-shift oscillators. (a) A three-section phase-shift oscillator. If the resistors in the phase shifting network are altered, care must be taken to preserve the correct biasing of the transistor. The transistor input and bias impedances modify the frequency of operation which departs from $1/[2\pi(\sqrt{6})CR]$. The potential divider in the collector circuit allows the voltage fed back to be adjusted for the best waveform; (b) an IC arrangement

Where the frequency of the oscillator voltage is to be variable, simultaneous adjustment of all the resistors or capacitors in a phase-shift network is not usually convenient. It is possible to use ganged capacitors consisting of three capacitors adjusted by the same spindle; such capacitors are constructed for radio purposes, however, and the maximum capacitance of any one section is not usually greater than 500 pF. Where this value is too small it is more convenient to use a Wien bridge oscillator in which the capacitor values are selected by switches and fine frequency variations obtained by using ganged variable resistors. Since only two resistors are involved, ganged components can be readily obtained.

The principles of a Wien bridge oscillator are shown in Figure 9.12(a). The voltages A and B are in phase at only one frequency given by $f = 1/2\pi CR$. If now an amplifier with an even number of stages is used in connection with the bridge, a Wien bridge oscillator results. The gain of the amplifier must make up for the attenuation in the bridge network.

$$B = \frac{Z_2}{Z_1 + Z_2} A$$

Therefore

$$\frac{A}{B} = 1 + \frac{Z_1}{Z_2}$$

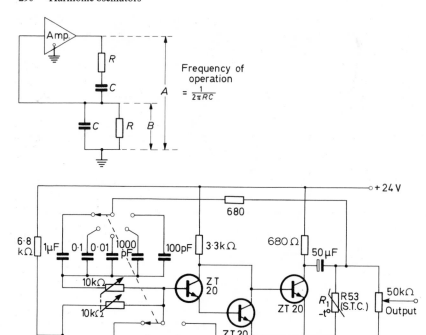

Figure 9.12 Wien bridge oscillator

where

$$Z_1 = R + \frac{1}{j\omega C}$$

$$Z_2 = \frac{R}{1 + j\omega CR}$$

Thus

$$\frac{A}{B} = 1 + \frac{(R + 1/j\omega C)(1 + j\omega CR)}{R}$$

$$= 3 + j\left(\omega CR - \frac{1}{\omega CR}\right)$$

A and B are in phase when the j-term is zero, i.e. when $\omega = 1/CR$. The frequency of oscillation is given by $f = 1/2\pi CR$. The attenuation is then 3. A and B are in phase because A/B is positive.

By using an even number of stages the output voltage is in phase with the input voltage. Connecting the bridge to the input of the amplifier as shown produces positive feedback at one frequency and oscillations at that frequency are sustained when the gain exceeds three.

Figure 9.12(b) shows a practical Wien bridge oscillator with variable frequency output. The first stage consists of a Darlington pair. The current gain and input impedance are high, which is desirable when operating this type of oscillator. The high input impedance ensures that the lower half of the Wien bridge is not upset by being loaded. To ensure that the voltage amplification is independent of frequency, negative feedback is introduced into the amplifier. A high initial gain is needed before the application of the feedback to make up for the fall that occurs when negative feedback is applied. A further advantage of the use of negative feedback is the low output impedance obtained; the loading effect of the Wien bridge is then minimized. The feedback is provided in this case by R_1 and R_2. A fraction of the output voltage is fed back in series with the emitter of the first stage.

By using a thermistor (S.T.C. type R53) for R_1, amplitude control is achieved. Any tendency for the output voltage to change is counteracted in the following way. A rise in output voltage causes an increase in the current through R_1 and R_2. The resistance of the thermistor, a small glass encapsulated bead type, decreases because of the inevitable rise in temperature due to the increased current. The feedback fraction thus increases which automatically reduces the gain; hence the output of the amplifier is reduced to almost its former value. Decreases in voltage output cause a rise in the resistance of the thermistor and a fall in the fraction of the voltage fed back to the emitter. A rise in gain results, which returns the amplitude of the output voltage almost to its former value. With the capacitor values given the frequency ranges covered are 15–200 Hz, 150 Hz–2 kHz, 1.5 kHz–20 kHz, 15 kHz–200 kHz, 150 kHz–2 MHz. The output voltage is about 1 V (r.m.s.). For a change in supply voltage of 4 V, the change in output voltage is less than 1 per cent and the change in frequency less than 2 per cent.

Figure 9.13 shows a version using an IC.

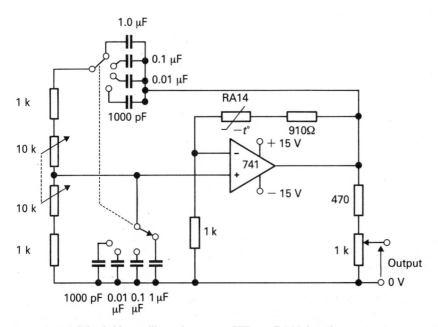

Figure 9.13 A Wien bridge oscillator that uses an ITT type RA14 thermistor

Low-frequency generators

When the required sine wave must have a very low frequency, say in the region of 1 Hz or less, the transistor circuits given so far are not satisfactory. The circuits must have a good response down to zero frequency together with satisfactory gain stability. It is almost impossible to arrange these with single transistor stages, and even with the more complicated circuit of Figure 9.12 the a.c. coupling in the last stage would prevent operation at very low frequencies. A solution to the problem of generating very low frequency sine waves can be obtained using analogue-computer techniques.

An analogue computer is an arrangement of d.c. amplifiers that can operate on voltages in a way that is analogous to the operations that could be performed by a mathematician using pencil and paper. Basically it is a machine for solving differential equations. We have already seen how ICs can be used to perform simple integration. A combination of two integrators and a sign reverser can be made to generate sine waves; since d.c. feedback amplifiers are used the sine waves can have a very low frequency.

Initially we note that the differential equation that has a sine wave solution is that representing simple harmonic motion, i.e.

$$\frac{d^2x}{dt^2} = -\omega^2 x$$

Consider now the arrangement shown in Figure 9.14(a). Here we have two cascaded integrators so that the output of the first integrator is proportional to the integral of the input voltage, e_i, and the output of the second integrator is

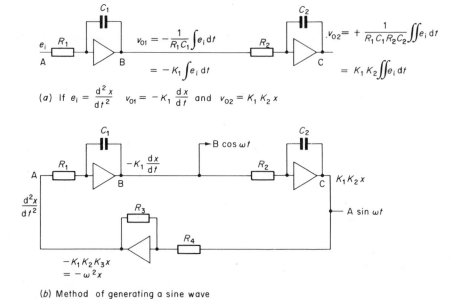

(a) If $e_i = \dfrac{d^2x}{dt^2}$ $\quad v_{01} = -K_1 \dfrac{dx}{dt}$ and $v_{02} = K_1 K_2 x$

(b) Method of generating a sine wave

Figure 9.14 Generation of low-frequency sine waves using operational amplifiers

proportional to the double integral of e_i. Hence if e_i is proportional to the second differential coefficient of x, then the output of the first integrator is proportional to dx/dt, and the output of the second integrator is proportional to x. It can be shown that a solution of the equation $d^2x/dt^2 = -\omega^2x$ is given by $x = A \sin \omega t$, hence if we can generate a voltage proportional to x, we have generated a sine wave; furthermore by arranging the circuit constants appropriately we can choose any practical value of ω and hence the frequency of the sine wave. Figure 9.14(b) shows how this is done. Assuming initially that a voltage analogous to d^2x/dt^2 is present at point A, the voltage at B is $-K_1 \, dx/dt$, and the voltage at C is $+K_1K_2x$, where $K_1 = 1/R_1C_1$ and $K_2 = 1/R_2C_2$. However, to obtain the voltage at A we must reverse the sign of the voltage at C and multiply it by a constant K_3 so that $K_1K_2K_3 = \omega^2$. The output from the sign reverser is then fed into the first integrator. Provided we are not concerned about initial conditions (i.e. that $A \sin \omega t$ must be a specified value, e.g. zero at time $t = 0$) then when the circuit is switched on, switching transients will initiate the action and very quickly the circuit will produce a sine wave at the point C. Additionally a cosine wave, i.e. a sine wave with a phase difference of $90°$, is produced at B.

To calculate the values of circuit components we may consider an example. Let us suppose that it is required to generate a sine wave having a frequency of $1\,\mathrm{Hz}$. This means that $\omega^2 = 4\pi^2$, hence $K_1K_2K_3$ must be equal $4\pi^2$. An intelligent initial guess can be made at the components in the integrators. Assume initially that $R_1 = R_2 = 1\,\mathrm{M}\Omega$ and $C_1 = C_2 = 1\,\mu\mathrm{F}$. Both these values can be obtained without any practical difficulty. Then $K_1 = K_2 = 1$ and $K_3 = 4\pi^2 = 39.5$; this means that $R_3/R_4 = 39.5$ therefore if $R_4 = 100\,\mathrm{k}\Omega$, R_3 must be $395\,\mathrm{k}\Omega$. The accuracy of the frequency generated depends upon the accuracy of the component values. For many purposes a value for R_3 of $390\,\mathrm{k}\Omega$ may well suffice. It is of course possible to make R_3 variable and adjust its value to give precisely the required frequency. It is not difficult to generate sine waves of different frequencies by having R_3 adjustable. The range of frequencies will, however, be limited since R_3 cannot assume all values from zero to infinity. However, by switching in different values for R_1, R_2, C_1 and C_2 a wide range of frequencies can be generated. The problem of maintaining the amplitude constant as the frequency is varied can be solved by feeding the output to a subsidiary feedback IC amplifier arrangement. By having a ganged control, adjustments of R_3 that result in changes of gain in the sign reverser can bring about compensating changes of gain in the subsidiary amplifier.

Exercises

1. By referring to Figures 9.4 and 9.5 show that the Colpitts oscillator will oscillate at a frequency given by

$$f_o = \frac{1}{2\pi \sqrt{\left[L_2 \dfrac{C_1C_3}{C_1 + C_3} \right]}}$$

2. Find expressions for the frequency of oscillation and the gain necessary to sustain oscillations for the circuit of Figure 9.12 when the capacitors and resistors in the frequency-determining arms are not equal.

3. Show how a four-section CR network can be used as a feedback path to produce oscillations of a specified frequency in an amplifier.

 Derive the conditions for maintaining oscillations of a sinusoidal waveform at a specified frequency. (Gain $= -18.39$; $f = \sqrt{0.7}/(2\pi CR.)$

Part 3

Digital electronics

Chapter 10

Fundamentals

The progress of modern industry depends to a large extent on our ability to introduce an increasing amount of auotmatic control into our industrial processes. Profitable fields for the introduction of automation techniques are the assembly of machine parts, the precision drilling and cutting of raw materials, the automatic sequencing of sorting, weighing, checking and storing of raw materials, and the automatic control of chemical and other processes. To an increasing extent the routine, and often mundane, work once performed by human beings is now being undertaken by machines. The digital computer is a good example of a sequencing machine that performs commercial and scientific calculations. We are already at the stage where machines are superior to man in many fields because of their ability to work to precise limits for long periods without fatigue. The speed of computer calculations is such that no human being could compete successfully with an electronic digital computer.

With all industrial control apparatus we must consider the problem of reliability. If, between two specified limits, a machine must distinguish accurately between two closely spaced states then the selection is not usually reliable. For example, let us suppose that a process depends upon measuring the collector current of a transistor. If the limits of the current were 0 and 10 mA, and within these limits we had accurately to distinguish between 5 and 6 mA, we would find that the control process was not very reliable one. Several factors, such as ageing of components and variations of the power supplies, could introduce errors into our system. For this reason control systems are designed to have only two possible states, namely 'on' and 'off'. The system has then to detect only when a signal is in the on or off state; it is easy to do this with a high degree of reliability.

A well-known on/off device for controlling industrial processes is the electromechanical relay. For over a hundred years relays have been the standard equipment used for routeing telephone signals. During the Second World War the design of switching networks for the selection and routeing of signals became very complicated. In an attempt to optimize a given system (i.e. use the minimum number of relays to perform a specified task) it was realized that the techniques involved were almost identical to those devised by the Irish mathematician, George Boole, for the symbolic representation of logic. His original efforts were directed to finding a symbolic algebra that could be used to represent arguments in philosophy and logic. His paper, published in 1847, dealt with the mathematical analysis of deductive reasoning and formulated relationships between the true and false propositions of his contemporary logicians. His ideas were not seen to have any practical application until 1938 when Shannon published a paper entitled 'Symbolic Analysis of Relay and Switching Circuits'. In the following years it was realized that digital computers were a practical possibility.

The starting point in the design of any digital system is the specification setting out the operations that are to be performed. The specification for each operation, or set of operations, is met by designing a suitable unit. Logical design consists of deciding how these units can best be interconnected to meet the specification. The design is efficient if it uses the minimum possible number of units. In the design of complicated systems a mathematical representation is convenient since it provides a shorthand by which the operational processes can be described. Such a shorthand can be very important because the design can follow mathematical lines and lead to simplifications that cut the cost of the installation. In many cases, these simplifications are by no means obvious from a consideration of the circuit configuration alone.

Boolean algebra is the mathematical method used in the design and optimization of digital systems. If the reader is going to use, design or discuss digital control systems, he must obviously learn the basic language first. He can then proceed to some of the specialized books that are entirely devoted to the subject.

Boolean algebra

In this form of algebra the symbols are borrowed from ordinary algebra, but we must rid our minds entirely of all the ordinary mathematical meanings. This form of algebra is concerned with the relationship between classes or sets, each class being represented by a symbol or letter of the alphabet. The symbol tells us nothing about the size or magnitude of the class, since size is irrelevant in Boolean algebra. We may make statements about any class, and these statements must be true or false. No half-truths or shades of meaning are allowed. This corresponds to our on/off concept. A switch is either on or off. We cannot have any intermediate state such as 'half-on' or 'nearly-off'. Each statement is given a truth value depending upon whether the statement is true or false. If it is true the statement is given a value of 1; if it is false we assign a value of 0 to the statement. In electronic logic systems the truth value is recognized as a voltage level in the circuit. Only two voltages are defined viz. the positive supply voltage and zero voltage. It is at our disposal to define one of them as logic 1 and the other as logic 0. When we define the positive supply voltage as logic 1, and zero voltage as logic 0, we are then said to be using positive logic. This is very convenient when npn transistors are used. The vast majority of integrated digital circuits use positive logic, and so throughout the text we shall assume that logic 1 is the positive supply voltage and logic 0 is zero volts. Logic 1 can therefore be referred to as the 'high' state and logic 0 as the 'low' state.

Logic systems are arrangements of electronic units that can be made to perform a specified function. When the end function (i.e. the task to be performed) depends only upon the states of the inputs, the logic is said to be combinational. We may, for example, want a lathe chuck to turn only when the workpiece is correctly positioned, a safety cage is closed and the machine has been switched on. All three input conditions must be satisfied before the lathe motor will turn. These conditions are known as the inputs. Such systems can be constructed by using special kinds of electronic switches. For reasons given later these switches are known as gates. Apart from power supplies and interfacing components, such as relays for example, gates are all that are required to build combinational logic circuits.

A second type of logic system is known as a sequential logic circuit. In this system the output is a function not only of the inputs at the time, but also of input conditions that have subsequently been removed. Apart from using logic gates, some form of memory must also be built into the system.

Combinational logic systems can be cconcisely described in terms of Boolean variables. Thus we may let the Boolean variable A present the statement 'The safety cage is closed'. If this statement is true then A = 1; if not, A = 0.

We may consider more than one statement simultaneously and decide whether the combination is true or false. Two, or more, statements may be combined to form one proposition by the use of the connectives AND and OR. For example let

A = the workpiece is correctly positioned
B = the safety cage is closed
C = the power switch is on

If each statement is true we write

A.B.C = 1

where the dot signifies the AND combination. Often, where no ambiguity arises, the dot may be omitted. The AND combination of two or more logic propositions is known as the 'logical product' (hence the dot) and is true only if every proposition is true.

It is possible to use the OR connective in two ways, hence there are two types of OR connective. For example, a child possessing 50p may buy an article costing 10p OR a different article costing 20p; he may, of course, buy both articles. Used in this way the connective is called an Inclusive-OR. Alternatively, the child possessing 50p may buy one article costing 40p OR a different article costing 30p, but he cannot buy both. Used in this way the connective is called an Exclusive-OR.

In logical propositions a combined statement using an Inclusive-OR is true if any one of the component or both statements is true. For example, if

C = rain is wet
D = snow is black

the statement 'rain is wet OR snow is black' is true because C is true (i.e. C = 1), although D = 0. We write this as C + D = 1

The use of the plus sign for the OR connective, rather than the AND connective, can be confusing for the beginner. Indeed, different symbols have been used by mathematicians and others, but control and electronic engineers have decided to adopt the dot and plus signs. The dot, representing a logical product (for the AND connective), and the plus sign representing a logical sum (for the OR connective), enable the familiar associative and distributive laws of algebra to be obeyed.

The Exclusive-OR is distinguished by using the symbol $\oplus$. Thus E $\oplus$ F = 1 if E = 1 or F = 1, but not both simultaneously.

If we wish to negate a statement we write a bar over the symbol representing the statement, for example:

G = snow is white

therefore

$\bar{G}$ = snow is not white

We thus have

$G = 1$, $\bar{G} = 0$ and $C + \bar{G} = 1$

i.e. it is true to say that 'rain is wet OR snow is not white' because one of statements is true.

A bar over the whole statement negates the statement; thus if

$A + \bar{C} = 1$ then $\overline{A + \bar{C}} = O$

Some of the logical identities given below are useful to those involved with logic systems.

$AB = BA$ $ABC = (AB)C = A(BC)$
$A1 = 1$ $AA = A$
$A0 = 0$ $A + A = 1$
$A(B+C) = AB+AC$ $A + 1 = 1$
$A + 0 = A$ $\bar{0} = 1$
$\bar{1} = 0$ $\bar{\bar{A}} = A$
$A\bar{A} = 0$

De Morgan's Theorem states that $\overline{AB} = \bar{A} + \bar{B}$ and $\overline{A + B} = \bar{A}\,\bar{B}$.

Any of the above identities can be proved by constructing a truth table. This type of table shows the logic state for every possible combination of input states. Examples are given in the next section.

Relay representation of the connectives

We commence our introduction to logic systems by considering the effect of opening and closing switches in simple circuits. In Figure 10.1, when the relay coil is energized, contacts are made to close, thus starting some system, say a motor that pumps liquid into a tank. Before the relay coil is energized two conditions must be satisfied, namely switch A AND switch B must be closed. Switch A may be a float switch controlled by the level of liquid in the tank, and switch B by a temperature-sensitive device. When switches A and B are open they may be said to be in the 0 state, and when they are closed they are in the 1 state. Both A and B must be in the 1 state for the relay to be energized. Thus A.B = 1 represents the position when the control system is set in motion. If A = 0 or B = 0 then the system will not operate. When A = 0 it is false to say that the liquid level is at a specified position. Similarly, B = 0 means that the temperature is not correct. A truth table is often used to enable us to see at a glance the output condition for any combination of states of A and B. This shown in Figure 10.1(a). There are only four different combinations possible with the two variables A and B. Only when A and B are 1 can the output (A.B) be 1.

In Figure 10.1(b) we have the OR combination. If either or both switches are in the 1 (i.e. on) state, the digital system is energized. As in the former case, we are assuming that with the relay not energized the contacts are normally open, i.e. in the 0 state; only when the relay coil is energized will the relay contacts close and come into the 1 state. The appropriate truth table is shown alongside.

Figure 10.1(c) shows the NOT combination. The relay contacts are normally closed, i.e. in the 1 state, when A is in the 0 state. On closing switch A (A = 1) then the output contacts are open, i.e. NOT closed.

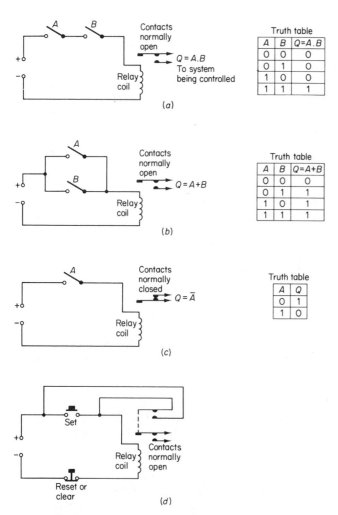

Figure 10.1 The use of the electromagnetic relays to realize the basic Boolean functions AND, OR and NOT. The useful MEMORY or HOLD circuit is also shown

Although not a logical connective, the MEMORY or HOLD circuit of Figure 10.1(d) is a useful, and basic, one in digital systems. In this operation we require that an impulse on the SET switch should energize the relay and that the relay coil should remain energized after the pulse has disappeared. This is achieved by having a second set of contacts on the relay wired across the SET switch. On closing the SET switch the relay is energized and both relay contacts are closed. Now when the SET switch opens there is still a path for the energizing current via the auxiliary relay contacts. In order to reset the system to its former state it is necessary to have a RESET switch, which is normally closed, in the position shown. On pressing the RESET switch the relay coil is no longer energized and the contacts open. There is now no longer a path via the SET switch or auxiliary contacts, and so the relay remains de-energized.

Transistor switches

Switches that depend upon electromechanical relays are suitable only for slow-speed systems; operating speeds of 1 to 10 ms are common. Because of their physical size, slow speed, vulnerability to wear in their mechanical moving parts, high power consumption, and somewhat unreliable performance after a period in service, electromechanical relays are being replaced in digital systems by semiconductor switches. By comparison, transistors are small, light in weight, and inexpensive; they have no moving parts, an almost indefinitely long life, and can switch at rates up to hundreds of megahertz.

An ideal switch is one which has zero resistance between its terminals when in the on position, an infinitely great resistance when off, and can switch from one state to the other in zero time. Clearly, under these conditions there can be no power dissipated in an ideal switch. The characteristics of a modern switching transistor are close enough to the ideal to make such a device an excellent logic switch. Constant research and development in transistor fabrication enable us to approach ever closer to the ideal switching characteristics.

Figure 10.2 shows a transistor connected in the common-emitter mode together with the corresponding characteristics and load line. We are not concerned here with those features which were important for amplifier action, such as linearity of characteristics and the precise position of the operating point. Instead we are interested in the behaviour of the transistor corresponding to points X and Y on the load line. When the input terminal A is connected to the zero voltage line no base current flows and the transistor is cut off. The operating point on the load line is then at X, and the voltage across the transistor, v_{CE}, is equal to the supply voltage. With modern silicon transistors the leakage current is so small that for all practical logic purposes the transistor is equivalent to an open switch. On transferring the input terminal to the positive supply line, base current flows. The magnitude of the base current depends upon the supply voltage and the base resistor R_B. Provided R_B is low enough, for a given supply voltage, sufficient base current can flow to saturate the transistor and the operating point moves up to position Y. The collector voltage falls to a low value, being little greater than 150 mV (this is some hundreds of millivolts smaller than v_{BE}); the transistor is now equivalent to a closed switch.

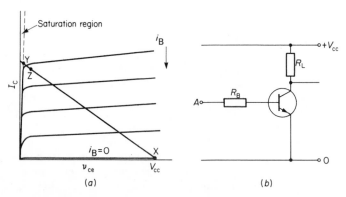

(a)   (b)

Figure 10.2 The use of the transistor as a switch

The speed at which the transistor can be switched is limited by charge storage effects. In the position corresponding to Y on the load line, holes are injected into the base via the base lead and, because of transistor action, a large number of electrons are injected into the base from the emitter region. In normal amplifier action the collector–base junction is reverse-biased to a comparatively large value; the collector–base voltage, v_{CB}, may be as much as one-third to one-half of the supply voltage. The charge carriers are therefore readily swept into the collector region. In switching action, however, when the transistor is heavily saturated, the collector–base voltage is not sufficiently large to remove the electrons quickly, and electron storage occurs in the base region. Before we can turn the transistor off, and reduce the collector current to zero, it is first necessary to sweep the stored excess charge out of the base region. The time taken to do this is one of the major factors that determines the switching time. Another major factor is the geometry of the base region. In a modern silicon high-frequency transistor, the base region is very thin and consequently there is little volume in which to store excess charge. The problem of sweeping out excess charge is thus alleviated.

An improvement in switching times can be achieved if the transistor is prevented from saturating when operating in the on state. We then avoid the problems associated with charge storage in the base region. Unfortunately it is not possible to avoid the saturation region by increasing R_B. While this latter course would restrict the base current, and hence the collector current, the spread of transistor characteristics in different samples of the same nominal transistor would make it impossible to arrive at a satisfactory design value for R_B. The prevention of saturation can be achieved, however, by connecting a diode between the base and collector, as shown in Figure 10.3(a). When the transistor is switched on, the collector current rises until $v_{CE} = v_{BE}$. At this point the diode is on the verge of conduction. A further rise in collector current makes $v_{CE} < v_{BE}$ and the diode conducts. Provided the diode voltage drop is less than that of the collector–base diode, the transistor cannot saturate. Further increases in the base current via R_B are bypassed through the diode. The transistor operating point then corresponds to the point Z in Figure 10.3(a). It will be appreciated that an ordinary silicon diode is unsuitable for this purpose since charge storage can take place in the depletion layer of its pn junction. The type of diode required for this application is known as a Schottky diode, a device that consists of a metal–semiconductor junction. In the conducting state the voltage drop in such a device is less than that in a silicon pn junction; additionally there is no charge storage, and hence switching times are

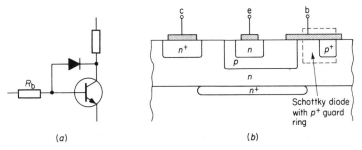

(a)   (b)

Figure 10.3 Schottky transistor – a non-saturating transistor switch. (a) Shows the circuit of the combined transistor and Schottky diode, and (b) shows how the circuit is realized using planar epitaxial techniques

reduced. In practice the use of a separate diode is unnecessary since, with planar epitaxial techniques, it is possible to fabricate the Schottky diode alongside the transistor. The construction is shown in Figure 10.3(b), where it will be seen that the diode is formed by extending the metal that makes contact with the base region. The diode itself therefore consists of the metal in contact with the n-type material of the collector. A guard ring of p-material effects the necessary isolation. The composite arrangement is often termed a Schottky transistor.

Logic circuitry terminology

Logic circuitry has, unfortunately, many associated jargon terms, some of which are explained in this section.

A logic switch is known as a 'gate'. This analogy is obvious enough in that an open gate allows the passage of information, whereas a closed gate does not. Similarly, a switch that is on will allow signal information to pass through it whereas a switch that is off will not.

Irrespective of the type of electronic circuitry that is used, all gates of the same class perform the same logic functions. In logic circuitry there are three basic types of gate, corresponding to the three Boolean functions AND, OR and NOT. The electronic circuitry that performs these functions is based on integrated circuit amplifiers that are operated as switches. Many gates combine the NOT function with the AND function. The result is a NOT AND gate known as a NAND gate. Similarly, we may design NOT OR gates; these are known as NOR gates. The various symbols used for these gates in logic circuit diagrams are shown in Figure 10.4 (see also Appendix 3).

Early electronic gates were based on circuitry that utilized discrete resistors and transistors (the so-called RTL or resistor–transistor logic gates). Later developments used discrete diodes and transistors (the DTL gates). For all practical

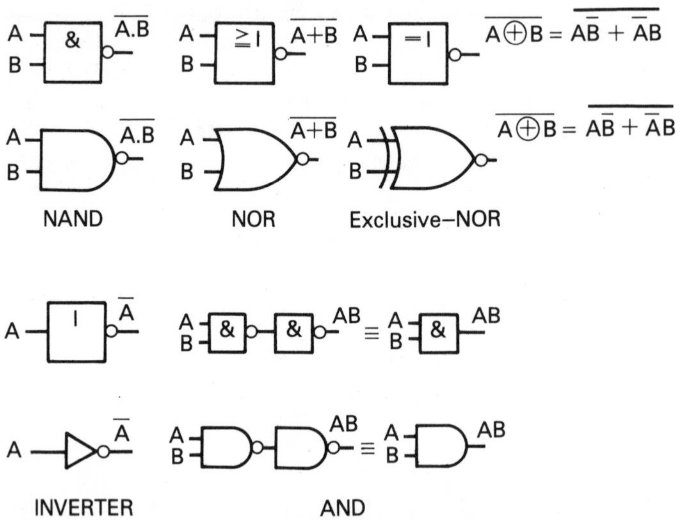

Figure 10.4 Logic symbols. Those at the top of each pair are to BS 3939 or IEC 117; those at the bottom of each pair are the American symbols to MIL-STD-806B. See also Appendix 3

purposes these gates have been superseded by IC logic gates. Two main systems are in common use today. They are the transistor–transistor logic (TTL) types, and those based on complementary metal-oxide semiconductor transistors – the CMOS types. The developments throughout the years have aimed at improving reliability, speed of operation, noise immunity, power efficiency, fan-in and fan-out capability.

Speed of operation has already been briefly discussed. Schottky transistors are available for high-speed operation.

Noise immunity is the term associated with the ability of a circuit to reject noise signals. Such signals arise, especially in factory locations, from the use of such production apparatus as electrical motors, thyristor control gear, electric arc welding equipment, etc. The noise consists of rapid variations of voltage having a spike or pulse-like waveform. Counting and control logic circuitry may well be affected by the presence of noise in the legitimate signal. If the circuitry has poor noise immunity, spurious counts are recorded or false control signals are given. CMOS gates have a greater noise immunity than TTL gates. Short spikes of voltage, known as 'glitches', may be generated within a logic system, especially in synchronous operation in which care has not been taken in the design of the circuit to produce correct timing of the logic operations.

The terms 'fan-in' and 'fan-out' refer to the number of input and output lines that may be associated with any particular gate. The terms originate from the time when the circuit symbol for a gate was a circle, with the input and output lines radiating like the spokes of a wheel. A set of input (or output) lines is reminiscent of the shape of a fan. The fan-out capability is the more important of the two. There is a practical limit to the number of logic loads that can be connected to a given gate. If too many loads are connected to the output stage, the driving gate is unable to deliver sufficient current to maintain the voltage across each load at the correct logic level. For example, a TTL gate must be energized from a 5 V supply. +5 V is defined as logic 1 and 0 V is defined as logic 0. Some tolerance is allowed on these voltage levels. Logic 1 level is any voltage between 2.4 V and 5 V; logic 0 level is any voltage below 0.8 V. If the output of a driving gate is intended to be at logic 1 the output voltage will be near to 5 V if the loading is light. When the number of loads is increased, however, the output voltage of the driving gate falls. When 2.4 V is reached we are connecting the maximum permissible number of logic loads to the driving gate. This number is the fan-out capability of the gate ($= 10$ for the 74 series TTL).

Modern logic systems use IC techniques in their fabrication, thus many hundreds of interconnected components are formed on one chip. The method of fabrication gives rise to three terms, viz. medium-scale integration (MSI), large-scale integration (LSI) and very large scale integration (VLSI). The terms indicate roughly the number of components that are fabricated simultaneously on single substrates.

In logic work the following identities are frequently used

$$\overline{AB} = \overline{A} + \overline{B}$$

and

$$\overline{A + B} = \overline{A}\,\overline{B}$$

Together, these identities are known as De Morgan's Theorem. The validity of the statements can be established by constructing a truth table, as shown in Figure 10.5. It should be noted in passing that $\overline{AB}$ does not equal $\overline{A}\,\overline{B}$.

A	B	$\overline{A}$	$\overline{B}$	AB	$\overline{AB}$	$\overline{A}+\overline{B}$	$\overline{A+B}$	$\overline{A}\,\overline{B}$
0	0	1	1	0	1	1	1	1
0	1	1	0	0	1	1	0	0
1	0	0	1	0	1	1	0	0
1	1	0	0	1	0	0	0	0

$$\overline{AB} = \overline{A} + \overline{B}$$
$$\overline{A+B} = \overline{A}\,\overline{B}$$

Figure 10.5 Truth table showing the proof of De Morgan's Theorem

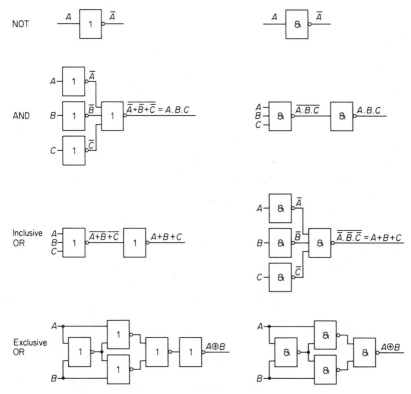

Figure 10.6 The basic logic function using NOR and NAND gates

Figure 10.6 shows how the various logic connectives can be realized using either NAND or NOR gates.

Transistor–transistor logic (TTL) gates

For high-speed logic work, TTL gates are now the usual choice. Many of the IC packages in the 7400 family of logic units are based on the gate circuit shown in Figure 10.7(c). Formerly, logic assemblies used RTL gates, but such gates incorporated many resistors and were not therefore suitable for IC fabrication.

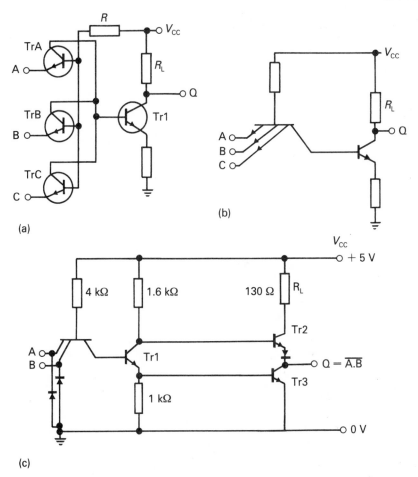

Figure 10.7 Transistor–transistor logic gate (a) Simple form (b) Multiple-emitter form (c) Multiple-emitter TTL NAND gate (¼ of 7400)

Some of the disadvantages of RTL gates were overcome by replacing the input resistors by diodes (DTL). In IC technology, however, it is cheaper to make transistors, and thus the circuit of Figure 10.7(a) evolved. The input transistors switch a single transistor, shown as Tr1. When the emitters A, B and C are connected to the zero-voltage line, the base–emitter junction of each input transistor is forward-biased, and all the current through R then passes, via the transistors, to the zero-voltage supply line. Under these conditions the base voltage of each input transistor is only about 0.6 V above zero, and hence the base–collector junctions are reverse-biased. The output transistor Tr1 is therefore 'starved' of base current and is thus cut off. This condition occurs if any one of the emitters of TrA, TrB or TrC is held at zero volts. If, however, A and B and C are connected to the positive supply rail, the base–emitter junctions are no longer forward-biased. The base–collector junctions of TrA, TrB and TrC thus all become forward-biased, and the current through R then passes into the base of Tr1. This

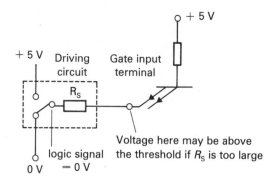

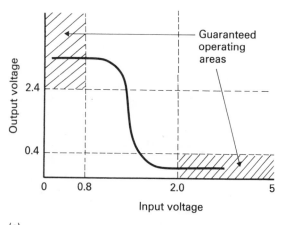

(e)

Figure 10.7 (d) Influence of the source resistance, R_s, of the driving circuit on the reliability of gate operation; (e) Transfer characteristic for the circuit of (c)

transistor becomes heavily conducting and the output voltage at Q falls. Using positive logic, therefore, the arrangement acts as a NAND gate.

The planar epitaxial technique enables an ingenious amalgamation of TrA, TrB and TrC to be achieved. Since all the bases are interconnected, as are all the collectors, it has been found possible to fabricate a single transistor having one base and one collector, but multiple emitters. A gate using multiple emitters may thus be called a multiple-emitter transistor logic (METL) gate.

The driving capability (i.e. the fan-out) of the circuits of Figure 10.7(a) and (b) is not high because the output impedance of the arrangement shown is too large to enable many following logic gates to be driven satisfactorily. Each logic load connected to the output has its own input capacitance; the total equivalent capacitor must commence charging via R_L at the moment when Tr1 is switched off. The charging time is quite long for practical values of R_L, and so the output voltage across the load does not rise to its logic 1 value as quickly as may be required. This mechanism limits the switching speed of the gate; conversely, for a given switching speed, the fan-out capability is limited.

These considerations have led to the development of the circuit shown in Figure 10.7(c). When one, or both, of the inputs are low, Tr1 is cut off. The base voltage of Tr2 therefore rises sufficiently to drive this transistor into conduction, and the output voltage across goes high. In this context, high is equivalent to a voltage of about 3.6 V, being 5 V minus a little more than two diode drops. Since the load capacitances are now being charged via a low-resistance path, the switching speed is fast. When the inputs are both high Tr1 is turned on. The base voltage of Tr2 falls, and that transistor is cut off. The voltage at the base of Tr2 is not only low enough to keep this transistor turned off, but also ensures that the diode connected into the emitter lead is reversed-biased. A very high impedance therefore appears between the +5 v supply rail and the collector of Tr3. Simultaneously, since Tr1 is turned on, its emitter voltage rises. The base voltage of Tr3 must therefore rise and Tr3 is then driven into conduction. Any load capacitance is discharged rapidly via the low-resistance path of Tr3, and the output voltage falls rapidly to about 200 mV.

Although normal operation involves only positive input signals, in practice negative input voltages may inadvertently be applied. For this reason diode limiters are used as input protection devices. The need for ever faster switching has led to the replacement of conventional bipolar transistors in logic gates by low-power Schottky transistors. The member of the 7400 family using these transistors can be recognized by the letters LS in the type number, e.g. 74LS00.

TTL gates require the supply voltage to be +5 V ±5%. Some care should be taken in the choice of power supply. In particular it is advisable that suppression circuitry be included to eliminate mains-borne interference. The +5 V supply line in the logic circuitry should be decoupled by mounting a small ceramic capacitor (0.1 µF) across the supply rails as close to the gates as possible. Ideally each IC should be decoupled separately, but it is usually satisfactory to make one decoupling capacitor serve a small number of ICs. The purpose of decoupling is to eliminate, or minimize, the adverse effects of switching noise generated within the logic system, and which may be present on the supply lines.

The influence of any external noise present on the input signal lines can be minimized by designing driving circuitry based on an understanding of the input requirements of a TTL gate. It is important to realize that the driving circuit are connected to the emitters of the multiple-emitter input transistor and therefore must be able to sink current satisfactorily when the input signal is low. If the impedance of the driving circuit is too high we may have the situation illustrated by Figure 10.7(d). Here, although the logic signal is zero, the voltage developed at the input emitter may be above the switching threshold of the gate. This threshold is determined by the collector potential of the multi-emitter input transistor, which in turn depends upon the v_{BE} values of Tr1 and Tr3. Typically, this threshold is about 1.4 V. Operating conditions that guarantee satisfactory performance are summarized in Figure 10.7(e). For TTL circuits, any input signal below 0.8 V is regarded as being low (i.e. at logic 0) and will give a guaranteed gate output voltage greater than 2.4 V. High input signals must exceed 2.0 V for guaranteed gate operation; the output voltage is then typically 200 mV when the loading is light. When the maximum number of logic loads are connected, however, the output voltage may rise to as much as 400 mV.

Maximum noise immunity is obtained when the input levels are removed as far as possible from the threshold levels. In the worst-case situation, when the input signal is 0.8 V the noise immunity is about 600 mV for a threshold level of 1.4 V. With this threshold level, high signals (which must exceed 2.0 V) also have a noise immunity of about 600 mV. These are not guaranteed noise immunity levels, however, because the threshold voltage differs for different samples of the same nominal gate. An important fact for the user to remember is that the maximum voltage that can be regarded as low is 0.8 V. If the driving circuit is itself a TTL gate its worst-case output voltage is 0.4 V maximum, and therefore the guaranteed noise immunity is only 400 mV. For high signals the minimum voltage delivered by a fully loaded TTL source is 2.4 V. The minimum input voltage that can be regarded as high by the following gate is 2.0 V. The guaranteed noise immunity under these circumstances is therefore also 400 mV.

We see that when the driving circuit is itself a TTL gate, the driving conditions are satisfactory. When the output of the driving gate is high, Tr2 acts as an emitter-follower, and the output voltage is two diode drops (1.2 V) below the collector voltage of Tr1. Since Tr1 is off, the output voltage of the gate is $V_{\rm CC}$ minus 1.2 V minus the small voltage drop across the 1.6 kΩ resistor, i.e. about 3.6 V. As many as 10 TTL logic loads may be connected to the output before the output voltage falls to 2.4 V.

Unused inputs should not be left unconnected; in the 'floating' state they are prone to pick up noise voltages. They may be connected to a used input, or alternatively connected to one of the supply rails. Unused active low inputs of all families may be connected directly to the zero-voltage line. Unused active high inputs of the LS TTL family may be connected directly to the positive supply rail, but the leads must be short and the supply rails adequately decoupled. Standard and Advanced low-power Schottky (ALS) TTL families require unused active high inputs to be tied to the positive supply rail via a pull-up resistor. Up to 25 unused inputs can be connected to a single resistor.

Standard 74 TTL gates dissipate 10 mW of power, can be operated at frequencies up to 35 MHz, and have a propagation delay time of 10 ns. (Propagation delay time is the interval between the arrival of an input signal and the completion of the appropriate change of the output voltage.) 74LS TTL has a power dissipation of only 2 mW, and can be operated up to 45 MHz with propagation delays of slightly less than 10 ns. Advanced low-power Schottky (74ALS) TTL has a power dissipation of 1 mW, can be operated at up to 50 MHz, and has a propagation delay time of 4 ns.

Emitter-coupled logic gates

A non-saturating circuit that uses conventional transistors is shown in Figure 10.8. The circuit is known as an emitter-coupled logic (ECL) gate, and represents perhaps the fastest switching circuit that is available at present. In this circuit, switching is achieved by switching current from TR4 by TR1, TR2 or TR3. The circuit is therefore a current-steering logic (CSL) arrangement. Since almost constant current is drawn irrespective of the logic state, the power supply requirements are eased and generated noise is almost non-existent. A further advantage is that two outputs, Q and $\bar{\rm Q}$, are available. The high cost of the gate, and comparatively large power dissipation, are disadvantages that must be suffered if switching speed of paramount importance.

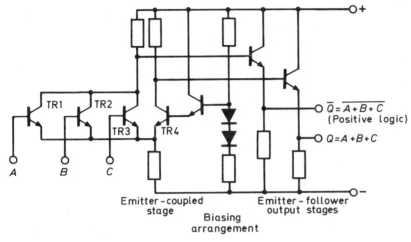

Figure 10.8 Emitter-coupled logic (ECL) gate

MOSFET gates

An examination of the drain characteristics of an FET shows that these devices can be used as switches in ways that are similar to those using bipolar transistors. With a resistive load we can construct a load line and see how the position of the operating point moves from an on position to an off position. In the off state, the current is very low (a few tens of picoamps) for a p-channel MOSFET. (The abbreviations MOSFET and MOST are interchangeable since they both refer to a metal-oxide semiconductor field-effect transistor.) For a bipolar transistor the off current is 10–20 nA. When on, a MOSFET voltage is comparatively high, being about 2 V compared with 0.2 V or 0.3 V for a bipolar transistor. A big advantage in using MOSFETs is that no driving current is required; many MOSFET gates can therefore be connected to a given output.

Figure 10.9(a) shows an NMOS inverter in which Tr2 acts as an active load (i.e. pull-up device) for Tr1. Predictions of the output voltage can be made using graphical data. Figure 10.10(a) shows the characteristics of Tr2. Since $v_{GS} = v_{DS}$ we can construct a line showing the relationship between the current through Tr2 and the voltage across it. At every point on the line, the drain–source voltage equals the gate–source voltage. This information can be transferred to the (identical) characteristics of Tr1. For any current through Tr1 and Tr2, the voltage across Tr2 must be subtracted from the supply voltage to give the corresponding drain–source voltage for Tr1. In the example shown we have assumed a supply voltage of 18 V. When the voltage across Tr2 is 15 V the voltage across Tr1 must be 3 V at the same current. When the voltage across Tr2 is, say, 10 V then that across Tr1 is 8 V. In this way a series of points can be plotted on the characteristics for Tr1 and a load line drawn. Had Tr2 been a resistor this load line would have been a straight line. We can now predict the output voltage for any supply voltage and logic state of A. In the example shown a supply voltage of 18 V is assumed. When the input to the inverter is high (i.e. A = 1) then $v_{GS} = 18$ V. The intersection of this characteristic with the load line shows that the output voltage will be 3 V. If the input is being

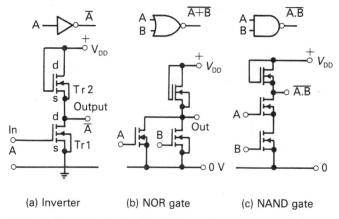

(a) Inverter   (b) NOR gate   (c) NAND gate

Figure 10.9 NMOS gates. In (b) and (c) depletion types are used as active loads. (Input protecting diodes are not shown)

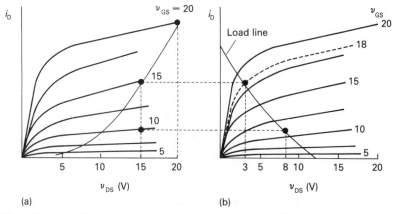

(a)   (b)

Figure 10.10 Predictions of output voltage from the inverter of Figure 10.9(a): (a) Characteristics for Tr 2; (b) characteristics for Tr 1

driven by another gate whose output is, say, 15 V then the output voltage would be just under 5 V with the characteristics shown in Figure 10.10(b).

NMOS gates are those that are use n-channel MOSTs. In Figures 10.9(b) and (c) n-channel enhancement devices are used in conjunction with a depletion type acting as an active load. These arrangements are found to have faster switching action than similar gates using enhancement types throughout. Where packing density is more important than speed, however, enhancement types are used throughout, since they are simpler to manufacture; on this score they are also superior to CMOS and TTL constructions.

CMOS gates

The gates of Figure 10.9 all dissipate power during the period that the load is carrying current, i.e. when the output is low. Power dissipation is a serious problem in LSI systems with their inevitable high packing density. This disadvantage can be overcome by using circuits that use both n-channel and p-channel enhancement MOSFETs. Since these types are mutually complementary, together they form complementary metal-oxide semiconductor gates, i.e. CMOS gates. Such gates dissipate power only during the transition periods; in the static periods no power is dissipated. As will be evident from the circuit diagrams shown in Figure 10.11 there is always a series arrangement in which one of the transistors is on, but the other is off; no current therefore flows through the chain during the static periods. Although the switching speed of these gates is slower than for comparable TTL gates, the low power dissipation is such an attractive feature that a whole family of CMOS gates has been developed. The 4000B range is a standard range now in common use. Functionally they perform the same logic tasks as their TTL equivalent, but require simpler power supplies and can operate with a wide range of supply voltages. Although the specification states that the supply voltage may be low as 3 V, the noise immunity and other performance parameters are not good at this low level. Supply voltages of about 5 V to 18 V are satisfactory. One of the major disadvantages of early forms of this type of gate was their vulnerability to static voltages on the input lines. To some extent the position has improved since the incorporation of protective circuitry in the input stages. Even with modern units, however, it pays to avoid excessive levels of static charge in handling and in storage. To this end the devices are often supplied with their pins embedded in black conductive plastic foam.

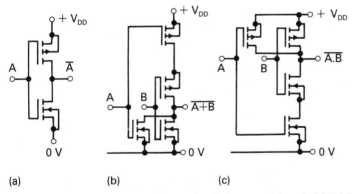

(a)   (b)   (c)

Figure 10.11 CMOS gates. (Input protecting diodes are not shown): (a) NOT gate; (b) two-input NOR gate; (c) two-input NAND gate

The manufacturers of TTL and CMOS families publish extensive literature on their products giving rules of operation, interfacing, etc. Any prospective logic circuit designer is therefore well advised to obtain copies of these publications.

In general, a logic system should be designed using only one type of logic unit. There are occasions when a wanted function may be available in one family, but the main system is using a different family. When mixing families care must be taken to ensure that voltage levels do not damage following units. Linking 5 V TTl directly onto 18 V CMOS is bound to lead to device failure. Voltage level-shifting circuits are therefore required; some examples of these are given in Figure 10.12.

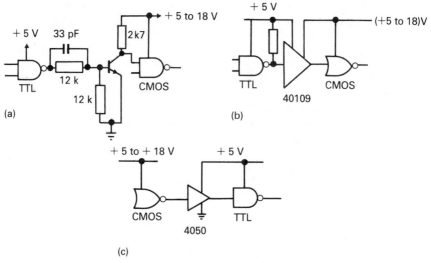

Figure 10.12 Level-shifting circuits for linking TTL to CMOS and vice versa

Tristate Logic Gates*

We shall briefly discuss the architecture of a digital computer in a later chapter, and content ourselves here with the knowledge that computer systems are complicated arrangements of logic gates and memories that operate in a sequential way. The interconnections of the logic units would be impossibly difficult if every logic block were to have its own output line to other parts of the system. The exchange of data between the central processing unit and the memory, as well as other peripheral units, therefore takes place along a set of common conductors known as a data bus. (A bus bar in electrical engineering is a bar of copper which is used to interconnect switches or other items of equipment.) In microcomputers it is common to have eight lines in the bus; eight bits of information can be transferred simultaneously. Sixteen-bit lines are also frequently used.

The type of logic gate discussed so far is unsuitable for driving bus lines. This is because both their high and low outputs are low impedance ones. Since the outputs of many gates must be connected to the same bus line, difficulties arise if some gate outputs go high while others remain low. The way out of the difficulty is to devise gates whose output lines can be at logic 0, i.e. the low voltage, or at logic 1, the high voltage, or can be open-circuited (i.e. have a very high impedance to ground). In this third condition a logic gate gives no defined logic output and some other gate connected to the line is then free to impose a high or low voltage as required. The term 'tristate' was devised by National Semiconductors, but is now in common use with other manufacturers products. Figure 10.13 shows a possible arrangement with a CMOS inverter. In addition to supplying a logic input, an ENABLE input must also be supplied. This ENABLE input may be high or low to allow the output information onto the data bus line; the choice of high or low depends upon the type

*Trademark of National Semiconductors

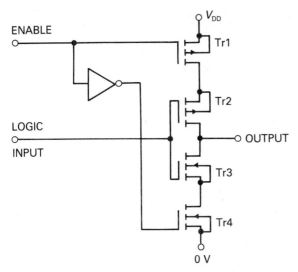

Figure 10.13 Output stage of a CMOS tristate logic gate

of logic unit used, and the manufacturer involved. In Figure 10.13, when the ENABLE signal is high both Tr1 and Tr4 are cut off. The output then acts as an open circuit (with an output resistance of 10^{10} Ω or more). When the ENABLE signal is low, Tr1 and Tr4 are both switched on and the gate then operates 'normally', giving low or high outputs at low impedance depending upon the input logic state.

Figure 10.14 shows a TTL tristate gate. When the ENABLE is high, diode D2 is reverse-biased, and the gate operates in the usual way. When the ENABLE signal is low, D2 conducts thus cutting off Tr1. Both Tr2 and Tr3 are then cut off and the output acts as an open-circuit.

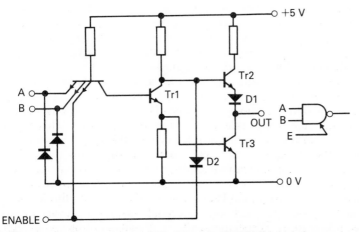

Figure 10.14 Tristate logic gate – TTL. When the ENABLE is high the gate acts as a normal two-input NAND gate. When the ENABLE is low the output is isolated from the rest of the system; irrespective of the input states, the output does not take up any particular state

One way of obtaining a high output impedance is to remove the active load of the output transistor, leaving the collector open-circuit. Such an arrangement is called an open-collector gate. When the output transistor is off, the output line is isolated from the supply lines (Figure 10.15(a)). Both NAND and NOR gates can be fabricated with open-collector outputs. Such outputs can be wired together to

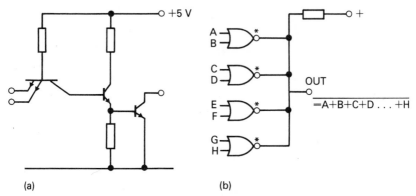

(a)   (b)

Figure 10.15 Open-collector gate that can be wired externally to provide easy expansion of fan-in capability: (a) Open-collector gate; (b) use of open-collector gates to form a multiple-input wired gate

provide an easy way of extending the fan-in capability. Figure 10.15(b) shows four two-input NOR open-collector gates wired together to form an eight-input gate. An external pull-up resistor is needed, and this makes the gate unsuitable for use in microprocessors system with their compact data bus arrangements. Noise problems with bus systems. and inferior speed of operation, make open-collector gates unsuitable for computer circuits.

Digital signals

The signals discussed in Chapter 2 were analogue signals in that their waveforms corresponded to the physical quantities that were converted into electrical form by suitable transducers. All periodic analogue signals can be synthesized by adding sine waveforms of the appropriate frequency, amplitude and relative phase. Sine waves are therefore the basic 'building bricks' of analogue signals.

Digital waveforms on the other hand all have the same basic shape. They are square waves whose amplitude is either zero, or some fixed voltage (e.g. +5 V in TTL systems) which can be defined as logic 1, or high. The basic digital waveform is a square wave with a 50 per cent duty cycle, i.e. a 1:1 mark-space ratio. This means that the waveform is high for the same time as it is low, and this corresponds to the string 0,1,0,1, etc. Before a square wave can convey information it must be modified (just as a sine wave must be modified, or modulated, before it can carry information).

In digital work, however, we are not at liberty to modulate the amplitude of the square wave. Information is impressed upon the digital waveform by varying the mark-space ratio according to some specified code. The pattern of 0s and 1s then conveys the message, which is then said to be encoded. Figure 10.16(a) shows the basic digital waveform while (b) shows part of a waveform which represents a

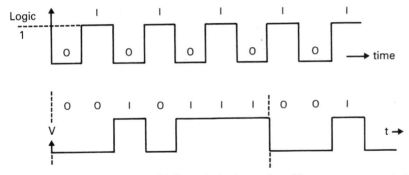

Figure 10.16 Digital waveforms: (a) Shows the basic waveform (b) represents an encoded signal

message. The region between the vertical dotted lines may represent a letter of the alphabet, a decimal digit or some other symbol. Examples of some codes, including the ASCII code, are given and explained in the next chapter.

The basic waveform shown in Figure 10.16(a) is the one used by digital computers to synchronize the multitude of operations within the system. It is this waveform that 'beats the time' and thus keeps all in order. For this reason it is often referred to as the clock waveform. The clock generator is one form of relaxation oscillator, the function of which is to produce a waveform in which a rapid transition takes place from the zero-voltage level to the high level, and vice versa.

Generation of rectangular waveforms

When we wish to produce oscillating voltages that have rectangular waveforms, it is usual to resort to two-state circuits in which there is an abrupt transition from one state to another. The multivibrator is the most commonly used circuit, which gets its name from the fact that square or rectangular waves are rich in harmonics. The basic circuit is given in Figure 10.17 in which two RC-coupled amplifiers are used, the output of one being connected to the input of the other, and vice versa. If we regard the arrangement as a two-stage amplifier (Figure 10.17(b)) we see that positive feedback occurs since the phase of the collector voltage of Tr2 is the same as that of the base voltage of Tr1. The coupling back of the output voltage of Tr2 to the input of Tr1 is such as to enhance any original disturbance that initiates the action. The circuit is thus rapidly driven into the condition whereby Tr1 is fully conducting while Tr2 is cut off. Such a condition is not permanently stable, however, because of the a.c. coupling via the capacitors; the condition is often referred to as being quasi-stable. The regenerative switching nature of the circuit is such as to drive the arrangement into its other quasi-stable state whereby Tr2 is fully conducting and Tr1 is cut off. As long as the supply voltage is present the circuit will continually oscillate between the two quasi-stable states; since there is no permanently stable state such a circuit is termed an astable multivibrator.

To assist in understanding the mode of operation of this circuit, the waveforms at different points in the circuit are shown in Figure 10.17 along with the circuit diagrams.

From capacitor-charging theory it is easy to show that the period of the square-wave output is $0.693 (C_2 R_3 + C_1 R_2)$ which, for symmetrical operation is

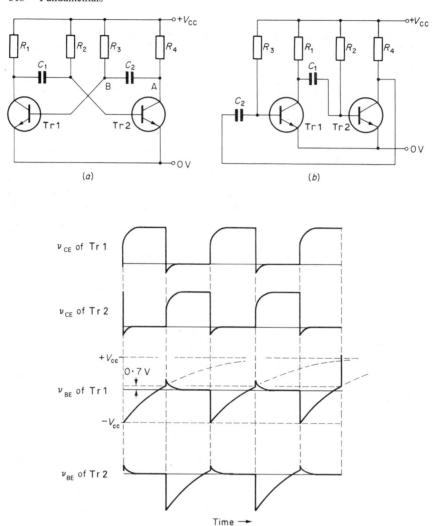

Figure 10.17 Astable multivibrator: (a) Basic cross-coupled astable multivibrator; (b) circuit of (a) to emphasize the amplifier arrangement with positive feedback

about 1.4 CR; the repetition frequency is thus about $0.7/CR$ when $C_1 = C_2 = C$ and $R_2 = R_3 = R$.

It is not often nowadays that we would wish to build a square-wave generator using discrete components. Better waveforms and compact circuitry are achieved using comparators.

Comparators are high-gain differential amplifiers built along the lines of IC operational amplifiers, but with circuitry suitably modified to perform a switching function. No negative feedback resistor is used, so only a tiny input voltage difference between the input terminals is enough to make the output voltage swing to almost $+V_{CC}$ or $-V_{CC}$. A reference voltage (V_{ref}) is applied to one of the input terminals (say the non-inverting terminal) and the voltage to be compared with the reference, v_{IN}, is applied to the other input terminal. Since the open-loop gain is very high (>100 dB), once v_{IN} exceeds V_{ref} by a very small amount then the output voltage swings rapidly to almost $-V_{CC}$. Conversely, when $v_{IN} < V_{ref}$ the output voltage is almost at $+V_{CC}$.

When ordinary operational amplifiers are used as comparators, slow rates of change of v_{IN} through the reference voltage gives slow rates of change of the output voltage. Digital circuitry is not usually very tolerant of slow rise and fall times. The position can be improved by the use of positive feedback as shown in Figure 10.18(a). The inevitable presence of noise on the input signal also produces difficulty in digital circuitry. As the signal passes through the reference voltage, superimposed noise can give multiple transitions at the output which causes glitches to appear in the digital equipment. (See the section on Schmitt triggers later in this chapter.) To minimize these effects special comparators are available in which the swing from $+V_{CC}$ to $-V_{CC}$ and vice versa is very fast (e.g. LM311N or NE 529). Internal positive feedback increases the transition speed. They also have open-collector output stages which allow the output voltages to be made compatible with other logic systems.

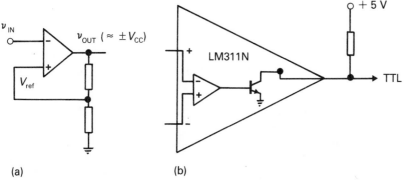

Figure 10.18 Comparators: (a) Improving the performance of an op-amp as a comparator by using positive feedback; (b) comparator with special output circuitry to make it TTL compatible

The generation of square waves can be achieved by using a comparator in the circuit of Figure 10.19. Initially we may assume that $v_{out} = V^+$ and therefore $V_{ref} = V^+/2$. The capacitor is initially uncharged, therefore $v_{IN} = 0$ at time $t = 0$; charging then commences. When the voltage across C has risen to almost infinitesimal fraction above $V^+/2$ the output switches rapidly to V^-. The capacitor then discharges until the voltage across it, i.e. v_{IN}, falls to zero, and then charges with the opposite polarity as the voltage continues to fall towards V^-. Once v_{IN} reaches a negative voltage marginally below $V^-/2$, the output suddenly shoots up to V^+ whereupon v_{IN} rises exponentially as before. From Figure 10.19 we see that the period of the output waveform must be twice the time it takes for v_{IN} to fall from $V^+/2$ to $V^-/2$.

During this time v_{IN} is given by

$$v_{IN} = \frac{3V}{2} e^{-t/CR} - V$$

where $V = |V^+| = |V^-|$.

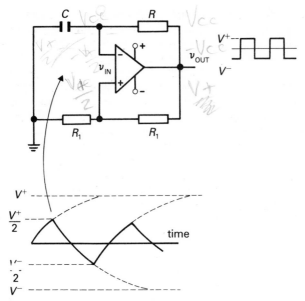

Figure 10.19 Use of a comparator to produce square waves

Taking $t = 0$ when $v_{IN} = V^+/2$, then after t_1, the half-period, $v_{IN} = V^-/2$, i.e. $-V/2$. Thus

$$-\frac{V}{2} = \frac{3V}{2} e^{-t/CR} - V \text{ i.e. } t_1 = (\log_e 3)CR = 1.1CR$$

The period of v_{out} is thus $2.2CR$, which gives a repetition frequency of $0.45/(CR)$.

Computers and microprocessor control circuitry require clock pulses of accurately known and stable frequencies. The usual way of generating the waveform is to use an oscillator that uses a quartz crystal. Bistable multivibrators (which are related to the astable type described above) are triggered at the appropriate time by the crystal output. Figure 10.20(a) shows one example of how this is done. Subdivisions of frequency are also available via decade dividers. Figure 10.20(b) shows how inverters may be used to generate clock pulses. Two inverters in effect form a positive feedback amplifier.

Telex and analogue signals

The encoding of the clock waveform to convey information is usually carried out with microprocessor and allied logic circuitry. In telex work, and long-distance intercommunication between computers, the transmitter consists of a typewriter keyboard which is connected to logic circuitry that encodes the message. A readable alphanumeric display is available to the operator who on pressing a key initiates the generation of a sequence of 0s and 1s that represents the corresponding character. At the receiving end the string of binary digits (known as bits) is decoded to produce the message. This message may be displayed on the screen of a visual display unit (VDU), but a printer also prints out the message giving what is called a hard copy.

Since the digital signals are subject to noise and distortion during transmission, it is a matter of some importance to reduce transmission errors to negligible proportions. Some method of error-detecting is therefore essential. The simplest of these methods is to use a parity check. An extra bit is added to the coded 'word' to indicate whether an error has occurred or not. The parity bit is either a 1 or a 0 so that the total number of 1s in the combined character-plus-parity-bit string is either odd or even. For odd parity the parity bit is 0 if the total number of 1s in the character string is already odd; if the number of 1s in the character string is even

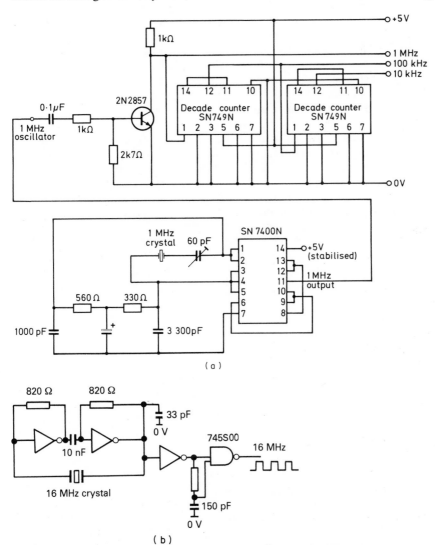

Figure 10.20 Square-wave generators in which frequency stability is maintained with the assistance of quartz crystals: (a) Use of ICs to yield square-wave pulses at standard repetition frequencies. The SN7490N (Texas) is an arrangement of four bistables and three logic gates that divide by 10; (b) clock-pulse generator for the BBC computer. Each inverter is part of the hex inverter package 74S04. Later circuitry divides the frequency to yield square waves to clock the microprocessor

then the added parity bit is 1. For even parity, the parity bit is 0 when the character string already contains an even number of 1s; the parity bit is 1 for an odd number of 1s in the character string. The parity bits are generated by the transmitter during the WRITE operation. A check is made at the receiver during the READ operation to ensure that the agreed mode of parity has been maintained. The receiver can then request a repeat transmission of those characters in which errors have been detected.

The system of error detection is obviously far from foolproof since if two errors occur within a string the error will escape detection. The error-detection mechanism can be improved greatly by using more sophisticated methods. For example, in radio telegraphy a 4-out-of-7 code is used. Every character pattern consists of seven digits, four of which are 1s, and the remaining three are 0s. Thirty-five different arrangements of four 1s and three 0s are possible, thus we can cover the alphanumeric range almost completely. If, for example, a string containing three 1s and four 0s (or five 1s and two 0s) were received, an error would be detected and a retransmission requested.

In systems where it is inconvenient, or impossible, to request a retransmission, error-correcting codes may be used. The forerunner of several ingenious error-correcting codes is the Hamming code, in which extra bits are added to each word so that when an error occurs it is possible to detect the position of the error in the character string. Suppose, for example, that denary digits (i.e decimal digits such as 4 or 9) were being transmitted as groups of four binary digits, then three other digits would be added to make a string of seven. The four binary digits, representing the denary digit, are taken in three groups of three. To each group of three digits a parity bit is added. Thus if the four 'information' digits were ABCD then the string transmitted would have seven digits, say ABCDRST. R is the parity bit for ABC, S for ACD and T for BCD. On reception, three parity checks are

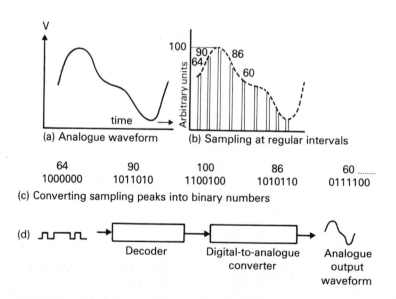

(a) Analogue waveform   (b) Sampling at regular intervals

64	90	100	86	60
1000000	1011010	1100100	1010110	0111100

(c) Converting sampling peaks into binary numbers

(d)   Decoder   Digital-to-analogue converter   Analogue output waveform

Figure 10.21 Digital transmission of analogue waveforms

made from which it is possible to detect if a transmission error has been made and, if so, which digit is in error. The receiving apparatus is then able to make a correction without further reference to the transmitter.

Now that the compact audio disc has become popular, many people realize that analogue signals can be encoded into digital form, and thus processed by digital equipment. The process involves sampling the analogue waveform at regular intervals and converting the peak voltage value of each sample into a denary number. These numbers can readily be encoded in to digital form, and transmitted as a digital signal. The process is called Pulse Code Modulation (PCM). The encoding is initially performed by an analogue-to-digital converter. At the receiver the digital information is then converted back into analogue form by digital-to analogue converters. Discussion of these converters takes place in the next chapter. Figure 10.21 shows in diagrammatic form the processes involved.

Both analogue and digital signals suffer distortion, and pick up noise, during the transmission process. One of the principal advantages of using digital signals is the ease with which the waveforms can be 'cleaned-up' at the receiving end. Figure 10.22 shows the noise pollution and waveform distortion that can affect digital

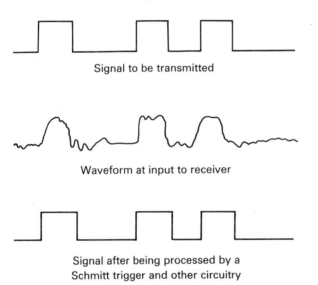

Signal to be transmitted

Waveform at input to receiver

Signal after being processed by a
Schmitt trigger and other circuitry

Figure 10.22 Recovery of data from a noisy and distorted digital signal

signals. Although the distortion and noise are quite severe, the pattern can be recovered perfectly by using a Schmitt trigger to clean the waveform. So far as the receiver is concerned these signals are therefore relatively immune from all types of electrical noise and distortion.

The Schmitt trigger

Although basically a voltage discriminator, and widely used as such, the circuit acts as a pulse shaper and will operate only on input signals above a predetermined voltage level. The circuit is thus very useful when we wish to discriminate between noise, or low-amplitude pulses, and pulses of larger amplitude.

A basic form of Schmitt trigger is shown in Figure 10.23(a). It is an emitter-coupled bistable which, because of the regenerative feedback common to multivibrators, may be regarded as a fast-acting switch, the action of which is precipitated when the input voltage exceeds what is called the 'trip' or 'threshold' level. Once initiated, the action proceeds irrespective of the waveform of the input signal; the only requirement is that the voltage level must remain above a certain critical value.

Initially, with V_{in} below the trip point, TR2 draws enough base current via R_1 and R_2 to be held in an on condition. TR1 is therefore off, because its emitter is maintained at the high voltage developed across R_4. Once V_{in} exceeds the voltage across R_4, plus about 650 mV, TR1 is turned on. The collector voltage of TR1 falls and a rapid regeneration action takes place in which the current through R_4 is diverted away from TR2 to TR1. TR2 is then cut off and the collector voltage rapidly rises to the supply voltage V_{CC}. As long as V_{in} exceeds the trip voltage the bistable will remain in this stable state. As the input voltage diminishes, a point is reached where a further cumulative action occurs and the trigger reverts to its initial stable state. The input voltage at which this occurs is less than the trip voltage. This phenomenon is called 'backlash' or 'hysteresis' and is due to the

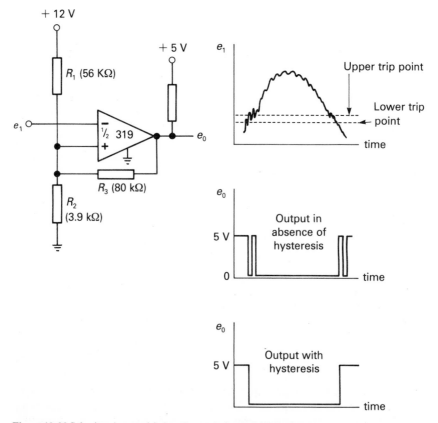

Figure 10.23 Schmitt triggers: (a) One form of trigger (b) Schmitt trigger with hysteresis. R_2 may be variable to adjust the value of the trip voltages

emitter-follower effect in TR1. During the initial rise of the base voltage of TR1, the emitter voltage is not affected, because TR1 is cut off. When TR1 is conducting, however, any subsequent fall in base voltage tends to be followed by a fall in emitter voltage because of the emitter-follower effect. Before TR1 can be cut off, and a cumulative action started, the base voltage must fall below the initial triggering level. For the circuit of Figure 10.23(a), with the values indicated, the hysteresis is about 1.8 V, the trip voltage being 8.5 V. The rise time of the output waveform is about 20 ns and the fall time is much longer, being approximately 30 ns.

For logic work, one of the serious disadvantages of the Schmitt trigger described above is the fact that the output pulse has a minimum value that is appreciably positive. Additionally, it is not possible to vary the trip voltage and hysteresis over wide limits without upsetting the operation of the circuit. These disadvantages can be overcome by using a high-gain IC operational amplifier. When used as such these IC amplifiers are often called 'voltage comparators' or 'level detectors'.

When operational amplifiers are employed as voltage comparators they are operated without feedback. The input signals are therefore subjected to the full open-loop gain. An extremely small difference between the input signals is thus sufficient to drive the output voltage to its full extent. With symmetrical power supply voltages, the output voltage is driven to almost $\pm V_{CC}$ depending upon the relative polarities of the input signals. A reference voltage is applied usually to the non-inverting terminal, and the varying voltage to be compared with the reference is applied to the inverting terminal. When the magnitude of a positive-going varying voltage becomes marginally greater than the reference voltage, the output voltage switches rapidly from about $+V_{CC}$ to $-V_{CC}$. For negative-going voltages that become marginally less than the reference voltage, the output states are reversed. (Comparators may be operated with a single supply voltage, but then the reference voltage must have polarity as the supply voltage.)

Conventional operational amplifiers suffer from two disadvantages when used as comparators. The transition speed of the output voltage is sometimes not fast enough for some applications, and, for logic work, the output voltages are not often TTL-compatible. For these reasons manufacturers produce ICs that are specifically designed for use as comparators; the delay in switching from one output voltage state to the other when input overdrive occurs is made much shorter than can be obtained with an ordinary op-amp. Comparators are also designed to have more flexible output circuits. The substitution of an open-collector output stage for the conventional push-pull arrangement allows external pull-up resistors to be used in connection with a separate voltage source. The output voltage can thus be made to be compatible with the rest of the logic circuitry. The LM 392, for example, is an 8-pin package that includes a precision open-collector comparator. The SN72710N is a high-speed comparator that controls the output voltage with inbuilt circuitry that includes a Zener diode. With a positive reference voltage the output voltage swings from $+3.2$ V to -0.5 V when the varying voltage exceeds the reference voltage, and is thus TTL-compatible. The LM319N is a dual high-speed comparator which is similar to the 710, but requires only a single rail supply; its open-collector output can interface directly with TTL systems and its response time is only about 80 ns. Since some comparators allow only a limited differential input range it may be advisable to use diode protection on the input circuits.

In logic systems a reference voltage of zero is to be avoided since it would not offer any noise immunity, and spurious triggering may result. A measure of noise immunity can be obtained by making the reference voltage larger than the peak voltage of any noise likely to be present. Trouble may still be encountered, however, when the noise is superimposed on a slowly rising input voltage. Under these circumstances, when the signal voltage is passing through the trip point, the noise plus signal voltage produces oscillations about the trip voltage level; as a result, spurious transitions of the output voltage occur. The solution to this problem is to introduce a controlled amount of hysteresis into the circuit. A popular way of doing this is to use positive feedback, as shown in Figure 10.23(b). Although a comparator with an open-collector output stage has been chosen (say an LM319), the principle of the circuit design can easily be extended to an op-amp or other type of comparator. With the arrangement shown, the output voltage is 5 V for signal voltages less than the trip voltage, and 0 V for signal voltages greater than the trip voltage. From a consideration of the circuit it will be seen that the trip voltage for rising input voltage waveforms is greater than that for falling input waveforms. The voltage difference between the two trip levels is known as the hysteresis. In the presence of hysteresis, spurious output voltage oscillations are avoided because once the first change in output voltage has been initiated by the rising noisy input signal, no further changes can take place provided the hysteresis is greater than the peak noise voltage.

Initial estimates of the values of the resistors R_1 to R_3 can easily be calculated from a knowledge of the voltages involved. In the example shown in Figure 10.23(b) it has been assumed that the required hysteresis is 0.2 V and that the upper trigger level is 1.0 V. The potential divider, R_1R_2, need not draw much current from the power supply. For the lower trip point, 0.8 V, the current through R_2 may be about 200 μA and hence $R_2 = 3.9\,k\Omega$. For the supply voltage shown the value of R_1 is 56 kΩ. The lower trip point is applicable for signal voltages above the trip point, and hence at this time the output voltage of the comparator is zero. It will be seen shortly that the shunting effect of R_3 can be neglected in this case. For signal voltages below the trip point the output voltage is 5 V and the trip point is required to be 1 V. The current through R_2 must rise to 250μA. Nearly all of the additional 50 μA will be supplied via R_3; since the voltage across R_3 is 4 V, the value of R_3 is 80 kΩ.

The monostable multivibrator (univibrator)

In the context of logic circuits the univibrator is used as a pulse lengthener. The Schmitt trigger is useful for squaring pulses with a poor waveform, but the duration of the pulse depends upon the time the voltage is above the trip point. When this time is not long enough to allow the logic circuitry to operate satisfactorily, or where very short duration pulses are to be counted, a square-wave of suitable duration can be produced by a monostable multivibrator. Figure 10.24(a) shows a circuit that uses npn transistors. Initially TR2 is on and TR1 is off. On arrival of an input pulse TR1 is urged to conduct. The consequent fall in collector voltage is transferred via C to the base of TR2. Multivibrator action then takes place and results in TR1 being sharply turned on while TR2 is sharply turned off. This is not a stable state however; C charges via the 2kΩ resistor and eventually the base voltage of TR2 rises to a value that turns on that resistor. A further cumulative action occurs resulting in the return to the original state. The duration of the output pulse depends upon the time constant. For the circuit shown the duration is about 1 ms when $C = 82\,nF$.

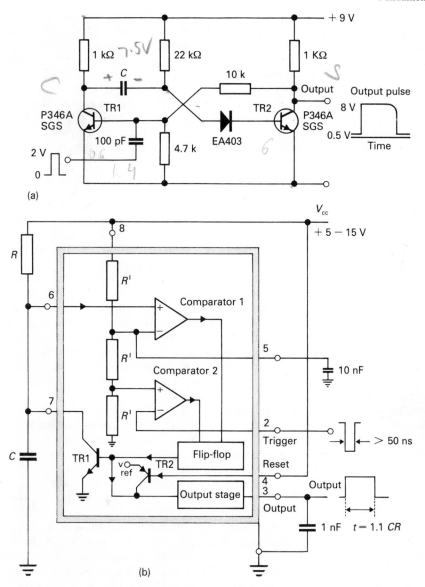

Figure 10.24 Monostable (one-shot) circuits: (b) shows the use of a standard 555 timer as a monostable. *C* should not be a ceramic or electrolytic type. Silvered mica, polystyrene and polycarbonate types are suitable. For larger capacitances a tantalum dielectric should be used

Monostable circuits are now readily available in integrated circuit form. The 74121 and 4047 are popular TTL and CMOS forms respectively. Such circuits are also known as one-shot circuits, since a single initiating pulse gives a single output pulse of longer duration. The length of the output pulse is determined by the values of an external resistor and capacitor. Some packages, such as the 74123 contain two monostables and also have facilities for retriggering and resetting.

The 555 timer is an arrangement of comparators and a bistable that can be used as a monostable. The internal circuit organization is shown in Figure 10.24(b) together with connections that must be made for monostable operation. Triggering occurs on the negative-going edge of the trigger pulse when the voltage level has fallen to below one-third of the positive supply voltage. The minimum pulse width required for triggering at about 25°C with a voltage trigger level of 0.3 of the supply voltage is 50 ns. Once the circuit has been triggered by a negative-going pulse the output remains high until the preset time has elapsed, even if further trigger pulses are applied during this period. The preset time is given approximately by $t = 1.1\, CR$ where C and R are the timimg components shown in the circuit diagram.

Number systems and counting codes

Many industrial processes depend upon counting events. Counting is also the central function involved in the operation of digital computers and digital instruments such as voltmeters and frequency meters. So far as people are concerned, the decimal system of counting is the one that is used for most purposes; such a system has its natural origin in the fact that we have ten digits on our hands. Electronic counting machines, however, work most reliably in a binary system because switching between two states is a well-defined action in properly designed transistor circuits. Unfortunately a problem arises at the interface between human beings and machines, hence we must devise code converters to translate from human language to machine language and vice versa. So far as counting is concerned the difficulty in translation results from the absence of a one-to-one correspondence between the characters of a binary code and those of the decimal code.

In the decimal system there are ten distinct and different digits (0, 1, 2. . . 9). For magnitudes greater than 9 the convention is to arrange digits in rows starting with the most significant on the left and concluding with the least significant on the right. The significance is determined by what is called the 'weighting' of a digit; thus arises the concept of the 'tens', 'hundreds', 'thousands', columns, etc. For example, $457 = 4 \times 10^2 + 5 \times 10^1 + 7 \times 10^0$. The number ten is called the radix, the numbers 0, 1, 2, etc. being the indices. Taken together, a radix and an index are termed the weight. Conversion of a decimal number to another expressed with any other radix is possible. Thus, had we all been born with only one hand it might well be that a radix of five would be the natural choice. A shepherd having a number of sheep might therefore say he had 232 sheep using the decimal system, but in a system using a radix of five he would say he had 1412 sheep, because for this system $1412 = 1 \times 5^3 + 4 \times 5^2 + 1 \times 5^1 + 2 \times 5^0$. The weightings, starting from the least significant end, are respectively 1, 5, 25, and 125. Using a pure binary code the weightings (again starting with the least significant) are $2^0, 2^1, 2^2, 2^3$, etc.; the magnitude of 232 in the binary system is 11101000.

The process of adding two binary numbers is similar to that for adding denary numbers. (Denary numbers are those numbers using a radix of 10, i.e. the familiar 'decimal' numbers.) We start with the least significant digit of each number. If the sum of these digits is less than 10 then we can use another digit to express this sum. For sums greater than 10 we use a single digit and then 'carry' a ten into the next column e.g. $7 + 5 = 2$ plus a 'carry' of 1 (ten) into the 'tens' column. In the binary system we have only two available digits, hence $1 + 0 = 1$ or $0 + 1 = 1$ but $1 + 1 =$

0 plus a 'carry' of 1 into the next column. A logic circuit designed to add two binary digits must therefore have a 'carry' line as well as the output 'sum' line. (See Chapter 11.)

Codes

It is possible to arrange sets of binary digits to represent numbers, letters of the alphabet or other information, by using a given code. Some of the important codes used in digital work are described below.

Binary decimal codes

The natural, or pure, binary code is not the best code for data transmission, for analogue-to-digital converters, or for instrumentation involving decimal readouts. For these purposes the simplicity of the pure binary code is abandoned. We retain the binary system of 0 and 1, but arrange for sets of binary digits (i.e. 'bits') to correspond to a decimal number or letters of the alphabet. The sets of bits are constructed by devising a code other than the pure binary code. When a correspondence with denary digits is involved, the codes are called 'binary decimal codes'. Representing any decimal number, we thus obtain a binary coded decimal (BCD) number. This BCD number consists of an arrangement of sets of bits, each set corresponding to a decimal digit. Any set must contain enough bits to enable the required number of unique and distinct arrangements to be recognized. For decimal counting we have to use four bits per decade since four bits yield 16 unique arrangements. This explains the importance of four-variable Karnaugh maps. Such maps will be shown to be useful for decoding BCD numbers and for code conversions involving BCD numbers. (For a description of Karnaugh maps, see Appendix 2).

Any arbitrary weighting may be selected for a binary decimal code. A well-used code is the 8421 BCD code. In this code the four digits have weightings of 8, 4, 2, and 1, respectively; thus the decimal digit 7 is represented as 0111 [i.e. $(0 \times 8) + (1 \times 4) + (1 \times 2) + (1 \times 1)$]. Using the 8421 BCD code, then ten decimal digits are represented by the first ten numbers of the pure binary code starting with 0000. A decimal number greater than nine is represented by sets of four bits each four-bit set representing a decimal digit. For example, the decimal number 192 is represented in the 8421 BCD code as 0001 1001 0010. Spaces have been left in the printing to show the four-bit sets; any counter, however, stores the information as 000110010010 since the electronic circuitry automatically distinguishes between the four-bit sets. It should be noted that although the sequence 000110010010 represents the decimal number 192 in the 8421 BCD code, such a sequence in the pure binary code represents 402 (i.e. $2^8 + 2^7 + 2^4 + 2^1$).

Since any arbitrary weightings may be used, it is understandable that codes other than 8421 BCD code have been investigated to see if any advantages could be gained by using different weightings. The 2421 BCD code is one such example. Examination shows that this weighting yields the same binary configuration as the 8421 code up to number seven. The decimal number eight is represented as 1110, and nine is represented as 1111. This code therefore makes for easy resetting of the bistables storing the equivalent of '9' to '0' because a single reset pulse may be applied to all bistables. Such is not the case with 8421 code.

A simplification in decimal arithmetic is possible if the nines complement is available; subtraction can then be reduced to an addition process. (A comparable simplification is possible in binary arithmetic by using the twos complement.)

The expression $(9 - y)$ is known as the 'nines complement' of y; for example, the nines complement of three is six. Thus

$$7 - 3 = 7 + 6 + 1 = (1)4$$

The '1' in the tens column is rejected by the electronic circuitry since no provision is made to 'carry it forward'. Obtaining the nines complement is therefore of some importance. A BCD code can be devised that makes it unnecessary to use additional circuitry for the generation of the nines complement. One such code is an Excess 3 code which is related to the 8421 BCD code. The configuration representing decimal 0 is defined as 0011 instead of 0000. Thereafter, the representation of any decimal digit x in the XS3 (Excess 3) code is identical with the representation of $x + 3$ in the 8421 code. Examination of the XS3 code shows that the nines complement of any digit is obtained by inverting each of the four binary digits in the set. Substantial economies are effected because the bistable storing any binary digit has two outputs Q and $\bar{Q}$. No further circuitry is therefore required to produce the nines complement.

For comparison, the three BCD codes discussed are given below;

BCD Code 8421	XS3(8421)	2421	Decimal digit
0000	0011	0000	0
0001	0100	0001	1
0010	0101	0010	2
0011	0110	0011	3
0100	0111	0100	4
0101	1000	0101	5
0110	1001	0110	6
0111	1010	0111	7
1000	1011	1110	8
1001	1100	1111	9

Examination of these BCD codes show that the transition from one decimal digit to the next digit often involves the simultaneous change of two, three or four bits. For data transmission purposes this can lead to very unreliable operation. Where, for example, the information is provided by a shaft encoder, it is impossible to operate two or more readout devices (e.g. photocells) at precisely the same time. Although the time differences are very small, such is the switching speed of modern bistables that non-synchronous operation leads to spurious counts. This difficulty has led to the invention of what are called Gray or reflected codes.

A Gray code is one in which the transition from one decimal digit to the next involves the change of only a single bit. The easiest way to generate a Gray code is to make use of a Karnaugh map. Figure 10.25 shows how this is done. A Gray code is sometimes referred to as a 'reflected' code. This is because if A, B, C and D and the decimal digits 0 to 9 are used as grid references, as shown in Figure 10.25, then the decimal numbers and corresponding binary representation may be plotted by filling in the squares. Certain reflections are recognized, e.g. in the section between 0 and 3, and 3 and 7. Many different Gray codes can be generated by using a different cell on the Karnaugh map as a starting point. Provided the transition from

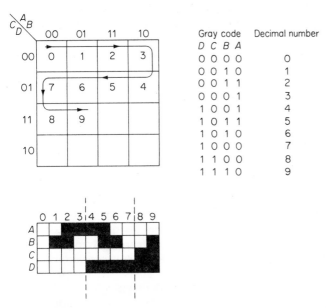

Gray code	Decimal number
D C B A	
0 0 0 0	0
0 0 1 0	1
0 0 1 1	2
0 0 0 1	3
1 0 0 1	4
1 0 1 1	5
1 0 1 0	6
1 0 0 0	7
1 1 0 0	8
1 1 1 0	9

Figure 10.25 The generation of one form of Gray code using a Karnaugh map. Plotting ABCD against the decimal numbers gives the pattern shown. Certain 'reflections' are evident (e.g. between 0 and 7 there is a form of symmetry about the line between 3 and 4). For this reason Gray type codes are sometimes called 'reflected codes'

one cell to the next is made in an orderly fashion, as exemplified in Figure 10.25, a Gray code results. Some Gray codes exhibit better reflections than others.

Gray codes are the ones used in shaft encoders. These devices give an output signal that consists of sets of binary digits. Each set corresponds to the angular position of the shaft relative to some datum direction. Essentially the shaft is attached to a disc upon which transparent and opaque areas are formed by a photographic process. These areas are arranged in concentric rings, each ring being provided with a photocell which detects the presence or absence of an opaque area. Sets of 0s and 1s are thus given out depending upon whether the photocells are illuminated or not. By choosing a Gray code only one photocell changes its state at any one time during the shaft rotation.

Octal and hexadecimal coding

For the reasons already given the basic machine code used in computers and digital equipment consists of sets of binary digits. When the information is being discussed or written by human beings, the long sets of 0s and 1s are confusing. Mistakes are easily made because the presence of a spurious 1 or 0 is difficult to detect. It is convenient therefore for human operators to use a radix other than 2. The choice of radix must be closely related to 2. The octal code has a radix of 8 (2^3). With this code long strings of 0s and 1s can be reduced to manageable form. The conversion of a binary number to an octal number is performed in a similar manner to that described at the beginning of the chapter. In the octal system eight digits are recognized (0, 1, 2, . . . 7) and the weightings starting with the LSB are 8^0, 8^1, 8^2, 8^3 and so on.

A code which is in very common use, especially in connection with microcomputers, is the *hexadecimal* code. This code has a base (i.e. radix) of 16. We therefore need only one symbol to represent a group of four bits. Since there are 16 different combinations of bits, 16 symbols are required; they are 0, 1 . . . 9, A, B, C, D, E, and F.

We thus have several codes with which to describe a given quantity. The following, for example, are all equivalent.

$$0100111000010011 = 19987_{10} = 47023_8 = 4E13_{16}$$

(To avoid ambiguity, small subscripts are used to indicate the radix.) In the above example we have a binary number expressed in denary, octal and hexadecimal form.

The ASCII code

Although the addition of binary numbers is central to the function of a computer, the latter would be of restricted value were it to be confined to arithmetic operations alone. Messages and information in a language that uses letters of the alphabet (e.g. English) and data of other kinds need also to be processed. Computers operate by coding letters of the alphabet, other symbols, and data into binary form. The computer can then operate on the binary 'strings' as though they were numbers. Computers must be able to communicate with other computers and digital equipment, so it is an advantage for all the digital systems to adopt the same code. The ASCII code is a suitable standard, and is now an internationally recognized code for the manipulation, storage and communication of data. The letters ASCII stand for American Standard Code for Information Interchange. It is a seven-bit alphanumeric code which covers letters of the alphabet and numbers as well as many other symbols and control characters. Figure 10.26 shows many of the binary codes together with their interpretations. The letters of the alphabet and other symbols are useful additions since human computer operators are happier with readable displays on a VDU (visual display unit) or via a printer; they also need to communicate with a computer via a typewriter-type keyboard. Decoding logic circuitry must therefore be devised that will allow humans to 'talk' to computers and vice versa.

The ASCII code is not the only one available for this purpose. An Extended Binary Coded Decimal Interchange Code (EBCDIC) has been developed by IBM for the use on their mainframe computers. It is an eight-bit code.

Decimal numbers	ASCII code in binary	in hex	Alphabetical characters	ASCII code in binary	in hex
0	011 0000	30	@	100 0000	
1	011 0001	31	A (a)	100 0001	41 (61)
2	011 0010	32	B (b)	100 0010	42 (62)
3	011 0011	33	C (c)	100 0011	43 (63)
4	011 0100	34	· ·	· ·	· ·
5	011 0101	35	· ·	· ·	· ·
6	011 0110	36	X (x)	101 1000	58 (78)
7	011 0111	37	Y (y)	101 1001	59 (79)
8	011 1000	38	Z (z)	101 1010	5A (7A)
9	011 1001	39			

Other symbols			Control signal meaning		ASCII code
:	011 1010	3A			
;	011 1011	3B	NUL	Null	000 0000
<	011 1100	3C	SOH	Start of heading	000 0001
=	011 1101	3D	STX	Start of text	000 0010
>	011 1110	3E	EOT	End of text	000 0011
?	011 1111	3F	EOT	End of transmission	000 0100
Space	010 0000	20	ENQ	Enquiry	000 0101
!	010 0001	21	ACK	Acknowledge	000 0110
"	010 0010	22	BEL	Bell (audio sound)	000 0111
#	010 0011	23	BS	Backspace	000 1000
$	010 0100	24	HT	Horizontal tabulation	000 1001
%	010 0101	25	LF	Line feed	000 1010
&	010 0110	26	VT	Vertical tabulation	000 1011
'	010 0111	27	FF	Form feed	000 1100
(	010 1000	28	CR	Carriage return	000 1101
)	010 1001	29			
*	010 1010	2A	[	101 1011	5B
+	010 1011	2B	\	101 1100	5C
,	010 1100	2C	]	101 1101	5D
-	010 1101	2D	↑	101 1110	5E
.	010 1110	2E	←	101 1111	5F
/	010 1111	2F			

Figure 10.26 Some of the ASCII (American Standard Code for Information Interchange) codes for numbers, alphabet lists and other common symbols

Chapter 11

Logic and digital circuits

Digital circuits and systems have now become so important that they dominate the field of electronics. The amount of money to be made from the manufacture and sale of digital equipment is so enormous that, understandably, the producers of digital integrated circuits spend a great deal of effort in devising those circuits and sub-systems needed to satisfy the demands of their customers. A vast array of gates and sub-systems are now available, so it is not surprising that many newcomers to the field are bewildered by the extensive range of circuits that can be bought, and the large amount of jargon that surrounds these products. This chapter is intended to introduce the reader who is new to the subject to some of the basic circuit arrangements that are used in combinational and sequential logic.

The ultimate sequential logic system is, of course, a digital computer. The reader will appreciate that in order to keep the cost and size of this book down to reasonable proportions, it is not possible to discuss in detail the vast subject of digital computers. Many books are now available that deal with this subject alone. It is hoped that this chapter will prove a suitable 'teeing-off' area for further studies in the computer field.

Combinational circuits

Circuits in which the output depends only upon the current state of the inputs are known as combinational logic circuits; they can be built using only logic gates and associated power supplies. Combinational logic systems are very easy to realize if the logic function is already known. Take, for example, the function

$$F = AB\bar{C}\bar{D} + A\bar{B}\bar{C}\bar{D} + A\bar{B}\bar{C}D + AB\bar{C}D + ABC\bar{D} + A\bar{B}C\bar{D}$$

By using a Karnaugh map (Figure 11.1(a)) we can minimize the function and obtain

$$F_{min} = A\bar{C} + A\bar{D}$$

(Karnaugh maps are briefly described in Appendix 2). Assuming that NAND gates are used, and that only A, B, C and D are present, it will be necessary to generate $\bar{C}$ and $\bar{D}$. The initial circuit then is as shown in Figure 11.1(b). Those gates marked with a cross are redundant, and thus the final circuit is as shown in Figure 11.1(c). Once experience is gained in reducing initial circuits to their final forms, the short cuts become obvious, and in many cases the final circuits can be drawn immediately.

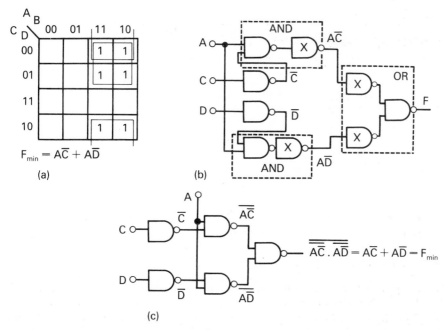

$$F_{min} = A\overline{C} + A\overline{D}$$

(a)   (b)   (c)

Figure 11.1 Combinational logic circuit to realize a given Boolean function

Adders

A useful practical combinational logic circuit is one that can perform the function of addition on two binary numbers. An arrangement of logic gates to perform the function of adding two binary digits is called a half-adder. The binary digits (usually abbreviated to bits – taking the 'b' from binary and the 'its' from digits) are presented to the input lines and the output line carries the sum; an additional output 'carry' line is also present. The circuit is given in Figure 11.2.

A full-adder is a logic circuit which has three inputs, viz. the two digits to be added plus a possible carry bit from an adder operating upon the adjacent less significant bits. Two outputs are available which are the sum and carry bits. Figure 11.3 shows the circuit arrangement.

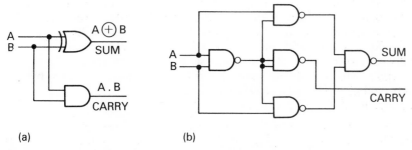

(a)   (b)

Figure 11.2 The half-adder: (a) uses an Exclusive-OR and an AND gate; (b) uses NAND gates to perform the same function

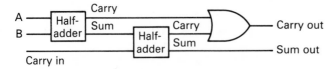

A	B	Carry in	Sum out	Carry out
0	0	0	0	0
0	0	1	1	0
0	1	0	1	0
0	1	1	0	1
1	0	0	1	0
1	0	1	0	1
1	1	0	0	1
1	1	1	1	1

Figure 11.3 The full-adder

Only rarely do we deal with numbers or data that are so simple that single-bit addition is sufficient. In practice a four-bit number is the minimum size we use, one reason being that it requires four binary digits to obtain sufficient different combinations (16 for four bits) to cover the 10 different denary digits. For many computer purposes the 16 different bit combinations of a four-bit number are too few to handle the large amounts of data to be processed. The usual word length in a microcomputer is eight bits. An eight-bit binary number has 256 different combinations of 0s and 1s. In jargon terms an eight-bit number is known as a *byte*, and a four-bit number is called a *nibble*. Both *nibbles* and *bytes* are not confined to numerical quantities; in computer work a given combination of 0s and 1s may represent a letter of the alphabet, as, for example, with the ASCII code. Since a string of 0s and 1s may represent both letters of the alphabet or numerical values, the string is said to represent an alphanumeric quantity.

The addition of two nibbles requires a half-adder and three full-adders; it may be that four full-adders are needed if carry information is also present at the inputs. The complexity of gates makes the arrangement an ideal candidate for IC fabrication. The 74LS83 IC, for example, is a TTL chip that will add two nibbles, as shown in Figure 11.4.

The numbers representing the data are carried by two sets of input lines. Other integrated circuits in the system may wish to interrogate these lines, which physically are copper conductors on a printed circuit board. As mentioned previously, these connectors are known collectively as data buses. In circuit diagrams it would be tedious and confusing to draw each individual line, especially if 8- or 16-bit words were involved. The data buses are therefore drawn as shown in Figure 11.5. Here two 74LS83s are used to add two 8-bit numbers. Each byte is divided into two nibbles. The least significant nibbles of each byte are added, as are the most significant nibbles. The combined result gives the sum byte.

Equality detector

Frequently in digital instrumentation and computers it is necessary to compare two multi-bit binary signals in order to detect equality between them; this is particularly

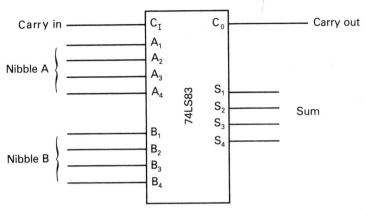

Figure 11.4 Full-adder for four-bit numbers

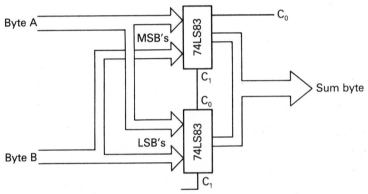

Figure 11.5 Method of adding two bytes using four-bit adders (MSB = Most significant bits; LSB = Least significant bits)

so when numerical data are involved. The principle of establishing equality depends upon comparing the corresponding bit-pairs in each signal. If all of these bit-pairs are equal simultaneously then the two signals are equal.

For a single pair of bits, say A_1 and A_2, then $A_1 = A_2$ when $A_1.A_2 + \bar{A}_1.\bar{A}_2 = 1$. Since

$$A_1.A_2 + \bar{A}_1.\bar{A}_2 = \overline{A_1.\bar{A}_2 + \bar{A}_1.A_2},$$

this will be recognized as the Exclusive-NOR function, and hence an Exclusive-NOR gate is all that is needed to indicate equality between the two corresponding bits. By using a separate Exclusive-NOR gate for each bit-pair in the signal, and connecting the outputs from these gates to the input of an AND gate, the output from the AND gate shows when the two signals are equal. The circuit diagram shown in Figure 11.6 shows how to detect equality between two four-bit signals.

Readers may care to devise another circuit, with three outputs, that compares two four-bit numbers, N_1 and N_2, that represent arithmetic quantities in the 8421-BCD code. The outputs must indicate when $N_1 = N_2$, when $N_1 > N_2$, and when $N_1 < N_2$ respectively. (This is Exercise 15 given at the end of the chapter.)

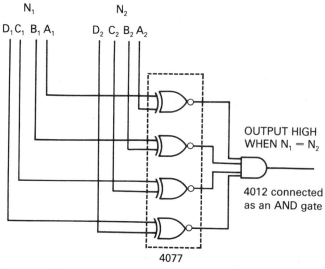

Figure 11.6 Equality detector for two four-bit binary signals

Decoders

Decoders, or code converters, are necessary interfaces between systems speaking different languages. We may, for example, wish to convert a binary number into readable form for a human being. We must therefore devise a logic system that will accept the binary digits at its input and deliver output signals that will energize a seven-segment display or the appropriate character on a printer, or cause the character to be displayed on the screen of a cathode-ray tube in a VDU. Many digital voltmeters give an 8421 BCD output in addition to the decimal display. It may be necessary to transmit the BCD information to a remote position. As we have already seen, the 8421 code is unsuitable for data processing and transmission. The problem at the interface between the voltmeter and data processing equipment is to convert the 8421 code into a more suitable one. The general method of designing the converters may be understood by taking a specific example. The following design procedure yields the logic circuitry necessary to convert the 8421 BCD code into the associated XS3 code.

The first step is to compare the truth tables of each code. If A, B, C, and D represent the four variables of the 8421 code, and A', B', C' and D' the variables of the XS3 code, then the comparison of the truth tables enables us to find a Boolean expression for A' in the terms of A, B, C and D that will make $A' = 1$. Similar expressions can be found for B', C' and D'. The results are summarized in Figure 11.7. The next step is to minimize these expressions by means of Karnaugh maps, as shown in Figure 11.8. The minimization yields four equivalent minimum Boolean expressions that can each be realized in logic form; the final code converter is shown in Figure 11.9.

The more common conversion or decoding functions are available in integrated circuit form. A BCD-to-decimal decoder is available as a 7442; a 7449 is a BCD-to-seven-segment driver. Before embarking on the construction of a decoder, the reader should first consult the list of available TTL and CMOS packages.

```
    8 4 2 1  (BCD)  XS 3
    D C B A         D' C' B' A'
0   0 0 0 0         0 0 1 1
1   0 0 0 1         0 1 0 0      A' = ĀB̄C̄D̄ + ĀB̄CD̄ + ĀBC̄D̄ + ĀBCD̄ + ĀB̄CD
2   0 0 1 0         0 1 0 1
3   0 0 1 1         0 1 1 0      B' = ĀB̄C̄D̄ + AB̄C̄D̄ + ĀB̄CD̄ + AB̄CD̄ + ĀBC̄D̄
4   0 1 0 0         0 1 1 1
5   0 1 0 1         1 0 0 0      C' = ABC̄D̄ + ĀB̄CD̄ + AB̄CD̄ + ĀBCD̄ + ABC̄D̄
6   0 1 1 0         1 0 0 1
7   0 1 1 1         1 0 1 0      D' = ĀBC̄D + ĀBCD + ABCD̄ + ĀB̄CD̄ + AB̄CD̄
8   1 0 0 0         1 0 1 1
9   1 0 0 1         1 1 0 0
```

Figure 11.7 The 8421 BCD code together with its associated XS3 code

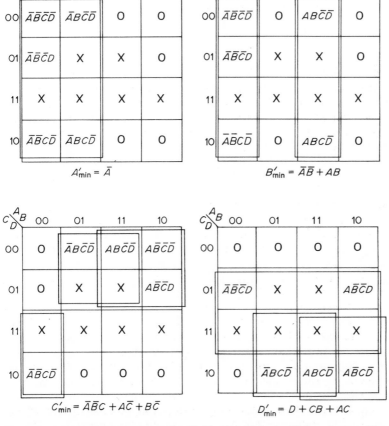

$$A'_{min} = \bar{A}$$

$$B'_{min} = \bar{A}B + AB$$

$$C'_{min} = \bar{A}\bar{B}C + A\bar{C} + B\bar{C}$$

$$D'_{min} = D + CB + AC$$

Figure 11.8 Reduction of expressions for A', B', C' and D' using Karnaugh maps. Those cells marked 'O' are not available for minimization since they are possible states in the 8421 BCD code. Those marked with a 'X' can be used since these combinations do not arise in the 8421 BCD code

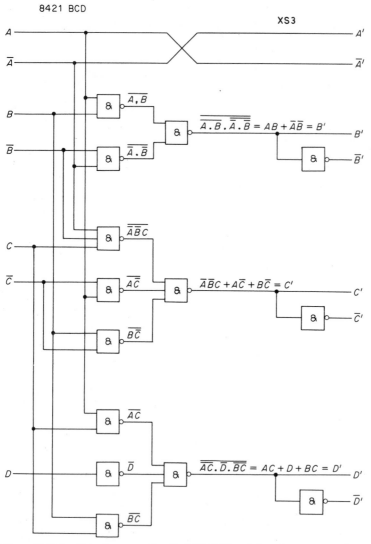

Figure 11.9 Translator for converting the 8421 BCD code into the corresponding XS3 code

Numeral displays

One of the most frequently used decoders in use today translates binary-coded numerical data into a form suitable for driving a numeral display that can be easily recognized by human operators. Three low-voltage readout devices are in current use; all depend upon the seven-segment display shown in Figure 11.10. The code converter may be designed with Karnaugh maps along the lines described earlier in the chapter. A map is drawn for each segment and the combinations of A, B, C and D representing the BCD digit are plotted and minimized in the usual way. For the 8421 BCD code, ICs are readily available for seven-segment decoding purposes (e.g. 7447, 7449 and 4543B.)

Hot-filament displays use a separate wire for each segment. When energized from the logic system, the wires glow in the same way as the filaments in any incandescent lamp. In spite of development, hot-filament displays consume a fair amount of power and have a limited life.

Crystal lamps that can emit light continuously for many years are being developed by most of the big firms, notably RCA in America. These lamps are made from synthetically grown gallium arsenide phosphide and gallium phosphide crystals. Light-emitting diodes (LEDs) made from these semiconductor compounds emit red, green or orange light when a small electric current is passed through them. For all practical purposes almost no heat is generated. Low power consumption, long life, fast operating speed, small size and extreme reliability are the advantages that are gained when these light sources are used. During operation, electrons are excited across the forbidden gap between the valence and conduction bands or between these bands and levels within the forbidden gap that are associated with impurities. For any given semiconductor material the energy changes are characteristic. For GaP and GaAsP the energy changes bring about the emission of visible light. Figures published by Bell Laboratories show that a GaP device, approximately 0.2 mm in size, produces a brightness level of about 5 mcd (or up to about 50 mcd for high-intensity types) when operated at 10 mA. Part of this brightness must be traded for area and contrast by using a suitable reflector.

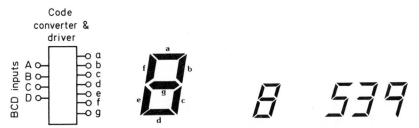

Figure 11.10 Seven-segment numeral display patterns used in connection with low-voltage numeral tubes that depend upon hot filaments, gallium arsenide phosphide diodes or liquid crystals. The basic arrangement is shown on the left; the pattern on the right shows how the number 539 appears

Much work is being undertaken at present to control the radiation pattern of the semiconductor wafer, and to select appropriate reflector materials and shapes. The LED can be used as a single lamp, or as a dot array to form numerals and letters of the alphabet, i.e. alphanumeric display. For numeral purposes the efficient seven-segment arrangement described above is often used.

Liquid crystal displays form the third group of indicator devices. Liquid crystals are organic substances that are termed 'mesomorphic', being intermediate between the solid and liquid state. Three types are identifiable: smectic (from the Greek, meaning soap-like); cholesteric, being derived from cholesterol; and nematic, meaning thread-like. It is the third type that is of special interest because such crystals can be controlled electronically to yield alphanumeric displays. In the undisturbed state, thin layers of this material appear to be quite clear and incident light passes through the liquid. For the displays used in electronic equipment the backing behind the liquid is black and so incident light is absorbed. The light itself consists of randomly oriented groups of molecules. The boundaries between the groups appear as snake-like threads when viewed under a microscope with crossed polaroids. When the liquid is agitated the molecules break away from the respective groups and a huge increase in the number of boundaries takes place. Incident light is then scattered throughout the liquid, and then reflected in much the same way as it is by milk (a mixture of fat and water). The liquid then appears to be white and diffuse. Increases in ambient light increase the contrast, and thus the display appears brighter; this is an important advantage over gallium phosphide displays. In electronic apparatus the turbulence in the crystal is achieved by applying a suitable voltage to the liquid. The presence of water as an impurity, together with the flow of ions, is essential to the mechanism. The turbulence is determined by the voltage level and by the number of ions present. For display purposes a thin ($\approx 50\,\mu m$) layer of liquid is sandwiched between a sheet of conductive glass and plastic film of Mylar. Where seven-segment numeral patterns are involved the plastic film has the appropriate parts cut away. Beneath the numeric pattern in the plastic film are areas of gold (or other conducting material e.g. glass treated with tin oxide) arranged in an identical pattern. Figure 11.11

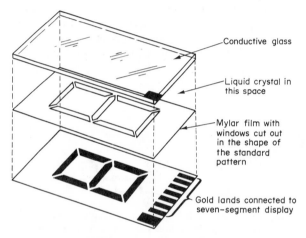

Figure 11.11 Exploded view of a liquid crystal numeral tube. The actul thickness of the liquid crystal film is about $50\,\mu m$. The whole sandwich is hermetically sealed. Gold end contacts are exposed and connect the pattern and conductive glass to the logic system

shows the essential features of this type of indicator. Those segments which are not activated stand out from the background as black areas; the segments to which an electric field is applied merge into the background colour of the mylar film. It is an easy matter to design the logic so that all the unwanted segments of a given digit are energized thus leaving the wanted undisturbed segments standing out as black areas.

Multiplexers

The term 'multiplexer' is used in telecommunications to denote the use of a common information channel for sending many messages. In frequency-division multiplexing, the frequency band transmitted by the common channel is divided into narrower bands, each of which constitutes a distinct channel. In time-division multiplexing, different channels are interrogated by intermittent sequential connections to the common channel.

Time-division multiplexing is frequently used by small computers and data acquisition equipment to interrogate some system that needs to be controlled, or from which data are required for analytic purposes. The signals from the various transducers in the system are switched sequentially, by a multiplexer, into the common input line known as an input port. Multiplexers of this type are based on transmission gates that have a very low 'on' resistance. Such gates are particularly easy to fabricate using CMOS technology. Figure 11.12(a) shows the circuit arrangement. When G1 is positive ($=V_{DD}$) the p-channel MOST (TR1) is off; if simultaneously G2 is held at zero voltage then the n-channel MOST (TR2) will also be off. There will thus be no transmission of information between the input and output terminals. Conversely, if G1 is at zero voltage and G2 is at V_{DD} then both transistors will be on and a low resistance path will be established between the input and output terminals. The 'on' resistance is as low as $30-100\,\Omega$ in some commercially available units. Figure 11.12(b) shows the addition of an inverter to provide the control signals to G1 and G2. Readily available ICs have two, four or eight of these bilateral switches on a single chip. The 4066, for example, is a common CMOS package which contains four single-pole single-throw switches. Transmission gate switches have many applications in analogue circuits, multiplexers and data acquisition systems. Digital-to-analogue converters, which are described later in the chapter, also use this type of switch.

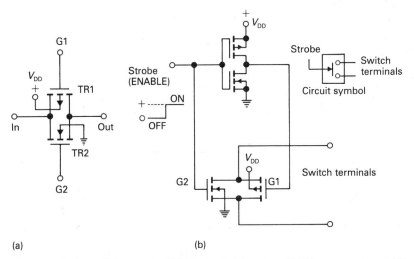

(a)

(b)

Figure 11.12 Transmission gate and bilateral switch based on CMOS arrangements: (a) Basic CMOS transmission gate; (b) CMOS bilateral switch

An array of switches can be operated by connecting their control lines to a control bus. A coded arrangement of 0s and 1s on the control bus thus allows information to be transmitted via selected switches. The control signal terminal may be called the STROBE input; this term is an alternative for ENABLE when several systems must be enabled sequentially. (The word 'strobe' – short for stroboscopic – comes from the Greek word meaning 'spinning'. Strobe lighting, for example, if a series of high-intensity light flashes that allow us to sample the position and appearance of a moving object at regular intervals of time.)

Time-division multiplexing can easily be achieved by altering the binary word on the control bus at regular intervals. We may wish, for example, to turn each of four switches in turn in order to sample the analogue outputs from a set of transducers. The sequence of words on the control bus would then be 0001, 0010, 0100 and 1000. The selection of one of four switches, as shown in Figure 11.13, requires a four-bit word. The selection can, however, be made more economically by using conventional gates in conjunction with a set of transmission gates, as shown in Figure 11.14(a). This arrangement selects one switch from four by using a two-bit code. Figure 11.14(b) shows a circuit that uses an IC multiplexer with an eight-line input capacity; in this case the selection process requires a three-bit code.

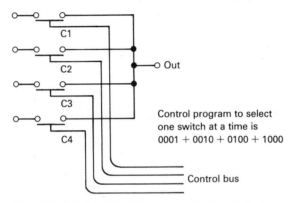

Figure 11.13 Each switch symbol represents the circuit of Figure 11.10 (b). With the given program, one switch is selected at any one time

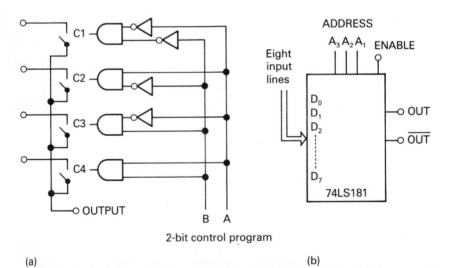

Figure 11.14 Multiplexers: (a) shows an arrangement of four transmission gates controlled by a two-bit program via selecting logic; (b) shows an IC version for eight input lines

Sequential logic circuits

In the circuits discussed so far, logic gates could be assembled to produce a given logic function at the output. The examples were of combinational logic in which the outputs depended only on the logic states of the inputs at the current time. There are many logic systems whose outputs are functions, not only of the inputs at a given time, but also upon logic states that were previously present, but are absent at the given time. Such systems are sequential logic circuits. Counters, digital computers and sequence-recognizing circuits are sequential logic systems. Take, for example, a circuit that has four inputs A_0, A_1, A_2 and A_3. If only one input is presented at any one time and the presentations occur in a random sequence, the problem may be to recognize when a given sequence, say A_1, A_0, A_3, A_2, appears at the input. A suitable circuit to perform the recognition cannot be made from logic gates alone; some form of memory must be incorporated. Memory is an essential part of all sequential logic systems.

Any device that can take up one of two stable states may be used as the basis of a memory system. At one time computer memories consisted of arrays of ferrite rings that could be magnetized or not. Today, especially with small home computers, magnetic tape may be used. Here the patterns of 0s and 1s are impressed upon the tape by magnetizing some parts of the tape with a tone of one frequency while the remaining parts are magnetized with a tone of different frequency. Floppy disks use the same principle except that the magnetizing medium (an oxide of iron) is spread on a flexible plastic disk; microcomputer systems can store 100 or 200 kilobytes of information on a single side of the disk. By using both sides, and a more sophisticated disk drive system, the storage capacity of the disk is greatly enhanced. The corresponding hard disk memories (the Winchester disks) are able to store many millions of bits of digital information.

Memories of more modest size rely on transistor bistable circuits. Bistables are referred to by computer people as flip-flops because the circuits remain in a stable state until they are flipped to a second state by a suitable triggering pulse. A resetting pulse can then make the bistable flop back into its original state.

The basic flip-flop, shown in Figure 11.15, consists of a pair of crossed gates. From an examination of the truth table shown in Figure 11.15 we cannot determine

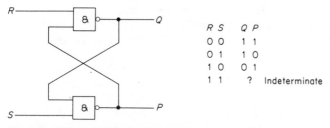

R S	Q P
0 0	1 1
0 1	1 0
1 0	0 1
1 1	? Indeterminate

Figure 11.15 Basic SR flip-flop and associated truth table

with certainty the state of the output when S (set) and R (reset) are both high. Here it will be appreciated that Q and P can be either 0 and 1 or 1 and 0 respectively, since both are possible stable states. A particular difficulty arises if the RS states go from 00 to 11 because then there is no way of knowing what the final state of the flip-flop will be. A race begins in which the faster gate of the flip-flop decides the final outcome. The way in which this unsatisfactory state of affairs is resolved is discussed below.

The clocked SR flip-flop

In addition to the ambiguity in the truth table for an SR flip-flop, a further disadvantage is evident in that such an arrangement is incapable of being operated synchronously with the rest of the system. This difficulty can be overcome by allowing the input gates of the memory unit to have two inputs, as shown in Figure 11.16. When the clock pulse is low the output cannot be altered from its previous state (Q_n) because the inputs to the cross-coupled gates remain at logic 1, whatever the states of R and S. The RS states can then be changed without upsetting the state of the flip-flop. When the clock goes high the changed output (Q_{n+1}) will go to a state, given in the truth table, determined by the RS inputs. Once again, however, a 11 arrangement is forbidden because it leads to an indeterminate output after the clock pulse has been received.

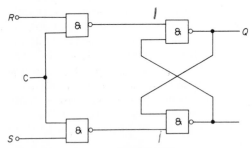

Figure 11.16 The clocked SR flip-flop:

RS	Q_{n+1}	(i.e. state taken up by Q after one clock pulse)
00	Q_n	Output remains in previous state
01	0	Output goes to 0 irrespective of previous state and stays at 0 even on receipt of further clock pulses
10	1	Output goes to 1 and stays there
11	?	Indeterminate state

In the late 1950s, M. Phister suggested a way in which the indeterminacy could be avoided. He showed in his book *The Logical Design of Digital Computers* (Wiley, 1958) that if the outputs were cross-coupled to the inputs, as shown in Figure 11.17 then the RS = 11 states could not occur. To distinguish this arrangement from the SR flip-flop, the letters S and R were changed to J and K. The choice seems to have been an arbitrary one; the letters do not stand for actual words as is the case with RS. Perhaps they were chosen merely to distinguish this flip-flop from others.

If we have J = K = 1, and assume that initially Q = 0, then $\bar{Q}$ = 1 and gate 1 is enabled while gate 2 is inhibited. On the arrival of the clock pulse all of the inputs to gate 1 become 1 and thus the output becomes 0. The bistable therefore changes state. We now have Q = 1, $\bar{Q}$ = 0 and consequently gate 2 is enabled while gate 1 is inhibited. On the arrival of the next clock pulse the output of gate 2 becomes 0, resetting the flip-flop to its original condition. The flip-flop exhibits a toggle action and divides by two. Seemingly, then, no trouble is encountered with 11 states for JK. Unfortunately such an arrangement cannot work, because the time it takes for

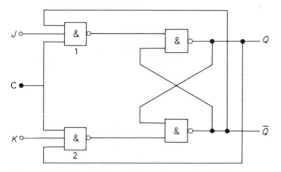

Figure 11.17 The clocked SR flip-flop modified in an attempt to avoid an indeterminate state:

JK	Q_{n+1}	
00	Q_n	Output does not change
01	0	Output goes to 0 and stays there
10	1	Output goes to 1 and stays there
11	$\bar{Q}_n$	Output apparently toggles, i.e. changes state for each clock pulse (but see text)

the transition from one output state to the other is much shorter than the duration of the clock pulse. The flip-flop will therefore oscillate, and at the end of the clock pulse period the state will be indeterminate. Although it is possible to introduce delays in the feedback lines, the resulting flip-flop would suffer a restriction. The delay must be longer than the duration of the longest clock pulse likely to be encountered. If shorter clock pulses are used there is no convenient way of reducing the delay when an IC construction is used. The maximum counting speed cannot then be realized.

The master-slave JK flip-flop

This type of flip-flop is easily the most versatile and popular flip-flop available today. In discrete component form its comparative complexity precluded its general use on economic grounds. With the development of IC techniques, however, the complexity is now no disadvantage. Since the JK flip-flop can function as an RS flip-flop, a clocked RS flip-flop, a T (toggle) flip-flop or a D (data) flip-flop, its popularity is understandable. Such versatility is a great advantage in the servicing of equipment since only one type of bistable is involved; this eases replacement and stock problems.

Several circuit arrangements have been proposed for the JK flip-flop, all of which overcome the oscillation problem. One such circuit is shown in Figure 11.18. In effect, the feedback signals are delayed by storing them for the duration of the clock pulse, together with J and K information, in a master flip-flop. The total information is gated into the master flip-flop on the leading edge of the clock pulse. The feedback from the Q and $\bar{Q}$ outputs ensures that the inputs to the master flip-flop never go from 00 to 11. While the clock is high the transfer gates are inhibited because a 0 on one transfer line is accompanied by a 1 from the corresponding output of the master flip-flop. Simultaneously, a 1 on the other transfer line is accompanied by a 0 from the other output of the master flip-flop. A HIGH clock pulse consequently inhibits both transfer gates, and transfer of information from the master flip-flop to the slave flip-flop is impossible. Once the clock pulse goes to 0 the input gates are inhibited. Each transfer line therefore goes

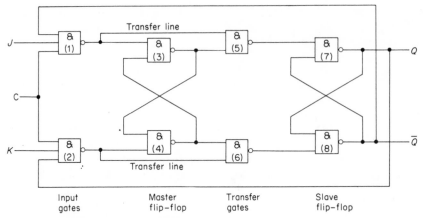

Figure 11.18 The master-slave JK flip-flop

JK	Q_{n+1}	
00	Q_n	No change
01	0	Goes to 0 if necessary and stays there
10	$\underline{1}$	Goes to 1 if necessary and stays there
11	$\overline{Q}_n$	Toggle action

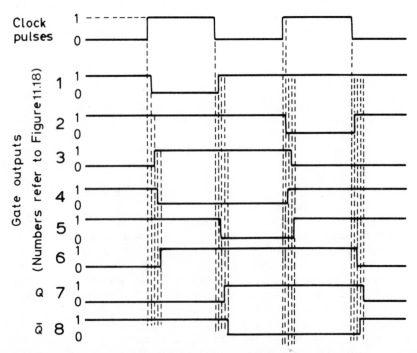

Figure 11.19 The timing diagram for the JK flip-flop shown in Figure 11.18. J = K = 1. The interval between two adjacent vertical dotted lines represents the delay time for a single NAND gate. The interval as shown has been deliberately exaggerated to clarify the changes that occur during counting operations

to 1 and the transfer gates are enabled. The information held in the master flip-flop is consequently transferred to the slave flip-flop. The output conditions change; although these changed conditions are fed back to the input gates, the master flip-flop is not upset because the input gates are inhibited. The arrival of the leading edge of the next pulse allows the new information to be fed into the master flip-flop. On the arrival of the trailing edge of the clock pulse, a transfer of this information to the slave flip-flop takes place and the outputs Q and $\bar{Q}$ return to their original state.

The timing diagram of the operation, shown in Figure 11.19, shows the state of the output of each gate throughout two clock periods. We see that the delay in entering the feedback information into the slave flip-flop depends only on the duration of the clock pulse. The speed of operation is therefore reduced as the clock pulse repetition rate is increased. The maximum operating speed is limited only by the delay time for the gates.

A popular circuit for the JK flip-flop is shown in Figure 11.20. Here the transfer lines, instead of being connected to gates 1 and 2 are joined and connected to the output of an inverter. The input to the inverter is fed with clock pulses. We have in effect two clocked RS flip-flops (one master and one slave); oscillations are prevented since the transfer of the information from the master to the slave cannot take place during the time that the clock pulse is HIGH. This is because the input from the inverter to the gates of the second RS flip-flop is LOW. Only when he clock goes LOW is the slave flip-flop enabled. The timing diagram is similar to that of Figure 11.19. It should be noted that the changes of the output takes place only on the trailing edge of the clock pulse.

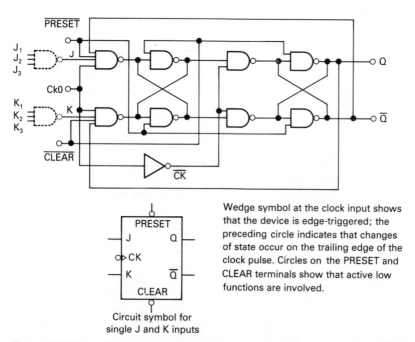

Wedge symbol at the clock input shows that the device is edge-triggered; the preceding circle indicates that changes of state occur on the trailing edge of the clock pulse. Circles on the PRESET and CLEAR terminals show that active low functions are involved.

Circuit symbol for single J and K inputs

Figure 11.20 Alternative form of JK flip-flop. By using additional NAND gates at the J and K inputs expansion of these inputs can be achieved. Some commercially available JK flip-flops have three J and three K inputs

All commercially available units have facilities for presetting the outputs to some required state prior to operation, and to clear the information after operation ready for a new start. In Figure 11.20 these terminals are labelled $\overline{\text{PRESET}}$ and $\overline{\text{CLEAR}}$. The significance of the negation bars is that these terminals are normally maintained at a high voltage, and are made to go low only when the PRESET or CLEAR functions are to be activated; these terminals are then said to be active low terminals. Units that do not use NAND gates may have PRESET and CLEAR terminals, in which case the logic levels are reversed, and the terminals are active high. Manufacturer's data should always be consulted prior to using the flip-flops.

D flip-flops

This type of flip-flop is used extensively in computer and digital control circuitry to 'capture' or latch on to data present on data lines or buses. Used in this way D (i.e. data) flip-flops are known as data latches, and are ideally suited for the temporary storage of transient information for later examination or manipulation.

JK flip-flops can be used, as shown in Figure 11.21, to form a D flip-flop. The output of the D flip-flop does not change until a clock pulse is applied. Once the input data is transferred to the output terminals it remains there, even when the

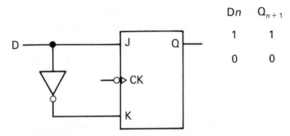

Dn	Q_{n+1}
1	1
0	0

Figure 11.21 Modification of a JK flip-flop to form a D flip-flop. D_n is the data state before the application of a clock pulse. Q_{n+1} is the output state after the clock pulse has been applied. Used in this way the JK flip-flop forms what is termed a data latch

input is removed. Not until the next clock pulse is applied is the new data transferred to the output of the flip-flop. The mode of operation may be understood by examining the truth table for the JK flip-flop given in Figure 11.18.

Where it is undesirable to have the output changes taking place on the trailing edge of the clock pulse, a suitable TTL circuit (e.g. the 7474 given in Figure 11.22) may be used. In this circuit, data transfer takes place at the leading edge of the clock pulse. This is an example of an edge-triggered flip-flop.

BCD counters

Counters that count in the pure binary code are easy to make with JK flip-flops. We simply arrange the flip-flops in a chain with sufficient units to accommodate the largest number likely to be encountered. All of the J and K inputs are tied to logic 1 (i.e. HIGH) and the clock terminal of a given flip-flop is connected to the Q output of the flip-flop immediately preceding it. Each flip-flop divides by two and hence pure binary counting can be achieved. The arrangement of flip-flops into a chain is, of course, not suitable for BCD counting. Four flip-flops are required to store the

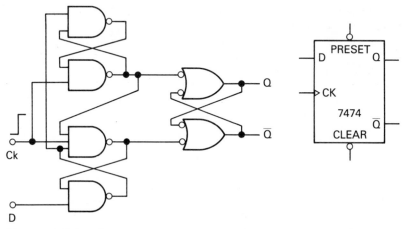

Figure 11.22 Edge-triggered D flip-flop, which triggers on the leading edge of the clock pulse. (The 7474 has two flip-flops in a single 14-pin package)

BCD equivalent of a decimal digit. For counting numbers greater than nine, sets of four flip-flops are required, each set representing the units, tens, hundreds, etc., respectively. The problem in BCD counting is to find a method of restoring any four-bit set to the state corresponding to decimal zero on the receipt of the tenth pulse. For any BCD set the states corresponding to the numbers 10–15 must be inhibited by the use of a suitable arrangement of gates, and the set must be made to return from the 9 condition to the 0 condition. RS flip-flops of the kind shown in Figure 11.16 can be used as the basis of a BCD counter provided suitable gating circuits are used to reset a four-bit set of flip-flops to the zero condition on the receipt of the tenth pulse. The design is not always easy because of the limitations of the RS flip-flop as a counting unit. As an easy introduction to the subject of BCD counters, let us consider the design of a ripple-through asynchronous counter to operate in the 8421 BCD code. The term 'ripple-through' arises because the pulses to be counted are fed into the input of the counter and then ripple through as the counting proceeds.

Examination of the truth table shows that the 8421 code is the pure binary code up to digit 9. This suggest that the starting point is a set of four flip-flops connected as for a binary counter. Once we move from 7 to 8, however, it is necessary to hold the B and C flip-flops at zero for both the ninth and tenth pulses, since on receipt of the tenth pulse all flip-flops must return to zero. The B and C flip-flops can be inhibited after the receipt of the eighth pulse by interposing an AND gate between the A and B flip-flops, as shown in Figure 11.23(b). By noting that $D = 0$ up to the count of 7, the $\bar{D}$ output 'enables' the gate for the first eight pulses. After eight pulses have been received $D = 1$ and $\bar{D} = 0$. No further pulses can be passed to the B and C flip-flops. By separating the S and R inputs of the D bistable, as shown in Figure 11.23(c), a further gate can be controlled by the D output. For counts up to eight this gate is inhibited because $D = 0$). On the count of eight, however, D becomes 1 and the gate is enabled; pulses are then routed past the B and C flip-flops. Flip-flop D is reset on the tenth pulse, whereupon $A = B = C = D = 0$. The four-bit set has therefore been reset in readiness for the arrival of the eleventh and subsequent pulses.

8421 BCD

D C B A	Decimal equiv
0 0 0 0	0
0 0 0 1	1
0 0 1 0	2
0 0 1 1	3
0 1 0 0	4
0 1 0 1	5
0 1 1 0	6
0 1 1 1	7
1 0 0 0	8
1 0 0 1	9

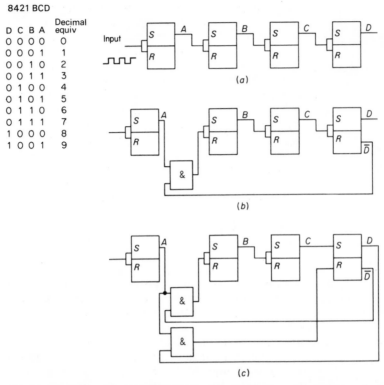

Figure 11.23 Design of an 8421 BCD counter. The SR flip-flops change state on the trailing edges of the respectively input pulses. (a) This arrangement cannot be reset on receipt of the tenth pulse; (b) this arrangement will count up to nine but the B, C and D flip-flops will be permanently inhibited thereafter; (c) circuit for resetting the bistables on the count of ten

The design of counters for other codes is by no means easy if we are restricted to the RS type of flip-flop. The reader can confirm this for himself by designing a counter for a 2421 BCD code using the procedure outlined above. Even though the 2421 BCD code is pure binary up to seven (assuming the least significant digits to be filled first), it is not too easy to see immediately the gating that is required.

One of the great advantages of the JK flip-flop is the ease with which counters can be designed. Unlike the design of counters using the SR flip-flop, we do not have to rely so much on intuition and ingenuity. Mapping techniques can be a help in difficult cases.

Asynchronous counters are those in which the individual flip-flops change state as required by their inputs, but do so independently from the rest of the system. Synchronous counters on the other hand are those in which all the flip-flops that are required to change do so simultaneously under the control of the clock-pulse generator.

The design of an asynchronous BCD counter may be illustrated by taking the example of an 8421 BCD counter. We first write down the truth table showing the states of ABCD corresponding to each decimal digit. A is the LSB and D the MSB of each of the four-bit binary equivalents of the respective decimal digit. We require four JK flip-flops for each bit, as shown in Figure 11.24. From the table given in Figure 11.23 we note that A must toggle (i.e. switch state) on the reception of each input pulse of the string to be counted. The A flip-flop is, in effect, a divide-by-two counter. The necessary JK code to ensure this is $J = 1$ and $K = 1$. The B bistable switches each time Q_A goes from 1 to 0; this bistable may therefore be clocked directly from Q_A. The JK code must be 1–1 for changes to occur. Examination of the truth table shows that changes should occur from 1_{10} to 9_{10} but not on the transition from 9_{10} back to 0. This can be achieved by energizing the J,K inputs from the $\bar{Q}_D$ output. Once 8_{10} is reached, the B bistable remains in its $Q_B = 0$ state until D is reset. The C bistable changes state each time Q_B goes from 1 to 0 and hence may be clocked directly from Q_B; the J and K inputs are permanently held HIGH. For the D bistable we note that a change of Q_D takes place only when Q_C goes from 1 to 0. We cannot clock D directly from Q_C, however, since resetting to zero on the arrival of the tenth pulse would not take place. Resetting can occur, however, by making $J = 1$, but only when $Q_B.Q_C = 1$. The counting circuit therefore is that shown in Figure 11.24.

In the above description the term 'clocking' has been used. It will be appreciated that a clock-pulse generator is not involved. The 'clock' pulses are those that are to be counted.

An alternative way of counting pulses is to treat the count input line as though it were from a clock-pulse generator and connect it to the clock input of every bistable. The counter bistables will then all change state simultaneously provided

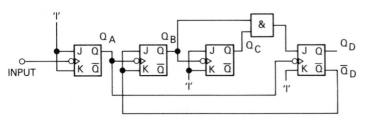

Figure 11.24 Asynchronous 8421 BCD counter that uses JK flip-flops

those that are required to change have the appropriate JK states. The counter is then operating synchronously. Synchronous counters can operate faster than the asynchronous equivalent since we do not have to wait for the information to ripple through the bistables. This ripple-through time limits the rate at which input pulses can be accepted by an asynchronous counter.

The design of synchronous BCD counters is not difficult especially if flip-flops with multiple inputs are provided for J and K. The truth table for the particular BCD code should be examined in connection with that appropriate to the JK flip-flop being used. As an example, the design of an 8421 BCD synchronous counter, using the truth tables of Figure 11.18 and 11.23, may be made using JK flip-flops in which the three J and K inputs each perform an AND function. The A flip-flop must toggle each time it is clocked. J and K must therefore be permanently HIGH. The B flip-flop toggles when Q_A goes from 1 to 0, but must be inhibited on the transition from 9_{10} to 0. One of the J and one of the K inputs must therefore be connected to Q_A. Connecting one of the J inputs to $\bar{Q}_D$ will guarantee toggling until 8_{10}, whereupon the JK states become 01; this means that Q_B goes LOW and stays there. C must toggle with changes of Q_B from 1 to 0, and hence all J and K inputs for the C bistable must be held HIGH. Q_D goes to 1 when A.B.C = 1. We therefore connect the three J inputs to Q_A, Q_B and Q_C respectively. If all three K inputs were held HIGH, Q_D would go HIGH on the receipt of the eighth pulse, but would revert to zero on receipt of the ninth pulse. To prevent this, two of the K inputs are held HIGH, the third being connected to Q_A. In this way Q_D remains at 1 until the arrival of the tenth pulse. Figure 11.25 shows the counting circuit.

In cases where it is not easy to use intuition in the design of counters, mapping techniques will enable a solution to be obtained. To illustrate the principles involved, the design of a 2421 BCD synchronous counter will be described. JK flip-flops and simple gating will be used.

The first step is to write out the appropriate table showing the present and next states of the four flip-flops along with the JK states necessary to give the correct outputs as the counting proceeds. Figure 11.26 shows the table. Karnaugh maps are then drawn for the J and K inputs of each flip-flop, as shown in Figure 11.27.

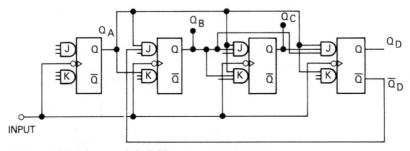

Figure 11.25 Synchronous 8421 BCD counter. The JK flip-flops used here each have three J and three K inputs. Each of the groups of three inputs performs an AND function. J and K inputs shown 'free' are connected to logic 1. The type used may be TTL 7472

Current state	Next state				
$Q_3 Q_2 Q_1 Q_0$	$Q_3 Q_2 Q_1 Q_0$	$J_3 K_3$	$J_2 K_2$	$J_1 K_1$	$J_0 K_0$
0 0 0 0	0 0 0 1	OX	OX	OX	IX
0 0 0 1	0 0 1 0	OX	OX	IX	XI
0 0 1 0	0 0 1 1	OX	OX	XO	IX
0 0 1 1	0 1 0 0	OX	IX	XI	XI
0 1 0 0	0 1 0 1	OX	XO	OX	IX
0 1 0 1	0 1 1 0	OX	XO	IX	XI
0 1 1 0	0 1 1 1	OX	XO	XO	IX
0 1 1 1	1 1 1 0	IX	XO	XO	XI
1 1 1 0	1 1 1 1	XO	XO	XO	IX
1 1 1 1	0 0 0 0	XI	XI	XI	XI

Figure 11.26 Table for the design of a synchronous 2421 BCD counter

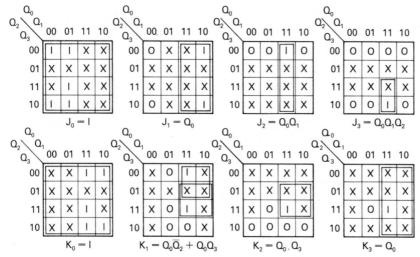

Figure 11.27 Karnaugh maps for use in designing a 2421 BCD synchronous counter

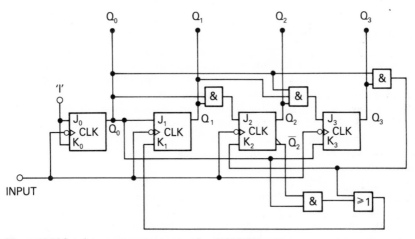

Figure 11.28 Synchronous counter to count in a 2421 BCD code

Minimum expressions are then obtained for each input from which the gating circuits are easily designed. The final circuit is shown in Figure 11.28.

Bounce suppression

In practical situations the pulses to be counted are often generated by mechanical switches. Unfortunately, such switches suffer from contact bounce. On operating the switch the contacts close, bounce apart and then close again. Usually several bounces are involved, lasting from about 1–10 ms. Although the periods between bounces are very short, such is the speed of modern counters that each bounce is counted separately, thus giving a spurious count. One way of avoiding this difficulty is to use an RS flip-flop as shown in Figure 11.29. On operating the switch the first

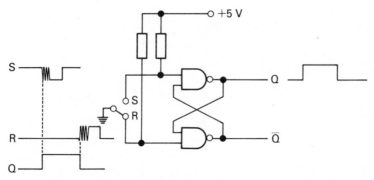

Figure 11.29 Circuit for eliminating the effects of contact bounce in mechanical switches

contact sets Q to 1; subsequent bounces are equivalent to applying multiple 'set' pulses and Q is not affected. When the switch returns to its original position the bistable is reset satisfactorily even though bouncing occurs.

Modulo-n counters

A counter that counts pulses from 0 to $(n-1)$, and then resets itself, is called a modulo-n counter. Thus a modulo-8 counter counts from 0, 1, 2, ... 7, and then resets to zero on the arrival of the eighth pulse. Modulo-n counters thus give one output pulse for every n input pulses. Such counters are useful in computer-controlled machine tools and in computers for incrementing or decrementing numbers held in chains of bistables known as registers.

Provided that asynchronous ripple-through counters are acceptable, the binary counters for modulo-2, 4, 8, ... 2^m present no problem. All that is necessary is to cascade m flip-flops for a modulo-2^m counter. Figure 11.23(a), for example, shows a modulo-16 counter. For this type of counter, resetting on the arrival of the nth pulse is automatic.

When synchronous operation is required, or for values of n that cannot be expressed as 2^m (where m is an integer), it is necessary to use gating circuits to effect the resetting.

Figure 11.30 shows the circuit for a synchronous modulo-8 counter. An examination of the truth table for this counter shows that the output from the B

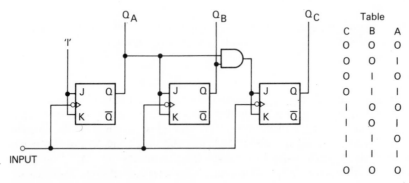

Table		
C	B	A
O	O	O
O	O	I
O	I	O
O	I	I
I	O	O
I	O	I
I	I	O
I	I	I
O	O	O

Figure 11.30 Synchronous modulo-8 counter

flip-flop changes state only when Q_A changes from 1 to 0. By using JK master-slave flip-flops, the JK states for the B flip-flop can be controlled by the Q_A output. Q_C changes state only when Q_A AND Q_B are both 1. This suggests the circuit shown.

A useful basic IC for modulo numbers up to 16 is the 74293. Figure 11.31 shows the schematic of this versatile counting IC together with the pin number connections. (An opportunity has been taken here to use the proposed new circuit symbol conventions; these new symbols, in so far as they affect the diagrams in this book, are briefly discussed in Appendix 4.) The divide-by-two section is merely a single flip-flop; the divide-by-eight has three available outputs, viz. the least significant bit (LSB), the next least significant (NLSB), and the most significant bit (MSB). All outputs are reset to zero when the voltage on both reset leads R0 (1) and R0 (2) is made to go HIGH. For any given modulo number, therefore, we need to examine the appropriate truth table and design the gating to make pins 12 and 13 go HIGH when a reset is required. Figure 11.32 shows, as an example, a modulo-5 counter.

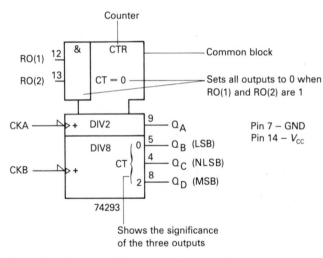

Figure 11.31 Integrated (TTL) circuit counting unit for use as a versatile counter. The + sign close to the clock inputs show that the counters increase as the pulses are received, i.e. they are count-up counters

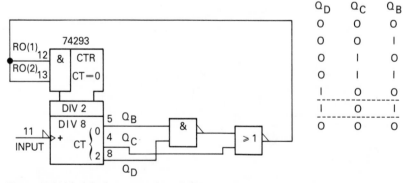

Figure 11.32 Modulo-5 counter using a 74293. The 101 output (shown between dotted lines in the state table) is a transient intermediate state

Reversible counters

So far, the counters we have been discussing increase their count by one on the receipt of an input pulse. When incremented in this way they are called up-counters. It is sometimes convenient, however, to subtract one from the count on receipt of every pulse, i.e. to decrement the counter; such counters are called down-counters.

Down-counters can be designed using similar methods to those employed for the design of up-counters. The state tables and JK codes need to be suitably modified. It will be found that the minimum expressions for the J and K inputs are identical in form to those obtained for the corresponding up-counter except that all flip-flop outputs are replaced by their complements. All that is necessary, therefore, is to interchange the connections to gates at each flip-flop replacing the Qs with $\bar{Q}$s, and vice versa. The principle can be appreciated by comparing the modulo-4-up- and down-counters shown in Figure 11.33.

The interchanging of the Q and $\bar{Q}$ outputs is achieved by designing a controlled gating circuit that effects the changeover when required. Figure 11.34 shows how this can be done for modulo-8 up/down counter.

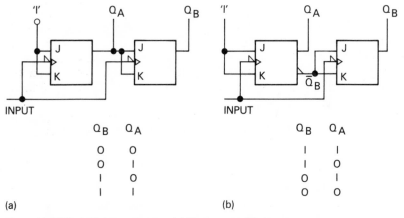

Figure 11.33 Up- and down-counters: (a) Up-counter; (b) down-counter

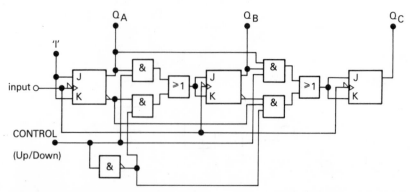

Figure 11.34 Modulo-8 up/down counter. It counts up when CONTROL is high, and down when CONTROL is low

Lock-out

A special difficulty arises in the design of counters that have redundant states. For example, any BCD counter is designed to count only up to 10, and hence six redundant combinations of flip-flop outputs exist. Normally these would not be encountered, but unfortunately in practical situations switch-on transients, and noise on the supply and signal lines may inadvertently switch the flip-flops into a redundant state. If the design of the counter is not satisfactory, continual cycling among the redundant states may occur; when this happens the counter cannot operate properly. The designer must incorporate circuitry to ensure that, after the reception of one or two pulses, the counter jumps back into a legitimate set of flip-flop outputs. Since flip-flops with resetting facilities are now readily available, it is usually prudent to include a gating system that resets the flip-flops to zero should any redundant state appear.

Timing

In all systems in which flip-flops and memories are associated with other logic units, timing of signals throughout the system is of the utmost importance if reliable operation is to be guaranteed. In particular, attention must be paid to the duration, shape and stability of the waveform of the clock pulses since the relative timings of the data, WRITE, READ and ADDRESS signals are all referred to these clock pulses. This is one reason why crystal oscillators (such as that shown in Figure 10.20(b)) are used as computer clock-pulse generators. Users must pay attention to the timing diagrams supplied by manufacturers when incorporating their products into sequential logic systems.

Figure 11.35(a) shows the idealized type of waveform diagram usually adopted by manufacturers. The straight-line approximations of the curved waveforms are

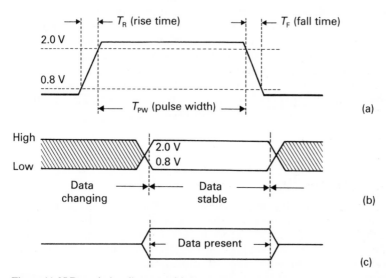

Figure 11.35 Data timing diagrams: (a) Clock pulse. Voltage levels are for TTL; (b) data timing; (c) data timing for outputs from tristate units

accurate enough for design purposes. Figures 11.35(b) and (c) show how data waveforms are usually represented. The hatched areas in (b) represent the periods during which the data may change (i.e. the logic outputs may go HIGH or LOW); the clear areas show the periods during which the data states must remain stable if reliable operation is to be guaranteed. Data waveforms for tristate logic outputs are usually represented as shown in Figure 11.35(c). No HIGH or LOW state is recognized until the data appears; thereafter, during the appearance, the data states remain stable.

In general, data must be present and stable for a setting-up period before triggering pulses are applied. Data must also be held for a minimum period after a trigger transition. For example, consider the operation of a master-slave

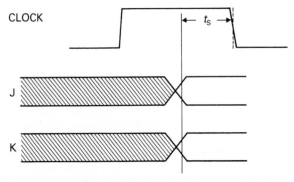

Figure 11.36 Timing for JK flip-flop. t_s is the set-up time for flip-flops that change their outputs on the trailing edge of the clock pulse. For 74LS76 $t_s = 20$ ns

edge-triggered flip-flop in which transitions of the output take place on the trailing edge (i.e. negative-going edge) of the clock pulse. When the clock goes HIGH the flip-flop is enabled, but if the data (i.e. the J,K state) is not present spurious operation will result. As soon as the data (say J = 1 and K = 1) is presented it takes time for the master flip-flop and input gate to establish their correct outputs. From Figure 11.19 we see that this time is $4t$, where t is the propagation delay time for a single gate; $4t$ is the set-up time, t_S. The data must be presented to the J and K inputs at least t_S before the trailing edge of the clock pulse appears. The situation may be represented diagrammatically as shown in Figure 11.36.

Shift registers

A bistable, such as a JK- or D- flip-flop, may be used to store a single bit. For computer operation, however, it is necessary to store complete words; 8-, 16- and 32-bit words are commonly used. It is therefore necessary to have a multi-bit store consisting of a set of bistables. Such a set is known as a register, and can be used as a form of memory for the temporary storage of information.

The digital information that is held in registers is frequently subject to arithmetic operations. These operations involve the addition of pairs of binary numbers. The least significant bit of each number is fed into an adder. On completion of the addition process, the next significant pair of bits is presented, and so on. To do this the two 'words' (i.e. strings of bits) are held in two chains of bistables known as shift registers. The function of each shift register is to move the whole word along the chain of bistables without altering the word shape.

One way of forming a shift register is to use a set of JK flip-flops cascaded as shown in Figure 11.37. The truth table summarizing the behaviour of the JK flip-flop is shown for convenience. Examination of the tables shows that if J = 1 and K = 0 then, after the application of one clock pulse, this data will be

J	K	Q_{n+1}	
0	0	Q_n	no change
0	1	0	goes to 0 if necessary and stays there
1	0	1	goes to 1 if necessary and stays there
1	1	$\overline{Q}_n$	toggle action

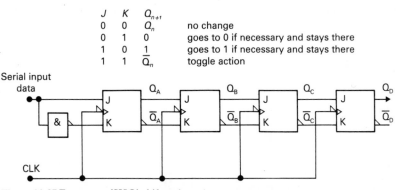

Figure 11.37 Four-stage (SISO) shift register that uses JK flip-flops

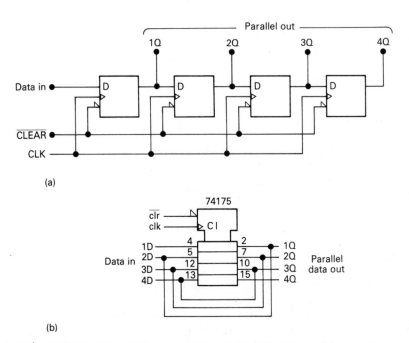

(a)

(b)

Figure 11.38 Use of data latches to produce a serial-in/parallel-out shift register: (a) Four-stage SIPO shift register that uses D-type data latches; (b) use of a four-bit parallel data latch (74175) to realize the SIPO shift register of (a)

transferred to the output of the flip-flop. By connecting the Q_A and $\bar{Q}_A$ outputs from this first flip-flop to the J and K terminals respectively of the next flip-flop, the data now programs the second flip-flop. On the application of the second clock pulse the data will be transferred from the output of flip-flop A to the output of flip-flop B; simultaneously the new data will be moved into flip-flop A. A third clock pulse transfers the data from B into C, from A into B and loads more new data into A, all simultaneously. We see, therefore, that each clock pulse shifts all of the data in the register one place to the right, taking in new data at the left, and 'dropping off' the data stored in the last flip-flop in the chain. Since the data word is presented to the input serially, and taken from the output serially, the circuit shown in Figure 11.37 is a serial-in/serial-out (SISO) shift register. Used in this way a shift

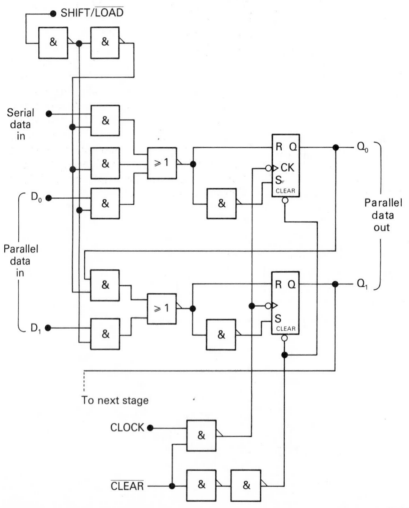

Figure 11.39 Logic circuit basis of a universal shift register. The two gates associated with $\overline{\text{CLEAR}}$ terminal ensure correct timing. The delay produced ensures that the clock is disabled before the $\overline{\text{CLEAR}}$ signal reaches the bistables

register can be used as a delay line. The delay time depends upon the number of flip-flops in the chain and the period of the clock pulses.

In many applications, especially in computers, the speed of data transfer can be considerably reduced if the output data is taken from the register in a parallel fashion. This merely involves having the outputs from each flip-flop available, as shown in Figure 11.38. Here we have a serial-in/parallel-out (SIPO) shift register. Figure 11.38(a) shows how four D-flip-flops can be used to realize a SIPO register. Figure 11.38(b) shows the use of a four-bit data latch (74175) to obtain the same result.

For the reasons given in the last pargraph, it is also desirable to be able to load the input data into the register in parallel fashion. Many commercial PIPO units also have provisions for serial inputs and outputs; very little extra logic circuitry is involved, usually no more than one or two extra gates. Such registers are versatile, and may be regarded as being universal, although it is not possible to shift the data word to the left as well as to the right; this facility requires a special IC.

Figure 11.39 shows the basis of a universal shift register. The mode of operation should be easy enough to deduce. The input switching arrangements are similar to those used in the up/down counter of Figure 11.34. Switching is necessary because the data on the parallel inputs must not be loaded while shifting is in progress. A couple of gates connected to a SHIFT/$\overline{\text{LOAD}}$ terminal control the loading process. When this terminal is HIGH the data input gates are DISABLED, while those gates necessary to connect the Q outputs of each flip-flop to the SET terminal of the next flip-flop in the chain are ENABLED. Conversely, when the voltage on the SHIFT/$\overline{\text{LOAD}}$ terminal is low the data gates are 'open' and allow loading to occur while the shift gates are disabled. Provisions are made for the serial input of data; the serial output is taken from the last flip-flop in the chain.

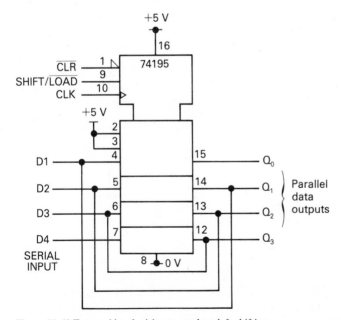

Figure 11.40 External hard-wiring to produce left-shifting

Parallel inputs and outputs require many pins if the number of bits in the data word is large. There is thus a practical limit to the capacity of a register within a single package. A 74195 is a readily available four-bit shift register that uses circuitry similar to that shown in Figure 11.39. Several of these may be grouped to expand the bit capacity of the overall system.

In computer work it is necessary to be able to shift the data word to the left as well as to the right. Figure 11.40 shows how this can be done with a 74195. To avoid hard-wiring external to the IC, a commercial unit designed for the purpose may be used. A 74198 is an example of an eight-bit bidirectional register; it is a 24-pin package.

Digital-to-analogue converters (DACs)

There are occasions, especially in control situations, when it is useful to convert digital output signals into their corresponding analogue forms. The ideal transfer function of a linear DAC may be expressed as

$$v_{OUT} = V_{ref}\left[\frac{B_1}{2^1} + \frac{B_2}{2^2} + \frac{B_3}{2^3} \dots \frac{B_n}{2^n}\right]$$

where V_{ref} is some reference voltage and $B_1 \dots B_n$ are the bits in the digital expression to be converted. B_1 is the most significant bit (MSB) and B_n is the least significant bit (LSB). In a four-bit number, for example, of the form we have already been using, the bits are D, C, B and A; D is B_1 and A is B_4 in the above notation. The maximum analogue voltage value for an n-bit number is $(2^n - 1)/2^n$; e.g. for a four-bit number, $v_{OUT} = 15/16$ths of V_{ref} when all the bits are 1.

It is possible to construct a DAC using an operational amplifier as an adder with multiple inputs. Unfortunately a large number of differently valued precision resistors are required. This range can be reduced to banks, each bank requiring the same four values of the input resistor. The disadvantage of this arrangement is that the reference voltage for each bank must be different and weighted.

A method of reducing the number of values needed for the resistors is to use a resistor ladder of the type shown in Figure 11.41. Here it will be seen that only two different values are required. The figure shows a five-bit number being converted, but, by extending the ladder network to the left, any length of digital 'word' can be accommodated. The number of bit inputs to the DAC determines the resolution of the converter.

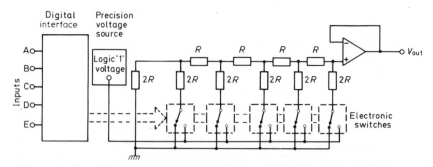

Figure 11.41 Schematic diagram showing the principles of operation of a digital-to-analogue converter

An analysis of the network can easily be made using mesh analysis and the usual conventions. To illustrate the principles involved a three-bit input is taken as an example, as shown in Figure 11.42.

$$-4Ri_1 + 2Ri_2 \qquad - V_{ref}(2^0) \qquad\qquad\qquad = 0$$
$$2Ri_1 - 5Ri_2 + 2Ri_3 + V_{ref}(2^0) - V_{ref}(2^1) \qquad\qquad = 0$$
$$2Ri_2 - 5Ri_3 \qquad\qquad + V_{ref}(2^1) - V_{ref}(2^2) = 0$$

where $(2^0) = $ LSB $(=0$ or $1)$; $(2^1) = $ NSLB $(= 0$ or $1)$ and $(2^2) = $ MSB $(= 0$ or $1)$.
From this set of three simultaneous equations we see that

$$-8Ri_2 + 4Ri_3 = 2V_{ref}(2^1) - V_{ref}(2^0)$$

and hence

$$-16Ri_3 = 4V_{ref}(2^2) - 2V_{ref}(2^1) - V_{ref}(2^0)$$

Since

$$v_{OUT} = V_{ref}(2^2) + 2Ri_3$$
$$8v_{OUT} - 8V_{ref}(2^2) = 16Ri_3 = -4V_{ref}(2^2) + 2V_{ref}(2^1) + V_{ref}(2^0)$$

Therefore

$$v_{OUT} = \frac{V_{ref}(2^2)}{2} + \frac{V_{ref}(2^1)}{4} + \frac{V_{ref}(2^0)}{8}$$

The digital number 110, for example, would, after conversion, yield the analogue voltage $(1/2 + 1/4) V_{ref}$, i.e. $0.75 V_{ref}$.

It will be noticed that the analogue output of the DAC goes from zero to a positive voltage using the arrangement shown in Figure 11.41. The operation under

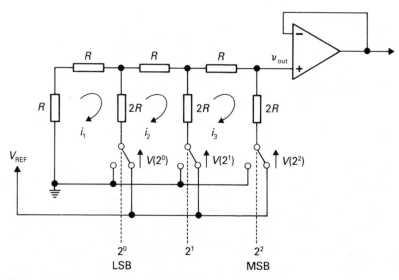

Figure 11.42 Three-bit D-to-A converter

these conditions is said to be unipolar (one polarity). A practical circuit of an eight-bit converter based on Ferranti components is given in Figure 11.43.

The ZN428 is a monolithic eight-bit converter with the R-2R ladder, a reference voltage source, transistor switches and a data latch all fabricated on a single chip. The transistor voltage switches are specially designed to have low off-set voltages (<1 mV). The data latches facilitate the updating from a data bus. They are said to be transparent when an ENABLE signal line is low, i.e. at logic 0; the data is then transferred direct to the switches. The analogue output from pin 5 is fed to the input of the operational amplifier ZN424P. Facilities are provided by external components that enable the DAC to be adjusted. The zero-adjust potential divider sets the zero when all of the input data bits are zero. With all the bits set to 1 the gain is adjusted until the output is full-scale minus 1 LSB. 1 LSB ≡ 19.5 mV for 5 V full-scale, thus with all the data bits at 1 the output analogue voltage should be adjusted to 4.9805 V. There are thus 255 increments of 19.5 mV throughout the scale; the resolution is therefore 19.5 mV. The output voltage range could be adjusted to 0–10 V if the supply is changed to suitable values.

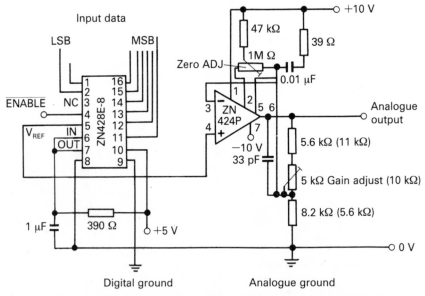

Figure 11.43 Practical circuit for an eight-bit unipolar digital-to-analogue converter. Semiconductor packages are Ferranti types. The full-scale deflection is 5 V. (If 15 V–0–15 V is desired to energize the ZN424P then resistor values shown in brackets should be used; all other values remain as shown. The full-scale analogue output voltage is then 10 V)

Where bipolar operation is required the circuit may be modified to that shown in Figure 11.44. Bipolar operation means that the output of the converter takes negative as well as positive values. A negative offset to the analogue output is achieved by setting the inverting input of the op-amp to a suitable fraction of V_{ref}. The usual procedure is to set all bits to zero with the ENABLE low and to adjust the offset until the amplifier output reads − Full-scale. Then with all bits set to 1 the gain is adjusted until the amplifier output reads + (Full-scale − 1 LSB). (1 LSB ≡ 39.1 mV for ± 5 V operation, thus the output should read 4.9609 V when all the input data bits are at 1.)

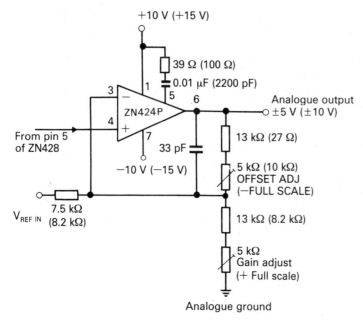

Figure 11.44 Modification to the op-amp section of Figure 11.43 to convert to bipolar operation. Component values in brackets are for 15 V–0–15 V supply voltages

Errors

In practice, DACs depart from perfection in several ways because it is impossible to ensure that the resistors of R-2R ladder network are all perfectly matched. The operational amplifier may also contribute to the errors. Figure 11.45 shows the transfer characteristic of an ideal three-bit DAC. For each of the input data codes there exists a discrete analogue output voltage. As the data bits progress from 000 to 111 the output voltage rises in a quantum fashion. Ideally, the discrete analogue voltages should lie on a straight line that passes through the origin. In practice, since the number of bits is much larger than three (say 8, 12 or 16), it is usual to draw the output characteristic as a continuous graph, but bearing in mind that the transfer characteristic is a discontinuous one.

Monotonicity

As the input digital code progresses in one-bit steps from 000 to the maximum number (111 in our example), the analogue output should increase by equal increments in staircase fashion. If this always occurs the converter is said to be monotonic. Errors in resistor matching or amplifier performance may cause the output to rise (or fall) above (or below) the ideal value. As the errors accumulate there will come a time at some stage in the conversion when the output drops with increasing bit count, as shown in Figure 11.46. The DAC is then said to be non-monotonic.

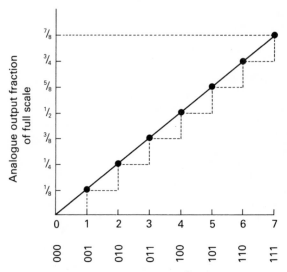

Figure 11.45 Transfer characteristic of an ideal three-bit DAC

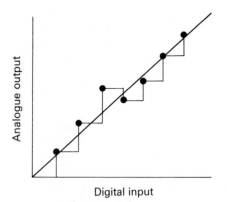

Digital input   **Figure 11.46** Non-monotonic DAC

Offset and gain errors

These errors are produced when the amplifier is not ideal or when the offset voltages in the transistor switches are not all identical. Figure 11.47 illustrates the effect of these errors.

Linearity errors

Offset and gain errors can be reduced to zero at the full-scale and zero points by trimming the variable resistors associated with the amplifier. Although this corrects the end points of the transfer characteristic, intermediate points still depart from their ideal positions. There is nothing that the user can do about this except to use a higher quality DAC. The difference between the actual output and the ideal output is usually expressed as a percentage of either the full-scale value or the LSB value.

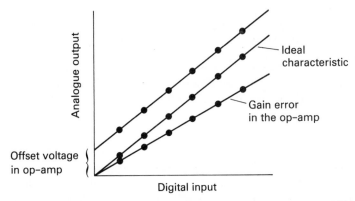

Figure 11.47 The effects on the analogue output voltage of gain errors and finite offset voltage in the op-amp buffer stage

Linearity errors less than $\pm$ 50 per cent of the LSB guarantee monotonic operation. It should be noted however that a DAC may have a linearity error in excess of 50 per cent and could still be monotonic.

The term 'differential non-linearity' is borrowed from the terminology associated with continuous functions. The slope of the line joining all the points of the transfer function of an ideal DAC should be constant. In a practical case the actual step height of the analogue output at a given point in the conversion may be greater or less than the ideal step height. If the difference is greater, the differential non-linearity is said to be positive; if it is less, then the non-linearity is negative. Positive non-linearities may not result in non-monotonic operation, but negative non-linearities in excess of 1 LSB always result in non-monotonicity.

Analogue-to-digital converters (ADCs)

These converters are the counterpart, or dual, of DACs; they convert a continuously varying analogue input voltage into corresponding digital output codes. The digital output code is usually a pure binary number corresponding to the value of the input voltage that produces it. Ideally, for each equal increment of input voltage, the binary code is incremented by 1 LSB. An ADC is usually adjusted so that transitions between codes occur at $\pm\frac{1}{2}$LSB on either side of the nominal analogue input for a particular code.

Many methods have been devised for achieving analogue-to-digital conversion, but we shall discuss four here.

Double ramp converters

This type of converter is also known as a dual-slope converter for reasons that will shortly be apparent. Although the conversion is slow compared with other types, the high resolution and accuracy make it a favourite for digital voltmeters and allied instruments. Usually, speed is not of paramount importance in this instrumentation.

The principle of operation of this type of converter is illustrated by Figure 11.48. The positive input voltage is fed to an integrator that has previously been zeroed. Negative input voltages may first be passed through a sign reverser using either a separate input terminal or, alternatively, automatic electronic circuitry, before being fed to the integrator. A high-impedance buffer stage is usually interposed to allow for adequate isolation between the integrator and the circuit delivering the voltage to be measured. Since the output voltage of the integrator is given by

$$v_{int} = -\frac{1}{RC} \int v_{in} \, dt$$

a given fixed input voltage produces a linear ramp with a negative slope. At the commencement of the integration, a clock-pulse generator feeds square waves into a counter. The integration proceeds until the counter is full, whereupon a signal pulse is generated that initiates a change in the control system. This change

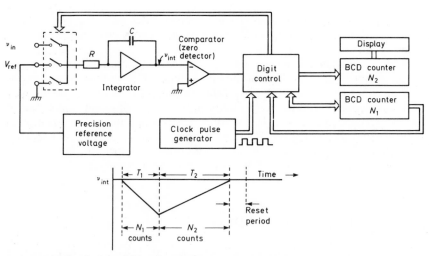

Figure 11.48 The dual-ramp (slope) analogue-to-digital converter

switches the input of the integrator, via the buffer stage, to a fixed negative reference voltage V_{ref}. If the time taken during the linear descent of the integrator output voltage is T_1 and the period of the clock-pulse generator is τ, then $T_1 = N_1 \tau$ where N_1 is the first count necessary to fill a given BCD counter. At the moment that the counter is full the integrator output voltage is

$$v_1 = -\frac{v_{in} T_1}{RC}$$

Once the input of the integrator is switched to $-V_{ref}$, the output waveform becomes a linear ramp with a positive slope. The integrator output voltage rises until it is zero; this zero voltage is detected by means of a comparator, the output of which is connected to the control circuitry. The time taken for the positive ramp to reach zero voltage is T_2. During this time, pulses from the clock are counted in a second BCD counter yielding a count of N_2; also $T_2 = N_2\tau$, hence $T_1/N_1 = T_2/N_2$. The height of the positive ramp is given by

$$v_2 = \frac{V_{ref}T_2}{RC}$$

From Figure 11.48 it is clear that $|V_2| = |-V_1|$, hence

$$\frac{v_{in}T_1}{RC} = \frac{V_{ref}T_2}{RC}$$

i.e.

$$v_{in} = V_{ref}\frac{T_2}{T_1} = V_{ref}\frac{N_2}{N_1}$$

Since N_1 is always a fixed count determined by the capacity of one of the counters, it is seen that the input voltage is proportional to N_2. The accuracy of the indication does not therefore depend upon the use of precision RC components for the integrator since v_{in} is proportional to the ratio N_2/N_1. Any errors on the negative ramp are therefore nullified by the same errors on the positive ramp. The accuracy of the measurement depends substantially on only one quantity, namely V_{ref}. A further advantage of using the integrator system (as opposed to others, e.g. a staircase-based system) is that noise on the input signal is averaged out and hence does not affect accuracy. The dual-ramp integrator is therefore the basis of many modern digital voltmeters. The displays with dual-ramp integrators are seemingly continuous since the cycle of 'negative ramp, positive ramp, reset and zeroing' can take place at rates approaching 10 per second. The conversion time of a Ferranti ZNA216E, for example is 160 ms.

Staircase ADC

In this type, an n-bit counter is incremented by a clock pulse generator via a logic control unit. The digital output of the counter is fed to a digital-to-analogue converter that gives a staircase output as described above. We thus have a steadily increasing analogue voltage (in step form) available. This voltage is fed to one of the inputs of a comparator; the voltage to be converted is fed to the other input of the comparator. When the staircase voltage from the DAC becomes marginally greater than the voltage to be converted, the output from the comparator suddenly changes state. This sudden change signals the end of the conversion, and the control logic then blocks the clock pulses from the counter. The count reached by the counter is the digital output equivalent of the input analogue voltage. Figure 11.49 illustrates the action. The control logic resets the system at suitable intervals ready for the next sampling.

Such converters are relatively slow since $(2^n - 1)$ clock pulses are required to drive an n-bit counter from zero to full-scale.

Successive approximation converters

A swifter way of achieving conversion than that described in the last section involves successive approximation. With this method, instead of incrementing a counter with single clock pulses, the MSB of a register is immediately set to 1, the

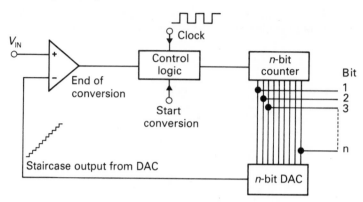

Figure 11.49 Staircase ADC

other bits remaining at 0. This causes the DAC to give an output of half the full-scale analogue value. By examining the truth table for any range of pure binary numbers from zero to n it will be seen that setting the MSB alone to 1 is equivalent to choosing the mid-point of the range. The half-range output of the DAC is fed to one input of the comparator and the voltage to be converted is fed to the other input. If the input voltage to be converted is in the upper half of the range the logic control unit keeps the MSB at 1; if the input voltage is in the lower half of the range the comparator gives the appropriate signal and the MSB is then set to 0. The control unit and register repeats the procedure for each bit in turn in order of decreasing significance until the LSB is set. The register then gives the appropriate output code. Figure 11.50 shows the configuration in diagrammatic form.

For a staircase ADC the conversion time depends upon the magnitude of the input analogue voltage. As we have seen, full-scale operation requires $(2^n - 1)$ clock pulses. The successive approximation converters require only n clock pulses for an n-bit register irrespective of the magnitude of the input analogue voltage. The Ferranti ZN432 10-bit successive approximation converter, for example, has a guaranteed conversion time of only 20 µs.

Another fast (20 µs conversion time) 10-bit successive approximation ADC is the AD573 from Analogue Devices. This package incorporates tristate output buffers and hence is particularly suited to interfacing with popular microprocessors. It is designed for analogue inputs of 0 to +10 V or ±5 V. The 20 µs conversion cycle is initiated by applying a positive pulse to the CONVERT terminals; action commences on the trailing edge of the CONVERT signal. Figure 11.51(a) shows the main organization of the device, while (b) and (c) show how to interface it with an Apple II peripheral connector and an 8085A microprocessor respectively.

Flash or parallel converters

A very rapid method of effecting analogue-to-digital conversion is to establish a potential gradient along a chain of resistors, as shown in Figure 11.52. To produce an n-bit converter each step in the chain must represent 1 LSB; we therefore require $(2^n - 1)$ equal increments. The chain of equal resistors is energized from a voltage reference source and each incremental voltage point is connected to a voltage comparator input. We therefore require $(2^n - 1)$ voltage comparators. All

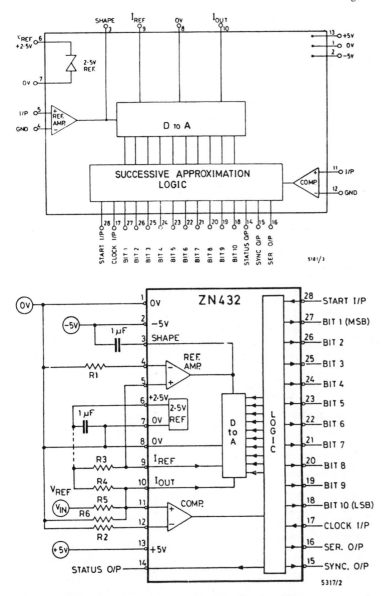

Figure 11.50 Block diagram of a successive approximation A/D converter together with typical external circuitry for a Ferranti type. For maximum input voltages of $\pm 2.5\,\text{V}$, $R_1 = 5\,\text{k}\Omega$, $R_2 = 1.25\,\text{k}\Omega$, $R_3 = 5\,\text{k}\Omega$, $R_4 = 2.5\,\text{k}\Omega$ and $R_5 = 2.5\,\text{k}\Omega$; for $V_{\text{IN}} = +2.5\,\text{V}$ the logic output is all 1s. When $V_{\text{IN}} = -2.5\,\text{V}$ the logic output is all 0s

of the other comparator inputs are connected together and brought to an input terminal. The input voltage to be converted is thus presented (or 'flashed') to each comparator simultaneously. At that point along the resistor chain where the input voltage exceeds the reference voltage, the appropriate comparator changes polarity. Since all of the comparator outputs are connected to a logic encoding IC,

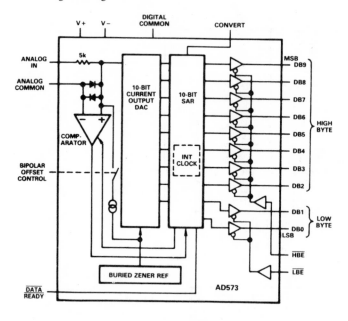

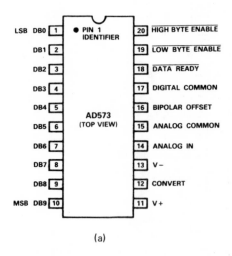

(a)

Figure 11.51 Ten-bit, fast A/D converter with interface circuitry to a microcomputer, and a popular microprocessor: (a) AD573 functional block diagram and pin connections;

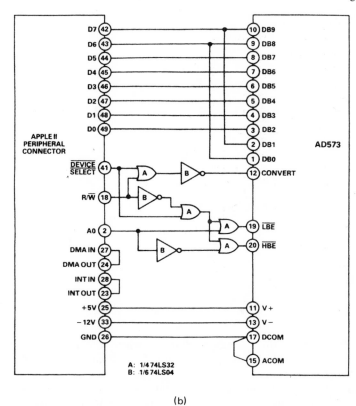

(b)

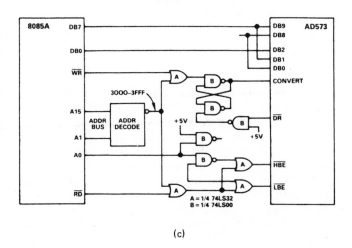

(c)

Figure 11.51 (b) AD573 interface to Apple II; (c) AD573–8085A interface connections

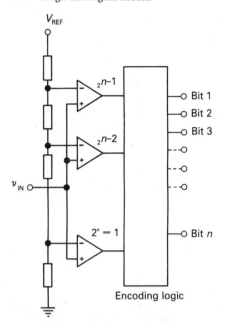

V_{REF}

2^{n-1}

2^{n-2}

v_{IN}

$2^0 = 1$

Bit 1
Bit 2
Bit 3

Bit n

Encoding logic

Figure 11.52 Flash (parallel) ADC

the latter detects the point along the reference chain at which the comparator changes state and delivers the appropriate output digital code. Usually a Gray code is used to reduce output errors.

Such converters are extremely fast, and conversion times as low as a few tens of nanoseconds are achievable in modern units. The main drawback is cost; a six-bit converter, for example, requires 63 comparators together with the precision resistor chain, voltage reference source and logic encoder.

ADC errors

Those analogue-to-digital converters that incorporate DACs in their circuitry are prone to similar errors as those encountered in DACs. If the reference DAC is non-monotonic then there will be missing codes in the analogue-to-digital conversion. Consider, for example, Figure 11.53 in which an eight-bit DAC exhibits non-monotonicity. The analogue output from the reference DAC falls at, say, step 4. If the analogue input voltage is less than the DAC output for step 3 then the comparator will stop the counter before 4 is reached; conversely, input voltages greater than 3 must also be greater than 4 so the comparator will not change state at 4 and therefore this code will be missing.

Quantizing error is due to the fact that for any digital output code there is a 1 LSB range of analogue input levels. From the ADC output code there is thus an uncertainty about the precise input analogue voltage equivalent to $\pm \frac{1}{2}$ LSB.

In operation, ADCs are trimmed so that the first transition of the output from 0 to 1 occurs when the input analogue voltage is at a level corresponding to $\frac{1}{2}$ LSB. As supplied, however, ADCs do not have the reference DAC adjusted for a $\frac{1}{2}$ LSB offset, and the first transition, in the absence of trimming, occurs for an input

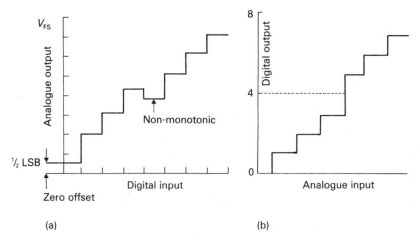

Figure 11.53 (a) Non-monotonic DAC used in an ADC gives rise to an ADC with a missing code, shown in (b)

voltage equivalent of 1 LSB. This error is usually combined with any DAC zero error and comparator offset voltage error; collectively these errors are known as the untrimmed zero transition error.

Gain error arises when the slope of the line obtained by joining the transition points is different from the straight line joining the zero and full-scale points. The difference at any transition point between the two lines is known as the non-linearity error.

The incremental size of input voltage to produce 1 LSB change in output is ideally $V/2^n$ where V is the full-scale value of the input voltage. This increment should produce 1 LSB change at all output values. The maximum difference between the actual increment to produce 1 LSB change and the ideal incremental value is known as the differential non-linearity. For differential non-linearity less than 1 LSB there will be no missing codes.

The *resolution* of an ADC (which is not the same thing as accuracy) is the number of output bits produced. The full resolution potential of the converter will not be realized in practice if there are missing codes and hence the *useful resolution* will be less than the *resolution*.

Several factors are involved in the choice of a suitable ADC. Accuracy, resolution, linearity and noise must be considered, but often it is the conversion speed which dictates the choice. Flash, or parallel ADCs have conversion speeds that are only 10 or 20 ns, but they require a good deal of electronic circuitry; consequently such converters can be very expensive. For more modest conversion speeds much cheaper ADCs are available. Very often in control systems the analogue input voltages vary slowly. For sinusoidal input signals the frequency range that can be handled by some ADCs is quite low. Take, for example, the ZN427E which is an eight-bit successive approximation ADC with a conversion time of 10 μs. For a 5 V input range 1 LSB 5 ≡ V/256 bits ≈ 20 mV per bit. For an error at the output of ±½ LSB the input voltage must change by no more then 10 mV in 10 μs, i.e. 1000 Vs⁻¹. For a sinusoidal input voltage given by

$$v = V_{max} \sin \omega t$$

then

$$\frac{dv}{dt} = \omega V_{max} \sin \omega t$$

therefore

$$\left(\frac{dv}{dt}\right)_{max} = 2\pi f_{max} V_{max}$$

For an input amplitude of 2.5 V

$$1000 = 2\pi f_{max}\, 2.5$$
$$\text{i.e. } f_{max} = 64 \text{ Hz}$$

We may improve the situation with respect to the maximum input frequency that can be accepted by using a sample-and-hold device.

Sample-and-hold circuits

It is becoming increasingly apparent that there are several advantages in using digital signals in communications, audio and video equipment. The information to be conveyed is usually contained originally in analogue signals. It is necessary, therefore, to digitize the signal information before it can be processed. For slowly varying signals the conversion time of available ADCs is short enough to allow instantaneous values of signal voltage to be converted rapidly into digital form without significant error. As we have seen above, however, the signal frequency does not have to be very high in audio or communication terms before difficulties arise. In order to allow sufficient time for instantaneous voltage values to be digitized accurately, we may use circuits that track the analogue signal, or sample it at suitable intervals. When it is necessary to effect a digitization of an instantaneous voltage value, we may need to 'hold' that value for a sufficient period to allow the ADC and digital circuitry to make the conversion. Figure 11.54 illustrates the situation; in the figure the periods for sampling and holding have been exaggerated in relation to the signal period in order to clarify the diagram.

The principle of sample-and-hold (SH) circuits involves charging a capacitor during the sample period so that the voltage on the capacitor is equal to the instantaneous analogue voltage at any given time. When a given capacitor voltage needs to be held for digitization the capacitor is disconnected from the charging source by an electronic switch. The voltage on the capacitor appears at a low-impedance output terminal by the use of a voltage-follower buffer stage. Similarly, the charging of the capacitor is achieved via the low output impedance of an input voltage-follower buffer. The electronic switch is conveniently an FET, as shown in Figure 11.54(b).

It will be noticed from the waveform diagram that the beginning of the HOLD output does not coincide with the fall in the control signal to logic 0. This is because it takes a finite time for the switch to operate and disconnect the capacitor from the charging source. This time difference is known as the *aperture time*. Provided the aperture time is short, the difference in analogue input voltage throughout this period is usually negligible. The aperture time varies with supply voltage and temperature, but for typical unit (LF 398) may be expected to be in the range

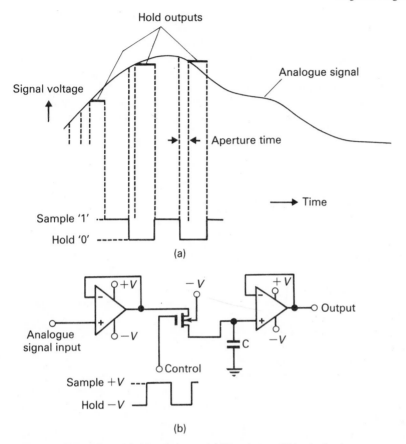

Figure 11.54 Sample-and-hold techniques: (a) Waveforms; (b) basic circuit

25–100 ns. We can see now that the addition of an SH circuit to an ADC can substantially improve the frequency performance of the overall arrangement. In the last section the example ADC could accept signals with frequencies up to only 64 Hz; there the rate of change was 10 mV/10 μs, i.e. 1000 Vs^{-1}. If we use an SH device with an aperture time of, say, 50 ns then the maximum rate of change of the input signal can be as high as 10 mV/50 ns, i.e. 2×10^5 Vs^{-1}. This allows an analogue frequency as high as 12.8 kHz to be handled, i.e. f_{max} has been increased by a factor of 200. In general, the improvement ratio is the conversion time of the ADC divided by the aperture time of the SH circuit.

Like all instrumentation circuits, the SH circuit is subject to error. During the sampling process there is a phase lag between the input signal and the voltage appearing across the capacitor. This is because the capacitor is charging not from a zero impedance source but from one with a finite resistance. At the time that the control voltage changes from SAMPLE to HOLD the capacitor will not have acquired the correct voltage, but some different voltage that was present at a time equivalent to the phase lag. This time difference is known as the *acquisition time* and is of the order of tens of microseconds depending upon the value of capacitor used. During the HOLD period the voltage across the capacitor remains constant only when there are no leakage paths, and the dielectric is perfect. In practice some droop appears (see also 'sag in CR circuits', earlier in the book); this droop contributes to the errors in digitization. A significant source of error involves the dielectric used in the capacitor especially in respect of dielectric absorption giving rise to a hysteresis effect. Polystyrene and polypropylene dielectrics are satisfactory, with mica and polycarbonate coming a poor second; ceramic types (used for decoupling IC supply lines, for example) are useless in this situation because of the high values of hysteresis involved. In connection with leakage it is advisable to use guard-ring techniques as suggested in Figure 11.55(b). This figure

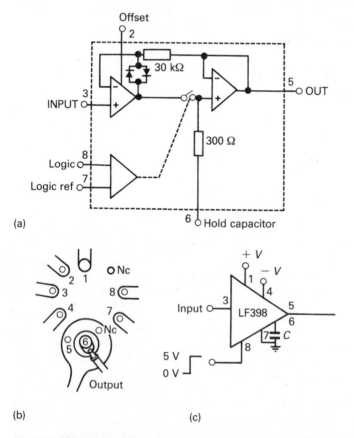

(a)

(b)   (c)

Figure 11.55 Practical sample-and-hold circuit: (a) Circuit of LF 398 sample-and-hold unit; (b) guard-ring technique for the HOLD capacitor. A 10-pin layout should be used; (c) typical connection. C is chosen from manufacturer's curves for the required HOLD step

also shows the circuit of the LF 398; from this the reader will see that a closed-loop configuration is being used. This arrangement usually gives better low-frequency tracking where speed of operation is not of the utmost importance. The open-loop arrangement of Figure 11.54(b) has a faster acquisition time.

Figure 11.56 shows an Analogue Device sample-and-hold IC together with the interface connections to their AD573 converter.

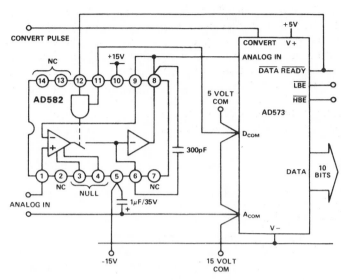

Figure 11.56 Sample-and-hold interface to the AD573

Memory systems

We have seen that by using TTL and CMOS gates it is possible to design and construct circuits capable of making decisions based upon the current state of the input signals. These combinational logic circuits may be used to perform such elementary tasks as checking door interlocks (say in a washing machine before a wash cycle starts), or they may be used to construct high-speed arithmetic units capable of adding or subtracting binary numbers. Combinational circuits are, however, unable to learn by experience. By using feedback lines and memory units, however, it is possible to construct a circuit that can take into account its past experience. Such sequential logic circuits provide the bases of digital computer systems in which the counting of events and the storage and retrieval of information are central functions. It is the size of the memory that dictates the ultimate complexity of the logic system that can be achieved. The more complex the system is, the more intricate and varied are the tasks that can be performed, i.e. the more powerful is the logic configuration. Systems that incorporate a few JK flip-flops, for example, are very elementary and can perform only the most rudimentary tasks.

The number of binary digits (bits) that can be stored in a memory is known as its capacity. By increasing the capacity of the memory to only very modest proportions, it becomes very difficult for human beings to design and assemble logic circuits from individual logic gates so as to exploit fully the memory capability. A different, and generalized, approach is required which involves assembling a basic set of logic units to form a digital computer. The computer is then 'programmed' to perform a specific task. The power and versatility of the computer depends largely upon the capacity of the memory incorporated.

One of the basic requirements of any computer system is some form of memory device. This memory must be capable of storing the complete data associated with a particular problem; additionally it must be able to hold information and instructions relating to the mathematical and logical operations that must be performed upon the data in order to achieve the desired result. The sequence of mathematical and logical operations is known as a program. (When the American spelling is adopted we are referring to a computer program, as distinct from the term 'programme' which has a wider meaning). It was shown earlier how several flip-flops could be connected to form a register. Such a register can be used as a memory to store a digital word. Clocked D-type flip-flops form the basis of many computer register-type memories. These types of memory cells, and the random access memory types (RAM) discussed below, are said to be volatile, because once the power source is switched off all of the digital information held in the bistables is lost.

As explained previously, bits in computers are arranged into *words*. Although 4 bits = 1 nibble and 8 bits = 1 byte, words can have any length; the word length is defined by the particular computer in use. Typical words may comprise 8, 12, 16 or 32 bits. Setting a given flip-flop so as to correspond to a bit value of 1 is called *writing*. Later it is frequently necessary to examine the state of the flip-flop in order to retrieve the stored information. We are said to interrogate, or *read*, the memory when information retrieval is required. Signals enable reading or writing to occur are often passed along the same line, which is called the read/write line. On circuit diagrams this line is marked $R/\overline{W}$ indicating that when the voltage on the line is high the computer is reading, and when it is low the computer is writing. Clearly, when large numbers of memory cells are involved it would be highly inconvenient, and very slow, to interrogate each cell sequentially until the required one was reached for reading. (This is one of the disadvantages of storing bits on magnetic tape. Each bit occupies a position along the tape, and all of the previous tape must be passed over the reading head before the required section is correctly positioned.) When many thousands of memory cells are fabricated on a single silicon chip it is obviously an advantage to be able to interrogate the cell array in a random fashion. Thus we can gain access to information in a given cell almost immediately. An array of cells in which this is possible is called a random access memory.

Random access memory (RAM)

Although the expression random access memory (RAM) is still widely used, the term read/write memory would be a better one. Another type of memory, described below, is one which can only be read. The read only memory (ROM) can be interrogated in a random fashion, so it too is a random access memory. To avoid confusion, the term RAM will be used for a read/write memory, and ROM for a read only memory.

Two basic classes of RAM are available, both of which are volatile. Figure 11.57 shows the circuit arrangements. Static RAMs are so called because once the information has been written into them the cells stay in that state until the power is turned off, or the alternative states are written in. Examination of the circuit diagram shows the familiar cross-coupled gate arrangement together with active loads, and read/write facilities. The latter consists of a word line which selects a suitable set of bistables ENABLING them so as to accept signals on the bit line. With the word line such that FETs T1 and T2 are conducting, information can be read or written via the bit lines. Packing density on the chip, speed of operation,

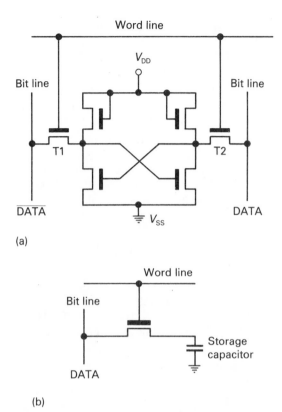

Word line

V_{DD}

Bit line

Bit line

T1

T2

$\overline{DATA}$

DATA

V_{SS}

(a)

Word line

Bit line

Storage
capacitor

DATA

(b)

Figure 11.57 Two forms of random access memory (RAM) cells: (a) Static RAM cell based on FETs; (b) dynamic RAM cell

and power dissipation are major considerations for the chip manufacturer. One of the advantages of static RAMs that use bipolar techniques is their high switching speed, but these types have relatively high power dissipations. As the demand for higher packing densities increases, some compromises have to be reached concerning the speed/power ratio. The FET types (as shown) are capable of higher packing densities and low power dissipation in the static state. CMOS is slower than TTL, but for many applications the speed is adequate.

Dynamic RAM devices use a small-valued capacitor as the storage element, buffered by an FET. Dynamic memories, because of the relative simplicity of the cell circuitry, are available with densities some four times (or more) greater than those of the static type.

Up to about 1985, 64 kilobit (65 536 bits) units were readily available; some manufacturers, notably in Japan, then started to run down production of 64K units in favour of the 256K DRAMs. Incidentally, manufacturers' literature refers to 64K ×1 units. This is to avoid ambiguity; the 1 refers to the word length, hence 64K ×1 = 64 kilobits. 16K × 4 is organized as 16K 4-bit words, i.e. 64 kilobits in total. (Readers should remember that 1K = 1024 in this context.) The mode of operation of the dynamic RAM (DRAM) involves the charging of a capacitor; inevitably the stored charge can dissipate. The logic state must therefore be 'refreshed' at suitable intervals (every few milliseconds), hence the term 'dynamic'. Sophisticated DRAM controller chips are now available which largely overcome the former problems associated with the refreshing mechanism.

RAM constitutes the main internal memory for a computer. RAM cells are also used for the registers in microprocessors. Banks of RAM chips are used as the main working memory for the computer.

Read only memory (ROM)

ROMs are required to provide a predetermined bit pattern of 1s and 0s at the data output lines for a particular address applied to the input. The basic storage element of a ROM need not be a flip-flop; usually storage is accomplished by the presence or absence of an interconnecting diode on a matrix grid of bit lines and word lines. Figure 11.58 shows the circuit of a simple ROM using a diode matrix. Programming is achieved during the final metallization process, which either completes or leaves open-circuit the connections to the anode of the diode. Manufacturing problems are thus eased since millions of identical units including complete diode arrays can be made in one batch run, but leaving the final metallization process to be added later when the customer's requirements are known.

The mode of operation may be understood by considering the circuit. Bit X will assume a logic state of 1 if either word line 'a' or 'x' is HIGH; this is because only two of the relevant diodes are connected. Word line 'a', however, is energized only when the address input is ABC = 000. Similarly, word line 'x' is energized only when ABC = 111. Hence

$$X = \overline{A}\overline{B}\overline{C} + ABC$$

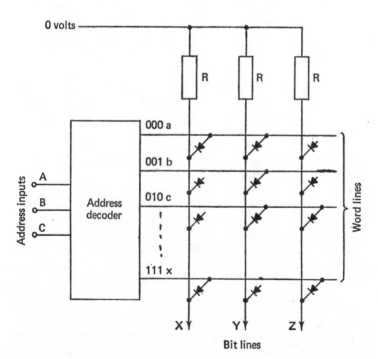

Figure 11.58 Read only memory (ROM)

Similarly, for Y and Z

$$Y = \overline{AB}\overline{C} + \overline{A}BC + A\overline{B}\overline{C}$$
$$Z = A\overline{B}\overline{C} + ABC$$

Although only a very small system has been shown here to illustrate the principle of operation, the use of ROMs becomes economic only in large complex systems where very many discrete logic gates would otherwise be required.

ROMs are best suited to combinational logic applications such as look-up tables for log, sin, cosine, code conversions, pattern generation, etc. With careful programming, however, they may be used for sequence generation. For example, in Figure 11.58 if the two outputs X and Y are fed back to the address inputs A and B, and input C is used as a clock input then the ROM may be programmed to execute any desired sequence of X, Y. Assuming that the contents of the ROM are as shown in Figure 11.59(a) the system acts as a two-stage binary counter. If the sequence starts at address $AB = 00$ and the clock input is LOW then $XY = 00$. These data outputs are fed back to the address inputs. Since the present address is equal to XY, the system is said to be in a stable state. When the clock input goes high, XY changes from 00 to 01, and no longer is $AB = 00$. Instead the address bits are forced to $AB = 01$. In this situation $XY = 01$, and hence the system changes to a second stable state (in which $XY = 01$). In like manner, the operating sequence of a system such as this may be derived (see Figure 11.59(b)).

Address location	Contents		Present address	Clock	Next address	Output sequence
ABC	XY		AB	C	XY	XY
000	00	Stable	00	0	00	00
001	01	Unstable	00	1	01	01
010	10	Stable	01	1	01	01
011	01	Unstable	01	0	10	10
100	10		10	0	10	10
101	11		10	1	11	11
110	00		11	1	11	11
111	11		11	0	00	00

Figure 11.59 (a) ROM contents for a two-stage binary counter (b) output sequence for the ROM counter

A more common use for ROMs is to hold the binary pattern which constitutes the microprocessor machine code instructions for the operating system of the microcomputer. Disk filing systems, and much computer software in the form of word-processing systems, database systems, etc. are held in ROMs.

The ROMs described so far are dedicated to one specific set of tasks; the programs are incorporated in the final stage of manufacture and once formed cannot be altered by the user. Unlike RAMs, the information held in ROMs is non-volatile, and hence not lost when the power supply is turned off.

When a large number of identical ROMs are required, the manufacturer uses a masking process to produce the necessary diode array that meets the customer's specification. These masks are used in the final metallizing process during which the necessary diode connections to the matrix lines are formed. Such devices are known as masked ROMs. One disadvantage in using masked ROMs, from the customer's point of view, is the delay between submitting the programming code to the manufacturer and the receipt of the prototype devices. One of the solutions to this problem is to use a programmable ROM. In this type of ROM all the diodes are connected to the matrix lines (i.e. the word and bit lines). Fusible links are placed in series with one of the connections to each diode. The fusible link can be 'burned-out' by the customer, using a sufficiently large current to do so. The customer is thus able to burn out selected links himself, thus producing the necessary stored program. In practice, special PROM 'programmers' are required for this purpose.

For those who wish to produce prototypes very swiftly, the most popular method is to use an erasable programmable read only memory, commonly referred to as an EPROM.

EPROM

These types of ROM use large arrays of MOSFETs as memory cells. No electrical connections are made to the gates. The gates are thus completely insulated from the rest of the devices, but may be charged by the application of a sufficiently large voltage pulse between the gate and source ends of the channel. Commonly available units require a voltage pulse of 21 V or 25 V depending upon the type. The so-called '5 V-only' devices operate from a single supply for read-only operation, but require a 21/25 volt pulse for programming purposes. For the duration of the pulse a large current passes from drain to source, and a temporary avalanche breakdown occurs in the dielectric that separates the gate from the channel. This causes a negative charge to accumulate on the gate. Once the programming pulse is removed, then, for all practical purposes the charge is trapped on the gate permanently; the leakage in normal usage is negligibly small. The FETs programmed in this way are then effectively on, and thus make the appropriate connection between the selected word and bit lines. The unprogrammed FETs are off and thus no connection is made at this point in the matrix.

The only way of releasing the trapped charge is to irradiate the gate with ultraviolet light. Obviously this cannot be done practically for an individual gate; the whole chip is irradiated via a small quartz window in the top of the containing package. The u.v. light must be intense and of the correct wavelength. Special u.v. fluorescent tubes are available for the purpose. Complete erasure may take up to 15 or 20 minutes. Once the erasure is complete, the device is then capable of accepting a new program.

Readily available EPROMs are sold under the generic part numbers of 2716, 2732, 2764, and 27128, for example, having densities of 16, 32, 64, and 128 kilobits respectively. These are organized into eight-bit words. Typical access times range from 550 to 250 ns; these are the times it takes to extract the data. During the periods that the EPROMs are active (i.e. being interrogated) the power dissipated in the chips is some 500–600 mW. In the standby mode the powers range from 130 to 200 mW. As packing densities increase, the power dissipation problems become serious. The figures quoted above are for NMOS devices. Because of their lower power dissipation, CMOS versions of EPROM are attracting considerable attention; their standby power dissipation is very small. CMOS gates are not so fast as other types, but frequently this is no disadvantage.

Electrically erasable programmable read only memories (EEPROMs) are becoming more readily available. Such devices avoid the necessity of generating and using an ultraviolet light source. At the moment (1985) such devices are expensive, but no doubt prices will fall as sales increase.

Problems of access

Having seen how thousands of memory cells can be assembled on a memory chip, the problem then arises as to how such cells can be organized so that access to the information can be made readily. We would never wish to have access to a single cell alone, since the information retrieved would be so limited. Within the memory, therefore, groups of cells are organized into registers, each register holding a word of information. As explained previously, the length of the word is dictated by the computer system. For many microcomputers a word length of 1 byte is used. Thus an array of 65 536 bistables (64 kilobits) has the capacity to hold 8 kilobytes of information, i.e. 8192 words. The memory cannot operate in isolation, but must interect with other parts of the system. It is obviously not feasible to have separate signal wires coming from the memory chips for every memory cell, and then to make individual connections to each part of the system that requires the information. We need to devise a logic system that will reduce to reasonable proportions the number of control and data lines connected to the memory chip.

Figure 11.60 shows the way in which any one of eight data bits may be selected using only three control lines. In general, if n control lines are used then any one of 2^n bits may be selected from memory merely by extending the number and configuration of AND and OR gates. Each block of decoding gates may be repeated for each output data bit. Thus eight blocks will give eight data outputs simultaneously. In this way any eight-bit register may be selected and its contents made to appear as eight-bit word on the eight data lines. The control lines are usually referred to as the *ADDRESS* lines. Physically, each line is a copper track on the printed circuit board (PCB). Collectively the tracks form the address bus. The bus lines are routed through the system and various other relevant parts are connected to them. In this way many devices are suitably interconnected. The data bus is a similar set of interconnections.

Readers will recognize Figure 11.60 as an example of a decoder. In this context the circuit is usually referred to as a multiplexer. From Figure 11.60 it will easily be seen how to increase the number of address lines. In most microcomputers 16-bit address lines are used. This means one of 64 kilobits can be selected. 16-bit address lines are usually associated with eight-bit data words. Each of the circuits of Figure 11.60, suitably extended to accommodate a 16-bit address bus, is repeated eight times, each complete circuit selecting one bit of a data word held in a memory register. (Integrated circuit technology is ideally suited to the repetition of blocks of circuitry.) Such a memory then has the ability to select one of 65 536 registers; the memory capacity is thus 64 kilobytes. Such a large memory is not (in 1985) economically available on a single chip; the computers therefore use banks of smaller capacity chips to perform the task. The BBC Model B microcomputer has sixteen 16K × 1 bit dynamic RAMs, thus making-up 32 kilobytes of RAM. (It also has 32 kilobytes of ROM.) The address and data lines are thus shared by the chips in order to achieve the required capacity.

Unlike the address bus which is unidirectional, the data lines must accommodate two-way traffic. All devices connected to the bidirectional data bus must be suitably

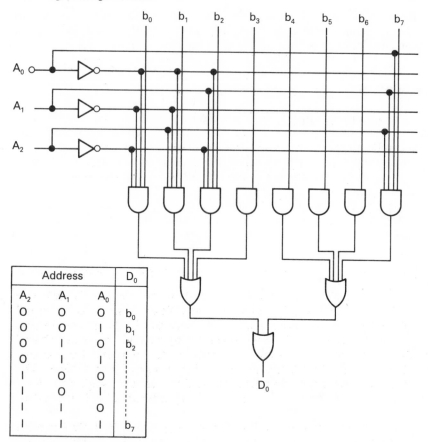

Figure 11.60 Method of selecting any one from eight bits using a three-bit address. (Connections for bits b_3 to b_6 not shown)

equipped with tristate logic, or other external circuitry to avoid confusion about which logic state a line is to assume. While a multiplexer must select a suitable bit from a large array, and energize a data line accordingly, the converse is also required, i.e. a data state on a single line has to be directed to a specified address in a large memory array. A circuit for doing this is called a demultiplexer. (The term 'demux' is often used as a shortened form of demultiplexer.) We will not here complicate the circuit by showing a demux for a large array; Figure 11.61 shows the principle involved in which data on a single line can be routed to a specific address by two control (address) lines.

Architecture of a small digital computer

Having seen how information in the form of binary digits can be stored and retrieved, we must now consider how such information can be controlled and manipulated in order to produce useful results. With such large-capacity memories available, there exists the potential to assemble complex and powerful logic

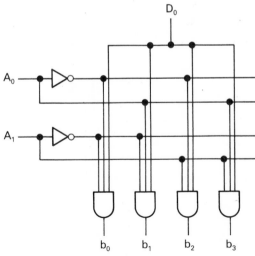

Address		D_0 connected to
A_1	A_0	
0	0	b_0
0	1	b_1
1	0	b_2
1	1	b_3

Figure 11.61 Data demultiplexer that routes data from D_0 to a specific bit line determined by the address

systems. These systems can perform tasks that involve manipulations and calculations with numerical data. Experimental scientific data can be collected and analysed in ways that would prove tedious, or impossible, for human beings. Real-time control systems can be controlled by these logic systems. Digital computers also have the power to store and manipulate alphabetical characters, since these characters can be represented by coded strings of binary digits. The business field is therefore open to the computer. Stock control, wage lists, bank accounts and data banks can now be handled very efficiently with the aid of computers.

The memory's full potential for information storage cannot be fully exploited unless additional logic circuitry is used to perform the tasks of data storing, retrieving and manipulation. Arithmetic and logical processes are performed by an arithmetic and logic unit (ALU). There also needs to be some sort of logic control unit which can route data between the ALU and the system's memory. Under the supervision of the control unit, data can be read from memory and stored in separate registers, or transferred to memory after manipulative tasks have been performed. Memory is used to store not only the data associated with a particular task, but also the sequence of instructions which must be performed upon the data.

Extra registers are included in the control unit to help in maintaining strict operation sequences. The program counter keeps a record of the address of the next instruction which is to be executed; the instruction register stores a copy of the instruction code so that the logic circuit of the control unit can decode it. A memory address register is one in which is stored the address of the data upon which the instruction must operate. As the complexity of the computer becomes greater, the power of the computer increases. More registers are then required in the central processing unit (CPU).

In addition to the memory unit, ALU and control unit, a computer needs to be able to communicate with the outside world; various input/output chips are therefore required. In addition, a clock-pulse generator to synchronize the whole system, and the power supplies, are needed. Figure 11.62 shows the fundamental architecture of a digital computer.

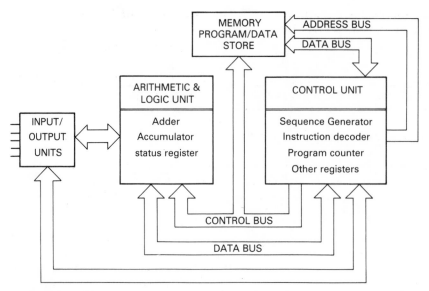

Figure 11.62 Architecture of a basic computer

Exercises

1. Outline the requirements of an ideal switch, and show how closely a junction transistor approaches these requirements.

   Describe the circuit of a simple TTL logic gate, and explain how it may be considered to perform either a NAND or a NOR function.

   Show how an Exclusive-OR function may be realized using NAND gates.

2. Explain, with the aid of a diagram, the main constructional features and operation of a p-channel enhancement-mode MOST. Show how p-channel and n-channel MOSTs may be combined to form a CMOS gate.

   Draw a circuit diagram of an Exclusive-OR gate that uses NAND gates. By writing the Boolean expressions representing the outputs of each NAND gate, prove that the Exclusive-OR functions can be realized by your configuration.

3. By the use of truth tables, or otherwise, prove the validity of De Morgan's Theorem, i.e. that (a) $\overline{AB} = \overline{A} + \overline{B}$ and (b) $\overline{A + B} = \overline{A}.\overline{B}$.

4. Show by means of Venn diagrams that (a) $\overline{AB} = \overline{A} + \overline{B}$ and (b) $\overline{A + B} = \overline{A}.\overline{B}$. Confirm your results with the aid of a truth table.

5. A car is parked on an inclined road. The factors that determine whether or not the car moves are

   A. the engine is running
   B. the clutch is engaged
   C. the brakes are on, and
   D. the ignition is switched on

   Devise the simplest logic system based on NAND gates that will indicate by means of a light bulb the conditions that govern whether or not the car will move. You may assume that the car will not move when the clutch is engaged, but the engine and brakes are off.

6. By the use of a Karnaugh map, or otherwise, reduce the following Boolean expression to its minimum form

$$F = \bar{A}\bar{B}\bar{C}\bar{D} + \bar{A}\bar{B}CD + ABC\bar{D} + \bar{A}\bar{B}C\bar{D} + AB\bar{C}\bar{D} + \bar{A}BC\bar{D}.$$

$$(Ans.\ \bar{A}\bar{B} + AB\bar{D})$$

7. Discuss the usefulness of Karnaugh maps and De Morgan's Theorem in the design of logic control apparatus.

   Show how the function

$$Z = \bar{A}\bar{B} + B(\bar{C} + \bar{A})$$

   may be generated by using (a) NAND gates and (b) NOR gates. Compare the two arrangements.

8. What is meant by the term 'BCD code'? Explain how such codes are constructed and why they are necessary in computing apparatus.

   Two counters operate, one in the 8421 code and the other in a 2421 code. Devise a logic circuit to detect coincidence of count in the counters.

   Devise a logic circuit that converts the 8421 code to a 2421 code.

9. A sequential circuit has four inputs x_1, x_2, x_3 and x_4 and a two-level clock waveform P, $\bar{P}$, P, $\bar{P}$ etc. Only one x input is presented at any one time. The x inputs occur in random sequence and change immediately following a change from $\bar{P}$ to P in the clock waveform. The sequential circuit is to recognize when the sequence x_2, x_3, x_4 appears in the input. Design a suitable sequence detector.

   Discuss possible industrial applications of sequence detectors.

10. Describe the structure of a MOST and give a brief account of its mode of operation. Show how such a device can be used as a logic gate, and proceed to make a critical comparison between MOST logic and the logic systems that use bipolar transistors.

11. Explain, with the aid of circuit diagrams, how a chain of JK bistable elements may be interconnected to form a shift register. Point out the differences between the mode of operation of a shift register and that of a ripple-through counter.

12. Design a synchronous counter to count in the Excess-3 8421 BCD code.

13. Design a modulo-9 counter and confirm that lock-out will not be a problem with your design.

14. Devise a logic circuit to replace the hard-wiring for the shift register shown in Figure 11.40 so that a single SHIFT RIGHT/SHIFT LEFT control line may be used.

15. N_1 and N_2 are two four-bit numbers representing arithmetic quantities in the 8421 BCD code. Devise a circuit that compares these numbers and gives an output when $N_1 = N_2$. Two other output are also required, one showing when $N_1 > N_2$ and the other indicating when $N_1 < N_2$.

Range of preferred values of resistors

Tolerance 20 %	Tolerance 10% (Silver)	Tolerance 5% (Gold)
10	10	10
		11
	12	12
		13
15	15	15
		16
	18	18
		20
22	22	22
		24
	27	27
		30
33	33	33
		36
	39	39
		43
47	47	47
		51
	56	56
		62
68	68	68
		75
	82	82
		91
100	100	100
	(E12 Series)	(E24 Series)

Scaling prefixes for electronic components

Prefix	Multiple	Symbol
tera	10^{12}	T
giga	10^{9}	G
mega	10^{6}	M
kilo	10^{3}	k
milli	10^{-3}	m
micro	10^{-6}	μ
nano	10^{-9}	n
pico	10^{-12}	p

BS 1852 resistance code

On circuit diagrams, and some manufacturers' components, examples of markings are

0.47Ω	is marked	R47
1Ω	" "	1RO
3.3Ω	" "	3R3
47Ω	" "	47R
$1k\Omega$	" "	1KO
$10k\Omega$	" "	10K

A further letter added indicates tolerance: $F = \pm1\%$, $G = \pm2\%$, $J = \pm5\%$, $K = \pm10\%$, $M = \pm20\%$; e.g. 3K3K represents a 3.3K resistor with a 10% tolerance.

Appendix 2

The Karnaugh Map

There are many occasions where diagrams can clarify a situation. For example, it is common to translate columns of data into graphical form, and to represent complex numbers by Argand diagrams. The corresponding graphical representation of a Boolean statement is a Venn diagram. Venn diagrams are topological maps, that is to say they represent those properties of a figure or shape that remain unchanged even when the figure is deformed. One well-known topological map is that showing the underground stations in the London area. Such maps are not intended to show the magnitude of the distances between stations, but merely show their interrelationship. Such an interrelationship is unchanged even if the map were to be deformed in some way. Figure A.2.1 shows some examples of Venn diagrams that

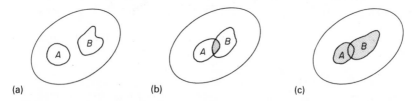

(a) (b) (c)

Figure A.2.1 Examples of Venn diagrams that represent Boolean statements: (a) Overall area represents the universe of discourse, e.g. human beings, A represents human beings with blue eyes, B represents human beings with brown hair; (b) shaded area represents those with blue eyes and brown hair, i.e. $A.B$; (c) shaded area represents those who have either blue eyes or brown hair (or both), i.e. $A + B$. Unshaded area represents those who have neither blue eyes nor brown hair, i.e. $\overline{A + B}$

represent Boolean statements. Since topological maps have no restriction on the shape, the Venn diagram showing two variables A and B may be drawn as shown in Figure A.2.2(a). Each subdivided area may be considered and a decision taken as to whether that area is or is not in A and B. Figure A.2.2(b) although of a different shape, is topologically identical to the diagram in Figure A.2.2(a). We see, therefore, that the top left hand area is in A and in B whereas the bottom left hand area is not in A and is in B. To each subdivided area we can therefore assign a Boolean product to present the situation.

When dealing with counters it is necessary to have four binary digits to represent a decimal digit. Maps dealing with four variables are therefore of great importance in digital work. When the areas of a four-variable Venn diagram are specially arranged, as shown in Figure A.2.3, the resulting diagram is known as a Karnaugh

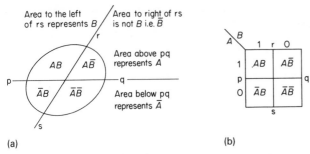

Figure A.2.2 (a) Shows a Venn diagram for two variables. Each subdivided area may be considered separately; it is then noted whether the subdivided area (or 'cell') is or is not in A or B; (b) is topologically identical to (a). The truth values 1 and 0 are used to indicate whether an area is, or is not, in A or B respectively

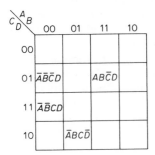

Figure A.2.3 The Karnaugh map. The truth values 0 and 1 for A, B, C and D respectively are chosen so that cells on the map are not only adjacent topologically but also in a Boolean sense (i.e. the Quine-McClusky theorem is satisfied). Each cell represents a four-variable Boolean AND expression

map. The value of such a map lies in the ease with which Boolean expressions can be minimized. The special arrangement is needed in order to incorporate the Quine-McCluskey theorem. Many maps are possible in which areas are actually adjacent, but to be adjacent in the Boolean sense the areas must represent Boolean products that differ by only one term. For example AB and $A\bar{B}$ differ by only one term, namely the B term. If AB and $A\bar{B}$ are represented on a map in adjacent positions, then they are adjacent, not only topologically, but also in a Boolean sense. Since $AB + A\bar{B} = A$ then the two areas may be combined into a single area representing A. As an extension of this idea let us consider four variables. The expressions $\bar{A}BCD$ and $\bar{A}B\bar{C}D$ are adjacent in a Boolean sense since the negation of only one variable (C) in one term yields the other term. They are also adjacent in a topological sense in the map shown in Figure A.2.3 and may, therefore, be combined into the single area $\bar{A}BD$. However, if we consider $\bar{A}BCD$ and $AB\bar{C}D$ then, although we could devise a map in which areas representing these Boolean expressions would be topologically adjacent, such a map would be of no value in minimizing Boolean functions because $\bar{A}BCD$ cannot be converted to $AB\bar{C}D$ by the change of only a single variable. This accounts for the special assignment of the values 0 and 1 to A, B, C and D in the Karnaugh map shown in Figure A.2.3.

Minimization using Karnaugh Maps

The technique of minimizing Boolean functions using a Karnaugh map is to plot each term of the function on the map. The result is then examined and areas are suitably combined so as to produce simpler expressions. The Boolean expression must be in the form of the sum of products before it can be plotted. If the original expression is not in this form it must be manipulated until it is. The sum of products means an arrangement of AND terms linked by the OR connective, e.g.

$$ABC\bar{D} + \bar{A}\bar{B}CD + AB\bar{C}D$$

If the original Boolean expression is the product of sums, i.e. a set of OR terms linked by the AND connective, for example

$$(A + \bar{B}).(B + C)$$

then De Morgan's Theorem must be invoked to effect the necessary transformation. An example of this transformation is given later.

Because of the importance of the four-variable map, we shall choose as examples those Boolean expressions that contain four variables. These expressions arise in connection with counting, and code conversion associated with decimal systems. The importance of counting is obvious enough in digital computing work, but counting also plays a vital role in many digital instruments (e.g. digital voltmeters and digital frequency meters).

Once each term of the expression has been plotted in its appropriate 'cell', inspection of the map is carried out to see how cells can be combined into larger areas. Figure A.2.4 shows some possibilities. Groups of two, four or eight cells

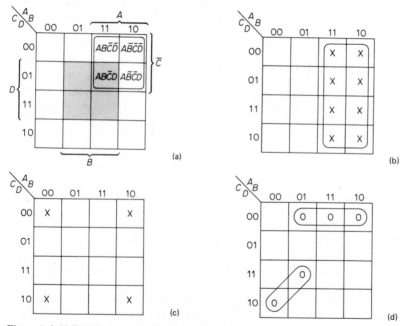

Figure A.2.4 Minimization using the Karnaugh map: (a) Combination of four cells to make area $A\bar{C}$, i.e. $ABC\bar{D} + A\bar{B}C\bar{D} + AB\bar{C}\bar{D} + A\bar{B}\bar{C}\bar{D} = A\bar{C}$. Shaded area shows another four-cell combination that represents BD; (b) combination of eight cells to make the area A, i.e. $ABC\bar{D} + A\bar{B}CD + AB\bar{C}D$ .. etc. $= A$; (c) cells marked with a cross are adjacent in a topological and Boolean sense. They may therefore be combined to represent $\bar{B}\bar{D}$; (d) forbidden combinations, since areas do not contain groups of 2, 4 or 8, or alternatively have no common boundary line

should be found that can be combined into the largest possible areas. Cells may be used more than once to form large groupings. Since the Karnaugh map is a topological one, it should be remembered that it can be rolled to form a cylinder. Topologically the vertical or horizontal edges of the flat diagram can be made to meet; they are therefore adjacent. Figure A.2.5 shows some further examples.

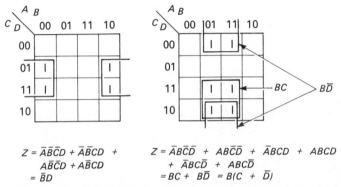

$$Z = \bar{A}\bar{B}\bar{C}D + \bar{A}\bar{B}CD +$$
$$A\bar{B}\bar{C}D + A\bar{B}CD$$
$$= \bar{B}D$$

$$Z = \bar{A}B\bar{C}\bar{D} + AB\bar{C}\bar{D} + \bar{A}BCD + ABCD$$
$$+ \bar{A}BC\bar{D} + ABC\bar{D}$$
$$= BC + B\bar{D} = B(C + \bar{D})$$

Figure A.2.5 Further examples of minimization using Karnaugh maps

For those who are approaching the subject of minimization for the first time, two difficulties may arise. The first is quite trivial. If the Boolean expression is a function of four variables A, B, C and D, then one of the terms may contain less than the four variables. Such a term is easily expanded by performing the reverse of factorization e.g.

$$AB\bar{C} = AB\bar{C}D + AB\bar{C}\bar{D}$$

It is unnecessary, however, to make the expansion if it is realised that a term such as $AB\bar{C}$ can be represented by an area on the map covering the cells $AB\bar{C}D$ and $AB\bar{C}\bar{D}$. We have already touched upon the second difficulty; if the expression to be minimized is in the form of the product of sums then, as it stands, the expression cannot be plotted immediately on a Karnaugh map. The difficulty can be resolved in the following way:

Let us suppose that the expression to be minimized is given by

$$Z = (A + \bar{B} + C + \bar{D})(\bar{A} + \bar{B} + C + \bar{D})$$

using De Morgan's Theorem

$$\bar{Z} = \overline{(A + \bar{B} + C + \bar{D})} + \overline{(\bar{A} + \bar{B} + C + \bar{D})}$$
$$= \bar{A}B\bar{C}D + AB\bar{C}D = Y \text{ (say)}$$

Y is in a form suitable for mapping. If we can find a minimum expression for Y this must be a minimum solution for $\bar{Z}$. Simple inversion then gives a minimum solution for $\bar{Z}$. In our case if we map $\bar{A}B\bar{C}D + AB\bar{C}D$, combining the areas gives $Y = B\bar{C}D$. (Although this is obvious from simple factorization, we are taking a simple two-term expression so as not to obscure the procedure by using more complicated expressions.) Since

$$Y_{\min} = B\bar{C}D = \bar{Z}_{\min}$$

$$Z_{\min} = \overline{B\bar{C}D}$$
$$= \bar{B} + C + \bar{D}$$

New logic symbols

Since logic systems were first introduced into industrial and other locations, manufacturers and international committees have proposed a multiplicity of symbol systems for use in circuit diagrams. Readers with some experience of logic circuits will, no doubt, use the system to which they have become accustomed during training and early work. Confused with the changes that have frequently been made in the past, they will probably be dismayed to learn of yet another proposed system. The International Electrotechnical Commission (I.E.C.) has been working on a new symbolic system that some claim, with justification, to be a powerful language in the field of logic circuit symbols. The complexity of modern logic circuits is such that some method is required of showing the relationship of each input of a logic circuit to each output without necessarily showing the circuit systems of the internal logic. Revision of the I.E.C. standards and those of the I.E.E.E. in the United States of America (Std. 91/ANSI Y32.14) are taking place, and will probably be complete and adopted by the time this book is published. Interested readers should be aware that revisions are under way, and be on the lookout for the definitive documents. In the meantime readers may care to consult a very good introduction and explanation of the new symbols given by F.A. Mann of Texas Instruments Incorporated in the *TTL Data Book for Design Engineers*, Vol. 2 (1983/84). In this publication he points out that dependency notation, showing how the outputs of a system are logically dependent upon the input states, is at the heart of the system.

The symbols themselves are of rectangular outline, to which is added general qualifying symbols. Figure A.3.1 shows the diagram of the general form. Some of

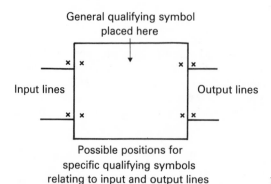

Figure A.3.1 General new logic symbol

the general qualifying symbols relevant to circuit diagrams in this book are given in Table A.3.1.

Associated logic symbols for input and output lines are given in Table A.3.2.

Some of the symbols that may be used inside the rectangular outline are shown in Table A.3.3.

Figure A.3.2 shows, by way of example, the symbols for a 2-input NAND gate, a 4-input OR gate and a NAND gate with open-collector output. Figure A.3.3 shows the symbol for a dual JK flip-flop.

When a circuit has an input function that is common to more than one element of the circuit, the symbol used is as shown in Figure A.3.4.

A common output arrangement is shown in Figure A.3.5.

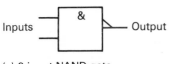

(a) 2-input NAND gate

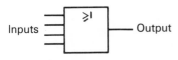

(b) 4-input OR gate

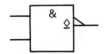

(c) 2-input NAND gate with open-collector output

Figure A.3.2 Examples of new logic symbols

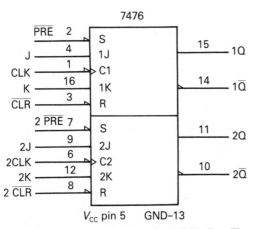

Figure A.3.3 Circuit symbol for a dual JK flip-flop. The numbers represent the pin positions on the IC package

Table A.3.1 General qualifying symbols

&	AND gate or function
≥1	OR gate or function
=1	Exclusive-OR
=	Logic identity
⎍	Schmitt trigger
X/Y	Coder or code converter. X is In; Y is Out; e.g. DEC/BCD and BIN/7-SEG
MUX	Multiplexer
DMUX or DX	Demultiplexer
Σ	Adder
P-Q	Subtractor
CPG	Look-ahead Carry Generator
π	Multiplier
COMP	Magnitude comparator
ALU	Arithmetic Logic Unit
SRGm	Shift register m = number of bits
CTRm	Counter m = number of bits
ROM	Read only memory
RAM	Random access memory
FIFO	First-in, first-out memory

Table A.3.2 Logic symbols for input and output lines

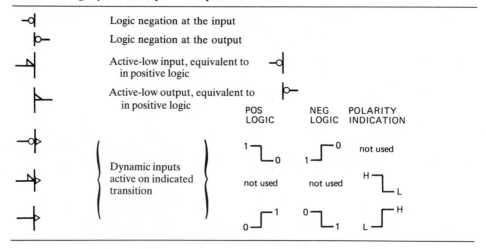

Table A.3.3 Some symbols used inside rectangular outlines

◊⊢	NPN open-collector output. Requires external pull-up device
⌂⊢	Passive pull-up output similar to above, but with built-in passive pull-up
▽⊢	3-state output
⊣EN	Enable input
J, K, R, S, T	Usual meanings associated with flip-flops, e.g. T = toggle.

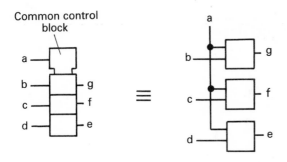

Example

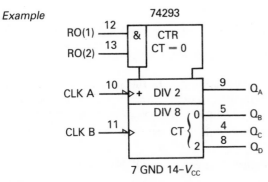

Figure A.3.4 Common input blocks

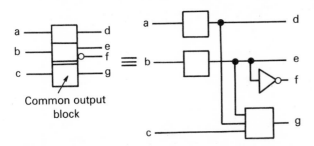

Figure A.3.5 Common output block. Qualifying symbols are required to denote logic functions

Appendix 4

Supplementary reading

The following books may be found useful as supplementary reading to this text:

(1) P. Horowitz and W. Hill, *The Art of Electronics*, Cambridge University Press (1980).

This is a very good book with a practical approach to the subject. It contains a wealth of down-to-earth detail showing the reader how to cope with actually assembling electronic circuits, as opposed to passing examinations in the subject.

(2) R. King, *Integrated Electronic Circuits and Systems*, Van Nostrand Reinhold (UK), (1983).

(3) G.B. Clayton, *Operational Amplifiers*, Butterworths (1979).

(4) A.C. Downton, *Computers and Microprocessors*, Van Nostrand Reinhold (UK), (1984).

(5) R. Meadows and A.J. Parsons, *Microprocessors: Essentials, Components and Systems*, Pitman (1983).

Index